INSECTS AND OTHER TERRESTRIAL ARTHROPODS: BIOLOGY, CHEMISTRY AND BEHAVIOR

TICKS

DISEASE, MANAGEMENT AND CONTROL

INSECTS AND OTHER TERRESTRIAL ARTHROPODS: BIOLOGY, CHEMISTRY AND BEHAVIOR

Additional books in this series can be found on Nova's website under the Series tab.

Additional E-books in this series can be found on Nova's website under the E-book tab.

INSECTS AND OTHER TERRESTRIAL ARTHROPODS:
BIOLOGY, CHEMISTRY AND BEHAVIOR

TICKS

DISEASE, MANAGEMENT AND CONTROL

MOGES WOLDEMESKEL
EDITOR

Nova Science Publishers, Inc.
New York

For permission to use material from this book please contact us:
Telephone 631-231-7269; Fax 631-231-8175
Web Site: http://www.novapublishers.com

NOTICE TO THE READER

Additional color graphics may be available in the e-book version of this book.

LIBRARY OF CONGRESS CATALOGING-IN-PUBLICATION DATA

Ticks: disease, management, and control / editor: Moges Woldemeskel.
p. cm.
Includes index.
ISBN 978-1-62081-136-8 (hardcover)
1. Ticks--Control. I. Woldemeskel, Moges.
SF810.T5T55 2012
571.9'86--dc23

2012005426

Published by Nova Science Publishers, Inc. ✝ New York

CONTENTS

PREFACE

Ticks and tick-borne diseases are among the major stumbling blocks to the development of livestock industry and entail heavy economic loses particularly in the tropics and subtropics. Ticks serve as vectors of several diseases and pose health hazards to animals and humans throughout the world. Attempts to control ticks and tick-borne diseases using different methods have been going on for several generations; however, ticks still cause insurmountable problems to the livestock industry and human and animal health. This book enlightens the reader on research and field experiences obtained from different parts of the world on the various chemical and biological approaches used in the control of ticks and tick-borne diseases. This book would serve as a valuable reference and guide for students, and researchers in biological and biomedical sciences and tick control authorities aimed at devising a sound tick control strategy.

Chapter 1 - Stress can be defined as a threat to homeostasis. Physiologic effects of acute stress include heightened awareness, increased heart rate, and muscle tone; all positive developments if the individual is being called upon to defend itself or escape. Chronic stress, however, can be detrimental to homeostasis. Ticks are external parasites and their blood-feeding behavior poses a significant economic burden to domestic animal agriculture worldwide. Direct negative economic effects of ticks on their hosts include loss of body weight and condition, lowered reproduction, and hide damage. These observed effects are due to a combination of a direct impact on homeostasis and lowered nutritional status via depression of intake. Additional hardships may result from disease transmission. Combating an external parasite burden may be considered a "cost of fitness" which incurs a drain on available energy. Protein-energy malnutrition has metabolic, endocrine and immune consequences especially with respect to parasitism. In this chapter, the effects of tick stress on grazing animals are discussed in relation to an interconnected series of physiological events involving multiple systems and indicators. Topics include: behavior, nutrition, metabolism, endocrine and immune factors, gastrointestinal function, and commensal bacteria. Recent research dealing with direct measurement of animal performance and metabolic indicators in cattle experiencing experimentally induced tick burdens will be discussed. Detection of tick stress via fecal analysis is presented as a means of bio-forensics.

Chapter 2 - The feeding period of hard ticks often exceeds one week and makes tick infestation a considerable intervention to host physiology. A broad range of inflammatory and immune reactions take place locally at the feeding site as well as systemically within the host body. Active modulation of the host immune response by tick saliva is thus required for the tick to complete its blood meal. Upon attachment, the tick inoculates its saliva containing a repertoire of bioactive molecules that help ticks to attach to the host, to overcome host hemostasis, and to prevent host pain and itching and the concomitant scratching. Furthermore, the tick must deal with the host innate immunity during primary infestation and with innate and adaptive immunity in subsequent infestations. The immunomodulatory properties of tick saliva and the roles of individual salivary components at the tick-host interface are described in this chapter, dealing with virtually all levels of nonspecific and specific host immune mechanisms. In addition, tick-borne pathogens, including the Lyme disease spirochete *Borrelia burgdorferi*, exploit the pharmacologic activities of tick salivary molecules to successfully maintain their life cycle. Identification of molecules responsible for this phenomenon, termed saliva-assisted transmission (SAT), can help in the design of an efficient transmission-blocking vaccine. Moreover, understanding of the immunity-based processes within the tick-host-pathogen interface opens a relatively new field for the development of drugs to treat various immunity-related diseases.

Chapter 3 - The tick *Ixodes ricinus (I.ricinus)* is the most multi-competent vector in Europe, responsible for significant diseases of humans and livestock throughout the northern hemisphere. Quantitative analysis of tick-borne disease risk and potential tick control depends on a full understanding of the tick's complex seasonal dynamics, for which a realistic model is an invaluable tool. The effects of individual environmental factors on population dynamics can be investigated in ways not easily possible by experimental manipulation of *in situ* populations. A population model for *I. ricinus*, based on a modified stage-classified Leslie matrix, gives a good fit to field data from UK sites and, after modification of a single parameter, also replicates divergent seasonal patterns in central Spain. This modification with respect to diapause, which is required before the model is more generally applicable across wide geographic areas, highlights gaps in the authors' knowledge of tick biology. The model is generally robust, but most sensitive to changes in parameters related to density-dependent mortality.

This model can be used to explore the relative impact of changes in climate, host density and acaricide-treated hosts on tick abundance and seasonality to identify the causes of apparent recent increases in tick populations, and to predict potential effects of future environmental change. The rise in temperature in 1989 across Europe could have increased the inter-stadial development rate of ticks, thereby reducing the cumulative effect of daily mortality rates and potentially raising population levels. An earlier onset of tick questing activity in warmer springs could increase contact between ticks and humans during recreation in the countryside in spring holidays. These tick population effects are more marked where initial background climate is more limiting, especially along latitudinal and altitudinal boundaries. At the same time, the observed increase in deer abundance is predicted to drive tick populations up, most markedly where deer are initially at low densities. Where host densities are already high, however, further increases may reduce the population of questing ticks as they find hosts more rapidly, even though the overall tick population (and therefore pathogen transmission potential) increases. Culling high-density deer populations could, therefore, initially cause an apparent increase in questing ticks, while adding more large hosts (e.g. sheep) would effectively reduce the number of questing ticks, and therefore the risk to humans. If such sheep were treated with acaricide, tick populations are predicted to decrease rapidly. Tick control in designated areas may be achieved by using sheep in this way as 'lethal mops', as occurred in the past.

Chapter 4 - Ethiopia has the largest livestock population in Africa, and lies in the tropics encompassing wide ranges of agro-climatic zones including the cool central highlands, wet tropical forest in the west and southwest and dry arid and semi-arid areas in the east, south and southeast. Therefore it would represent various climatic zones and livestock production systems in tropical Africa. Ticks are the main ectoparasites infesting domestic and wild animals and are among the major stumbling blocks in livestock production in Ethiopia and tropical Africa. Ixodid and argasid ticks collected from cattle, sheep, goats, camels, equines and a few from wild animals in Ethiopia are described. In this chapter, forty-four tick species in 9 genera recorded in different agro-ecological zones of the country and their possible role in disease transmission are briefly described. The identified ticks infesting domestic and wild animals belonged to the genus *Rhipicephalus* (15 species), *Hyalomma* (9 species), *Amblyomma* (8 species), *Haemaphysalis* (4 species), *Aponomma* (2 species), *Boophilus* (2 species), *Ornithodorous* (2 species), *Ixodes* (1 species), and *Argas* (1 species). Understanding the distribution, seasonal dynamics and the host range on various tick species infesting domestic and wild animals is very important to envisage and implement sound and effective tick- and tick-borne disease control strategies.

Chapter 5 - Canine babesiosis is an important worldwide, tick-borne disease caused by hemoprotozoan parasites of the genus *Babesia*. It is a small pleomorphic intraerythrocytic parasite that can cause erythrocyte destruction and hemolytic anemia. The parasite is transmitted through the bites of ixodid ticks, such as *Rhipicephalus sanguineus*. Definitive diagnosis of canine babesiosis is based on medical history and clinical signs together with the identification of *Babesia* spp. within infected erythrocytes, positive serologic results, or detection of amplification of nucleic acid extracted from blood or tissues. Subjects with high mortality can be identified using several prognostic factors. Combination of babesicidal agents and antibiotics, and supportive cares are integrated as treatment program for canine babesiosis. Tick (vector) control is the key to prevent canine babesiosis. This chapter describes the current information regarding epidemiology, pathogenesis, and pathophysiology of the clinical manifestations associated with tick-borne canine babesiosis, and the methods of diagnosis, prognostic factors, treatments and prevention of canine babesiosis.

Chapter 6 - Cutaneous dermatophilosis is a severe skin disease of domestic animals caused by the bacterium *Dermatophilus congolensis* (*D. congolensis*). Although the disease is caused by *D. congolensis*, several risk factors are associated with the pathology and pathogenesis of the disease. Tick infestation is among the major risk factors, which play a very important role in the establishment, transmission and progression of the disease. Dermatophilosis and tick infestation entail heavy economic losses in animal production due to treatment costs, management of the disease and condemnation of skin and hides in various parts of the world, particularly in the tropical countries. In this chapter, the role ticks play in the pathology and pathogenesis of dermatophilosis and the effect of tick control in the management of the disease will be covered.

Chapter 7 - Tick resistance to acaricides is an increasing problem in tropical Africa. It poses a real economic threat to the livestock and veterinary pharmaceutical industries. New acaricides are extremely expensive to develop. Therefore, the present acaricides should be seen as an ever-diminishing resource, which should be protected by all means possible. To alleviate tick resistance problems through resistance management regimens, it is very important to determining the pattern of resistance to ticks in a given area. A field survey was carried out at selected communal and commercial farms in the Eastern Cape and Northwest

Provinces of South Africa to monitor the levels of field ticks (B. decoloratus, R. evertsi evertsi, R. appendiculatus and A. hebraeum) resistance to Amitraz, Cypermethrin and Chlorfenvinphos.

On the communal farms high levels of tick resistance were detected to Cypermethrin as well as partial resistance to Chlorfenvinphos whilst no resistance was detected against Amitraz. On the commercial farms, however, ticks were equally resistant to Amitraz, Cypermethrin and Chlorfenvinphos. The populations of B. decoloratus on these farms had developed higher levels of resistance to the test acaricides than the populations of R. evertsi evertsi, R. appendiculatus and A. hebraeum. Higher level of tick resistance to Amitraz was observed on commercial farms than on communal farms. However, there was no significant difference in tick resistance to Chlorfenvinphos and Cypermethrin at both the commercial and communal farms. It was summarized that inappropriate use of acaricides on the commercial as well as the communal farms might have resulted in higher tick resistance to the currently available acaricides.

This chapter describes the levels of tick resistance to acaricides at selected commercial and communal farms in South Africa, compares the *in vitro* adult and larval test methods and discusses the acaricide management strategies, which may increase the lifespan of the presently used acaricides.

Chapter 8 - Different approaches have been used in Mexico during several years for the control of the cattle tick *Rhipicephalus (Boophilus) microplus,* the vector of two very important diseases of cattle: babesiosis, caused by the haemoprotozoans *Babesia bovis* and *B. bigemina* and anaplasmosis caused by the haemotropic rickettsia *Anaplasma marginale*. The most frequently used methods in Mexico include strategic acaricidal baths, which when used improperly have generated tick populations resistant to some acaricides, use of anti-tick commercial vaccines, and the more recently proposed use of entomopathogenic fungus for biological control of ticks. These methods utilized in Mexico have shown various degrees of success. In this chapter the efficacy of the aforementioned methods of tick control and their drawbacks are discussed. Furthermore, the tendencies for the use of alternative new control procedures, for example the use of the fly *Megaselia scalaris* as a biologic control method as well as the practice of integrated control strategies is described.

Chapter 9 - The tick *Rhipicephalus (Boophilus) microplus* is a bovine ectoparasite that causes severe economic losses in herds in tropical and subtropical areas due to parasitism and disease transmission. Tick control is currently based mainly on the use of chemicals. However, alternative strategies are required due to the ability of ticks to develop resistance, the demands of consumers for chemical-free foods, and the environmental impact of chemical acaricides. In accordance with this scenario, the use of fungi to biocontrol ticks is an efficient and important tool. The entomopathogenic and acaricidal fungus *Metarhizium anisopliae* is the most studied fungus for the biocontrol of *R. microplus*. Various studies on different aspects of the use of *M. anisopliae* to control ticks and other pests have been published during the decade. Such knowledge is crucial in the search for more efficient strains and for optimizing new formulations for large-scale use of this efficient, economic and environmentally safer form of tick control. Here The authors focus on the biocontrol of the tick *R. microplus* based on the use of *M. anisopliae*, including the different fungal isolates, their virulence, the biochemical and molecular aspects of the pathogenesis and the control strategies.

Chapter 10 - Ticks are vectors of various protozoan and bacterial diseases. Ticks transmit infectious diseases such as Lyme disease, Japanese spotted fever, Rocky Mountain spotted fever, Colorado tick fever, and others. It is significant to check for the presence of a tick bite, because immediate and precise treatment may reduce the morbidity and the mortality due to tick-borne diseases. The management of tick-bite should not be limited to the treatment of only the tick-bite lesions. Physicians need to examine and consider the possibility of consequent hazardous tick-transmitted infectious diseases. As the distribution of ticks is dependent on the environment, it is recommended for physicians to know the types of ticks prevalent in their surroundings, the time ticks bite their victims, and the types of tick-borne disease transmitted by a given tick species prevalent in the surrounding. Recently, dermoscopy is used for detecting and examining the tick-bite lesions and the involved biting tick. It also helps to determine whether or not the tick has been completely removed from the lesion. Dermoscopy is useful for easy initial classification of the tick species, which will suggest the strategy for treatment. Now, the term entodermoscopy is proposed for the new field connecting entomology, infectiology and dermatology.

In: Ticks: Disease, Management and Control
Editor: Moges Woldemeskel
ISBN: 978-1-62081-136-8

Chapter 1

THE PHYSIOLOGY OF TICK-INDUCED STRESS IN GRAZING ANIMALS

Douglas R. Tolleson,[1,*] Pete D. Teel,[2] Gordon E. Carstens[3] and Thomas H. Welsh, Jr[3]

[1]University of Arizona, School of Natural Resources and the Environment, Arizona, US
[2]Texas A and M University, Department of Entomology, Texas, US
[3]Texas A and M University, Department of Animal Science, Texas, US

ABSTRACT

Stress can be defined as a threat to homeostasis. Physiologic effects of acute stress include heightened awareness, increased heart rate, and muscle tone; all positive developments if the individual is being called upon to defend itself or escape. Chronic stress, however, can be detrimental to homeostasis. Ticks are external parasites and their blood-feeding behavior poses a significant economic burden to domestic animal agriculture worldwide. Direct negative economic effects of ticks on their hosts include loss of body weight and condition, lowered reproduction, and hide damage. These observed effects are due to a combination of a direct impact on homeostasis and lowered nutritional status via depression of intake. Additional hardships may result from disease transmission. Combating an external parasite burden may be considered a "cost of fitness" which incurs a drain on available energy. Protein-energy malnutrition has metabolic, endocrine and immune consequences especially with respect to parasitism. In this chapter, the effects of tick stress on grazing animals are discussed in relation to an interconnected series of physiological events involving multiple systems and indicators. Topics include: behavior, nutrition, metabolism, endocrine and immune factors, gastrointestinal function, and commensal bacteria. Recent research dealing with direct measurement of animal performance and metabolic indicators in cattle experiencing experimentally induced tick burdens will be discussed. Detection of tick stress via fecal analysis is presented as a means of bio-forensics.

* Correspondence:E-mail address: dougt@cals.arizona.edu.

Introduction

Stress can be defined as a threat to homeostasis, some event which the body perceives as harmful and thus requires defensive action. The concept of stress in biological terms was first proposed by Selye [1]. His observation that rats receiving frequent injections developed (among other symptoms) enlarged adrenals regardless of what the injections contained, led others to determine that the hypothalamus, the anterior pituitary, and the adrenals function to produce hormones responsible for the physiological actions now known as the "stress response". In general, this response involves processes which prepare the body for "fight or flight". Included among these processes are: tachycardia, vasoconstriction in the skin, vasodilation in skeletal muscle, and temporary/short-term cessation of "non-essential" processes such as digestion or immune function.

The endocrinology of stress involves catecholamines released from the adrenal medulla as a result of sympathetic nervous system stimulation and, a feedback driven cascade of hormones produced by an interrelated group of tissues known as the hypothalamic-pituitary-adrenal (HPA) axis. Briefly, upon perception of a stressor, the hypothalamus secretes corticotrophin releasing hormone (CRH) which acts upon the anterior pituitary gland's corticotroph cells to stimulate adrenocorticotropic hormone (ACTH) synthesis and secretion. The adrenal cortex subsequently releases glucocorticosteroids (e.g. cortisol and corticosterone) in response to ACTH. Some of the physiologic effects of acute stress were previously alluded to: heightened awareness, increased heart rate and muscle tone, all positive developments if the individual is being called upon to defend itself or escape. Chronic stress, however, can be detrimental to homeostasis. Classic work by Sapolsky [2] with baboons has illustrated the long-term effects of stress on overall health and well-being. Immunity is, of course, a significant player in this process.

The Immune System

One of the most exciting and rapidly expanding fields in the life sciences is immunology. Though driven primarily by interest in combating disease, research into the immune system has also helped expose the intricate connections among various systems in the body. Disciplines such as psychoneuroimmunology, for instance, deal with the actions of such neuroendocrine tissues as sympathetic neurons and HPA hormones on metabolism and the immune system [3]. In addition to indirectly affecting immune function by potentially shunting resources away from immune tissues during a stressful situation, direct effects of glucocorticoids on various immuno-active cells have been demonstrated [4]. Mononuclear cells contain receptors for glucocorticoids [5]. Glucocorticoids inhibit major histocompatibility II expression [6] and tumor necrosis factor production [7] in macrophages. Cortisol [8] reduced lymphocyte production in swine. Hormones of non-adrenal origin have direct effects on immune tissues as well. Growth hormone and insulin-like growth factor 1, for instance, in addition to their better-known metabolic functions, also expedite maturation of lymphocytes in bone marrow [9] and thymus [10]. The communication between the immune system and other physiological systems is bi-directional. Cytokines, the chemical messengers produced by immune cells, influence many metabolic functions; for example protein synthesis and muscle maintenance. Interleukin-1, Interleukin-6, and tumor necrosis factor-α are among the cytokines considered to be "pro-inflammatory". Insulin-like growth factor-1, both circulatory and intramuscular, is reduced by elevated concentrations of these cytokines [11]. Thyroid hormones are important regulators of basal metabolic rate. Pro-inflammatory cytokines diminish the response of thyroid-stimulating hormone to thyrotropin-releasing hormone [12], thus affecting peripheral thyroid hormone levels and ultimately basal metabolism.

Stress and the Immune System

Venkatraman and Pendergast [13] in a review of the effects of stress on the immune system state:

> ...stress leads to a proportional increase in stress hormone levels and concomitant changes in several aspects of immunity, including the following: high cortisol; neutrophilia; lymphopenia; decreases in granulocyte oxidative burst, nasal mucociliary clearance, natural killer cell activity, lymphocyte proliferation, the delayed-type sensitivity response, the production of cytokines in response to mitogens, and nasal and salivary immunoglobulin A levels; blunted major histocompatibility complex II expression in macrophages; and increases in blood granulocyte and monocyte phagocytosis, and pro- and anti-inflammatory cytokines. In addition to providing fuel for exercise, glycolysis, glutaminolysis, fat oxidation and protein degradation participate in metabolism and synthesis of the immune components. Compromising, or overusing, any of these components may lead to immunosuppression.

Specifically, chronic stress suppressed Immunoglobulin G titer to keyhole limpet hemocyanin in hamsters [14]. Matalka [15] reviews evidence that acute stress induces the production of pro-inflammatory cytokines (i.e. promotes a T_h1 response), while chronic stress causes dysregulation in immune function by shifting the cytokine pattern in favor of a T_h2 response. Chronic heat stress reduced white blood cell counts and antibody production in laying hens [16], while acute cold stress enhanced the innate immune response in growing chicks [17]. Beef steers grazing endophyte infected fescue have been reported to exhibit decreased phagocytic activity and expression of major histocompatibility complex II [18]. However, Rice et al. [19] found that antibody titers to Conconavilin A tended to be higher and were higher ($P < 0.05$) against sheep red blood cells in cattle grazing infected fescue versus those consuming the non-infected variety. The effect of "stress" on immunity is thus not a "one size fits all" phenomenon. Rook [20] states that, feedback from cytokines on the HPA axis helps the body to control the immune response. Cortisol plays an important role in regulating immunity and inflammation, but as with many other vital physiological functions a delicate balance must be maintained; either too much or not enough can be harmful.

Gastrointestinal Tract

The gastrointestinal tract (GIT) is a hollow tube extending from the oral cavity to the anus. The function of the GIT is digestion, absorption, and excretion of food residues. Basic architecture varies among species but in mammals generally the components are: the mouth, esophagus, stomach, small intestine, cecum, large intestine, and associated structures such as the liver, pancreas, and gall bladder. Since the small intestine is the primary area for immune activity in the GIT, the discussion in this section will focus on the immunological activities of this organ.

The small intestine is composed of three sections: the duodenum, jejunum, and ileum. It is the longest section of the GIT (~ 50 m in bovines, 4 m in canines, 6 m in humans) and the major site of nutrient absorption. The duodenum, jejunum and proximal ileum are the sites where most digestion and absorption occur [21]. On cross-section, from the serosa to the lumen, the small intestine consists of the longitudinal and circular outer muscular layers, followed by the sub-mucosa and finally the mucosa lining the lumen. The mucosa is the "front line" for both nutrient absorption and immunity. The absorptive surface area of the small intestine is greatly expanded by: 1) folding of the sub-mucosal layer (*valvula conniventes*), 2) the presence of finger like outward projections known as villi upon these folds, 3) and lastly, the villi are in turn lined with epithelial cells possessing a luminal cell surface covered by microvilli. The microvilli are also referred to as the brush border. The net result of the above convolutions and projections is a surface area of approximately 250 m^2 in the adult human [22]. The columnar absorptive cells, enterocytes, line the lumen of the small intestine. These are replaced every 3 to5 days by new cells produced from the simple tubular glands between the villi called crypts of Lieberkuhn. The second most numerous cells found here are goblet cells, important for their mucus secreting functions.

Digestion beyond that which occurs in the upper sections of the GIT is carried out by specific enzymes on the cell surface of the enterocytes, in a layer of glycoproteins known as the glycocalyx. The net result is a population of molecules of a size and configuration appropriate for absorption. Absorption is often accomplished by nutrient dependent mechanisms coupled to Na/K transport, an energy consuming process responsible for much of the net energy of maintenance in animals. Undigested and unabsorbed feedstuffs are passed along the tract through the colon, and ultimately excreted through the anus along with resident microbes, various digestive secretions and sloughed epithelial cells [23, 24].

As important as the digestive functions of the GIT are to the animal, they are by no means the only important functions. The body is exposed to many potential antigens through ingestion of food and associated microbes. The gut epithelium is often described as being in a constant state of controlled inflammation. As such, the commensal microbial population exists in a delicate balance with the body's immune defenses.

Immune Function of the Gastrointestinal Tract

The GIT provides three barriers against infection: 1) physical structure, 2) innate immune system, and 3) adaptive immune system. Each is interrelated and dependent upon the other two. The physical barrier [25] consists of primarily the mucosa and associated layers. The epithelia, as previously mentioned, line the lumen of the tract and come into immediate contact with microbes and digesta. Tight junctions between cells here allow passage of water and ions but effectively block diffusion of macromolecules and microorganisms. The glycocalyx of columnar epithelial cells and mucus from goblet cells help shield the inner layers from direct microbial attachment and invasion. Paneth cells are found at the base of the luminal crypts and have both secretory (IgA) and phagocytic (lysozyme) functions. Also found in the epithelia are membranous or microfold (M) cells [26], which have specialized adaptations for transporting antigens. The lamina propria is the site of Peyer's patches (discrete clusters of lymphoid follicles), and in the sub-mucosa are numerous primary follicles as well as active germinal centers. The Paneth cells, M cells, and Peyer's patches are thus not only part of the physical barrier to infection, but also play an active part in the innate immune barrier.

Innate Immunity in the Gastrointestinal Tract

Innate immunity occurs in the gut associated lymphoid tissue. This somewhat amorphous collection of cells is designed to not only defend against intrusion by potentially harmful microbes, but perhaps equally significant also, to not attack helpful organisms. Several cell types and mechanisms within cells exist to accomplish this unique balance. Nicoletti [25] has recently reviewed the origin, structure, and function of M cells. These special cuboidal epithelial cells can be found adjacent to enterocytes but possess a distinct morphology. The M cells exhibit considerably less microvilli than do other epithelial resident cells and the characteristic thick glycoclayx is noticeably absent. Such adaptations facilitate active phagocytosis of large particulate matter within their area. This feature obviously serves to eliminate invading bacteria by internalizing and presenting them to macrophages, but also exposes the body to exploitation by certain microbes which use this route to gain entry to the underlying tissue [27, 28]. Another interesting feature, which differentiates M cells from prototypical intestinal epithelial cells, is the deep invagination of the basolateral membrane. Intraepithelial pockets are thus formed which create intimate contacts between the M cell and specific antigen presenting cells (APC) which home to this area. Such modification of the basolateral membrane also results in shortened transit for endocytotic vesicles within the M cells.

In addition to being interspersed throughout the gut epithelia, the M cells form areas of high concentration (> 50%) immediately above the Peyer's patches. These areas are referred to as follicle-associated epithelium [29]. Kiyono and Fukuyama [26] have presented a thorough review of the origin, structure, and function of the Peyer's patch. The Peyer's patches are dome-shaped structures within the lamina propria, somewhat analogous to the deep cortex and follicles of a lymph node but with no afferent or efferent lymphatic vessels [30]. They are in fact clusters of these follicles. Similar in form and function to lymph nodes, they are thus sites of antigen encounter for activation and induction of B and T cells and have areas of concentration of each type of lymphocyte. Immigration of naïve B and T cells from primary lymphoid tissues occurs via high endothelial venules which lace the lamina propria. Amplification of both effector and memory cells is another similar function of Peyer's patches to that of the lymph nodes, resulting in enhanced IgA secretion and cytotoxic killing. Antigen presentation occurs via macrophages and dendritic cells (DC) which have been in close contact with the invaginated basal membrane of M cells. These APCs, especially DC are an integral link between innate and adaptive immunity.

Dendritic cells are APCs of bone marrow origin and are pivotal in directing the immune response in the gut. They exert a disproportionately large effect on T cell function: one DC may interact with 300 to 1000 T cells [31]. Two major functions of DCs include antigen acquisition and lymphocyte activation. Gut DCs acquire antigen through: 1) connection with M cells, 2) phagocytosis of apoptotic endothelial cells, 3) bacteria which enter the lamina propria through damaged epithelia, and 4) insertion of dendrites between tight junctions into the intestinal lumen. These cellular intelligence operatives are thus exposed to all the extra-cellular environment has to offer. Highly adaptable and sensitive, DCs are capable of not only discriminating sub-molecular variation in antigens, but also of modifying the reaction and differentiation of lymphocytes as a result. For instance, lipopolysaccharide (LPS) from Escherichia coli stimulates IL-12 and subsequent Th1 expansion upon interaction with DCs, while Porphyromonas gingivalis derived LPS does not stimulate IL-12 and directs a Th2 response [32]. Such fine specificity requires the involvement of precise pattern recognition receptors (PRR). Pathogen associated molecular patterns, or PAMPs, are recognized by a family of cell surface PRRs called Toll-like receptors (TLRs). First reported by Beutler [33] and many others since, TLRs are specific for several highly conserved microbial molecules, moieties, and structures. Double stranded RNA (TLR-3), unmethylated DNA (TLR-9), flagellin (TLR-5), and the previously mentioned LPS (TLR-4) are noteworthy examples [34]. The response to TLR recognition of microbial invasion is an intracellular signaling cascade including such molecules as TRAM, TRIF, MyD88, IRAK-1, TRAF-6 and ultimately the transcription factor, NF kappa beta. The products of genes activated by NF kappa beta include the pro-inflammatory cytokines TNF, IL-1, and IL-6 thus stimulating innate immunity. In addition, NF kappa beta also functions in the expression of MHC, IL-2, Il-12 and IFN gamma: all of which are molecules important to the adaptive response [35].

Adaptive Immunity in the Gastrointestinal Tract

One may not think of the adaptive immune system as a barrier in the sense that the physical structure of the GIT and its associated cells of the innate immune system are, but nonetheless, it is. The adaptive system must be bypassed, broken through or compromised to allow a sustained infection to occur. In this system, lymphocytes of both the B and T variety are selected by antigen and activated. This process is then followed by expansion of the immune cells and hopefully, elimination of the infection. Finally, these cells undergo negative feedback and return to a "resting" state [36]. This entire procedure is done in concert with the previously described cells of the innate immune response. Mesenteric lymph nodes and Peyer's patches are sites of antigen encounter for lymphocytes in the gut. Antigen recognition and proper co-stimulation result in up-regulation of cytokines such as IL-2, IL-4 and IFN gamma which in turn regulate differentiation of naïve T cells to T_h1, T_h2, and CTL subsets. Decreased expression of CCR7 by T cells and a concomitant increased expression by B cells cause migration out of their respective areas in lymphoid tissue towards each other at the follicular edge. The B cell/T cell interaction culminates in B cell maturation, and differentiation into effector or memory cells. Secretion of IgA is the primary function of GIT B cells. All three barriers must interact to ward off the constant potential infections in an environment so conducive to microbial growth. As alluded to earlier, one mystery of this system is that it must also recognize and not fight off bacteria that are at the least not harmful, and very often beneficial, to the host.

INTESTINAL MICROBES

> Constructive conjunctive symbiosis is an association between two different organisms, with bodily union between them, which is of benefit to the physiologic processes of one (or both) [a] of them (Merck Source 2005 [a] Authors' addition)

The so called commensal bacteria, those that live within the GIT without harming the host are, perhaps more accurately described by the definition given above. Human beings are in fact, by number of cells, about 90% microbial with 10^8 to 10^{12} bacterial cells/g of intestinal content. In fact, an extension of the Human Genome Project, i.e. the Human Microbiome Project has been initiated to help characterize and understand this important aspect of health and well-being [37]. The gut microflora have even been described as a metabolic organ [38]. Fermentation by hindgut microbiota provides volatile fatty acids, of which butyrate is the primary metabolic substrate for intestinal epithelia [39]. Acetate is ultimately used by peripheral tissues and propionate provides a substrate for hepatic gluconeogenesis. Energy from plant structural carbohydrates, otherwise not available to animals, is obtained in this manner. Coprophagy is practiced by some animal species, allowing the re-capture of nitrogen from bacterial growth in the hindgut. Additionally, the B vitamins biotin, folic acid, and pantothenate, as well as vitamin K are synthesized by resident bacteria. Aside from these more well-known functions, intestinal microflora also participate in many important physiologic interactions with the host.

Research with gnotobiotic or germ-free animals has elucidated numerous benefits derived from intestinal prokaryotes. Such work has indicated that short-chain fatty acids of microbial origin stimulate intestinal blood flow, epithelial proliferation, and gene expression. Germ-free animals exhibit decreased GIT development and epithelial turnover as compared to their microbe inhabited cohorts. Intestinal bacteria stimulate healthy development of gut immune function. For example, as opposed to typical lymphoid tissue, the normal state of Peyer's patch follicles includes chronic germinal centers; the result of constant stimulation by enteric antigens [40]. As a member of the common mucosal immune system, a fully functional gut has implications for immunity in other parts of the body. Bacteria are also known to achieve cell-cell communication both between and across species and thus regulate not only the activity of the colony but favorably influence reactions from the host [41]. A recent discourse by Sperandino [42] reviewed the subject of bacterial cross-talk with the endocrine and neuroendocrine systems. Bacterial growth is stimulated by mammalian hormones and cytokines [43]. Blalock [44] has referred to the immune system as the 6th sense. Could it be that resident microbes are more than just opportunistic bystanders in this process? Constructive conjunctive symbionts, perhaps?

One way that the balance between distress and eustress can be manifested is through the population of GIT microbes. As previously discussed, "good" bacteria are key players in a healthy gut; they deny space to potentially harmful species, digest structural carbohydrates, and provide essential vitamins. An imbalance in the composition of the microbial species in the GIT is implicated in such maladies as irritable bowel syndrome [22]. Stressed animals may exhibit marked differences in diet selection and intake, passage rate, digestion, and absorption. Some bacteria are substrate specific, others are generalists. Changes in any one of the previous factors could result in hindgut substrates very different from that found in non-stressed animals, and accordingly, very different microbial populations.

In his review, Tannock [45] cites numerous monogastric animal studies in which various stressors induce changes to normal microbiota. Tannock [45] also states that two bacterial populations are particularly sensitive to diet-induced stress: *lactobacilli* usually decrease under stressful conditions while *Escherichia coli* usually increase. Maternal separation in rhesus monkeys resulted in lowered fecal *lactobacilli* and increased susceptibility to infection [46]. Pigs subjected to early weaning, acute cold, or mixing with non-littermates exhibited increased fecal shedding of *E. coli* compared to controls [47]. Long-term transport in cattle has also been reported to increase fecal shedding of *E. coli.* [48]. Similarly, transport-induced stress resulted in increased shedding of *Campylobacter spp.* in broiler feces [49]. Bacteria constitute 30 to 50% of fecal material in ruminants and up to 90% in poultry; monitoring changes in the fecal population of bacteria in response to stress could provide a method to quantify stress in free-ranging birds and animals.

Fecal Near Infrared Spectroscopy

Near infrared reflectance spectroscopy (NIRS) involves the detection of light in the near infrared (NIR) band (~800 to 2500 nm) that has been reflected by a substance of interest. When irradiated by the light, organic bonds, primarily CH, OH, and NH, begin to vibrate at characteristic frequencies corresponding to those found within the NIR band. Light at wavelengths of similar frequencies are absorbed by the bond, dissimilar wavelengths are reflected. This occurrence results in a biochemical "snapshot" of the material, via summary of the relative population of bonds present and their detected absorbance and reflection of NIR light. The use of NIRS is widespread in agriculture and natural resource management and one specific application, fecal NIRS, has particular utility for investigations into the nutritional ecology of herbivores [50, 51]. Fecal analysis has obvious advantages as a non-invasive means of obtaining information from free-ranging animals. Samples are relatively easy to obtain and this method allows the animals to go about their daily routines and not be subjected to movement, gathering, and handling. The dry matter in feces is largely bacteria, undigested food, sloughed endothelial cells, digestive secretions and metabolized endocrine products [23]. Thus, feces are a repository of physiological information, fraught with opportunity for scientific exploitation. Quantifying nutritional parameters seems a logical primary application of the NIRS technology and indeed, diet quality in several species of livestock and wildlife including cattle [52-55], goats [56], sheep [57, 58], white-tailed deer [59], and elk [60, 61] has been determined. Coates [52] and Tolleson et al., [62] have employed fecal NIRS to predict dry matter intake in ruminants. Dietary tannin content [63] as well as intake of pine needle [*Pinus ponderosa,* 64], leafy spurge [*Euphorbia esula*, 65] and mountain big sagebrush [*Artemisia tridentata*, 66] have been determined with this technique. The use of NIRS allows the researcher to bypass traditional chemistry once a calibration equation is established, thus NIRS is a faster, less expensive technique in the long run.

In addition to diet quality, fecal NIRS has been used to discriminate between fecal samples according to physiological characteristics such as sex and species in cervids [67] or sex and reproductive status in cattle [68-70] and sheep [71]. Discrimination between samples from pen-fed animals with and without an external parasite burden has been reported [72], and Teel et al. [73] have similarly used NIRS of feces to distinguish between samples from grazing cattle with divergent parasite loads. The mechanisms by which physico-chemical differences induced by stress or physiological status are expressed in feces, is not yet known. Stress-mediated changes in digestion, absorption, immunity, and microbiota all seem possible. Monitoring any or all of these via blood or feces would fall under the aforementioned field of bio-forensics.

One hypothesis for the divergent fecal NIR spectra noted above is that tick-induced stress causes changes in the endocrine/immunologic/metabolic homeostasis of the animal which in turn affects host/microbial relationships in the hindgut. Stress responses in the animal could alter intake, diet selection, passage rate and absorption. Additionally, the gut-associated innate immune system could be altered by endocrine changes due to stress. Soderholm et al. [74] reported increased gut inflammation in chronic stressed versus non-stressed rats. Additionally, degradation of the brush border membrane occurred in surgically stressed rats [75]. The responses listed here all have an endocrine or immune system connection and in fact, direct communication between host and bacteria has been shown to occur through stress related endocrine factors [42]. Tick related stress should also result in disrupted homeostasis and thus alter the host/bacterial ecology. Research conducted in free-ranging animals is needed to better understand and quantify the effects of tick-induced stress on microbial ecology of the hindgut.

Tick Attachment and Feeding

The tick life cycle consists of four stages: 1) egg 2) larvae 3) nymph 4) adult. A single blood meal is taken during each of the latter three stages. The majority of ixodid ("hard tick") species attach and blood-feed on a separate host for each of these three stages with an intervening period off-host for molting or egg laying. These are known as three-host ticks. Some tick species exhibit a two-host life cycle in which fed-larvae remain attached and molt on the animal, then feed as nymphs before dropping off to molt to adults. Other tick species complete all three blood-feedings and associated molting and mating on a single host, known as the one-host life cycle. In most species, mating of adults takes place on the host. Oviposition by the female occurs after detachment, and the cycle starts again. It is obviously the attachment and feeding by ticks that have detrimental effects on the host.

Attachment is initiated by insertion of the hypostome. This appendage has many hook-like projections which inhibit removal of the attached tick. Additionally, ixodid ticks secrete a cement-like substance produced by the salivary glands which helps to hold the feeding tick in place and prevent leakage of the blood meal. This cement consists primarily of proteins and glycoproteins but also contains lipids [76]. After attachment, formation of a feeding lesion occurs. Small vessels rupture and drain blood into the lesion. Normal response to such an injury, namely platelet formation, plasma coagulation cascade, and vasoconstriction, would suffice to stop the hemorrhage. Tick saliva, however, contains anti-haemostatic factors affecting all three processes. Platelet aggregation inhibitors found in tick saliva include apyrase [77], moubatin [78], disagregin [79], variabilin [80], and prostaglandins [81]. Anticoagulation factors in tick saliva operate through inhibition of either serine protease factor Xa, or thrombin [82]. The most well-known anticoagulant present in tick saliva is tick anticoagulant peptide (TAP) [82]. Anticoagulation serves purposes in tick feeding other than the obvious role of maintaining blood flow through the feeding lesion; it also keeps blood circulating through mouthparts and gut tissues. Ticks concentrate the blood meal and excrete approximately 33 to 50% of the fluid intake back into the host [82] through the salivary glands. Vasodilation also serves to promote blood flow at the feeding site. Prostaglandins, in addition to inhibition of platelet aggregation, are potent vasodilators. When prostaglandin synthesis is impaired, engorged tick weights can be reduced up to 30% [83].

Tick-Induced Immuno-Suppression

In addition to anti-haemostatic factors, tick saliva also contains many other pharmacologically active compounds which suppress inflammation and the immune response, both acquired and innate [84]. All ixodid species possess immunoglobulin-binding proteins [85]. As the name suggests, these proteins bind to host antibodies, preventing the subsequent binding to epitopes within the tick. Antibodies are then passed through the tick digestive system and returned to the host. Extracts from tick salivary glands have been demonstrated to suppress human natural killer cell [86] and mouse interferon α [87] activity. Decreased natural killer cell function could also lead to reduced production of interferon, a component of the cytokine cascade involved in directing the T helper cell response [88]. Immunomodulation by the tick is obviously self-serving, but it is also the mechanism which facilitates transmission of tick-borne diseases.

Saliva activated transmission (SAT) factors [89] are secreted into the skin lesion during feeding. Though SATs are produced by both infected and non-infected ticks [84], and function to affect the reaction of skin in the immediate area of the feeding lesion, they also facilitate pathogen transmission to the host and between ticks. For instance, a virus may enter the skin at the feeding lesion and go undetected due to the aforementioned effects of tick saliva constituents. At the feeding sites in the skin, the virus may infect Langerhans cells and be carried to a lymph node where naive lymphocytes are activated [90] and return to the skin, attracted to tick affected areas. At these locations, uninfected ticks are exposed to the virus. Vector mediated effects on pathogen transmission and immunomodulation by ectoparasites would also be expected to affect overall immune function in the host. If these effects can be adequately characterized and quantified, they could be utilized in the development of bioassays to determine the effects of ticks and or other stressors on livestock.

The Effect of Ticks on Indicators of Stress and Metabolism

The animal performance, endocrine, immune, and metabolic responses of cattle with and without an *Ambyomma americanum* tick burden has been recently examined [91]. Although tick treatment of growing cattle fed moderate level of nutrition did not significantly affect the overall performance or feed efficiency during the study, the tick-treated animals did exhibit more outward signs of stress, i.e. head tossing, vocalization and grooming, as compared to the non-treated animals. Platelet counts and blood protein concentration were higher in tick-treated versus control subjects. Monocyte counts were also higher in tick-treated subjects compared to controls on day 10 of the study. There was a tendency for some animal performance indicators to be affected by tick stress. For instance, average daily gain was numerically lowest between days 14 and 21, just after the intensive tick feeding and engorgement period. Additionally, within the tick treatment group, the cumulative number of replete female ticks at day 14 was negatively correlated with body weight gain between days 14 and 21. There was also a 12 percentage point reduction in mean proportional dry matter intake for tick treated cattle from day 9 to day 13 of the *A. americanum* feeding cycle.

Tick treatment in this same study resulted in numerically lower liver growth hormone mRNA as compared to control animals. Overall, liver IGF1 gene expression was lower in tick-treated than control animals and a treatment by day interaction for liver IGF1 gene expression such that IGF1 mRNA was reduced in tick-treated versus control cattle 35 days after initial exposure to ticks. It was interesting to note that the effects of tick treatment on the liver somatotropic axis occurred well past the acute effects of an external parasite burden, indicating a carryover effect that could be cumulative with prolonged tick infestation. It should also be noted that these cattle were relatively well-fed in a controlled environment. The effects of a tick burden on host metabolism could thus be expressed for much longer than currently thought and would likely be exacerbated by environmental factors such as extreme temperatures or a poor nutritional environment. Warmer drier conditions lead to lower nutritional environments for grazing animals [92]. Pre-weaned beef calves infested with *A. americanum* gained less than controls and this difference in performance was more pronounced during a drought year [93]. Within the natural range of *A. americanum* (south east and south central US), cattle exposure to ticks often occurs coincident to summer droughts.

Similar results were observed by Australian workers [94, 95] in cattle exposed to *Boophilus microplus,* for which both direct effects on cattle metabolic and immune systems (e.g. lower packed cell volume and increased gamma-globulin) and "anorectic' effects (e.g. lower intake and weight gain) were reported. Dairy cows infested with this same tick species produced less milk with less butterfat than did control cows [96]. In contrast, Willis et al. [97] found elevated heart rates in growing beef steers infested with *A. americanum* but no effects on intake or any indicators of metabolism in blood. Differences in tick species, host sex, age, and nutritional status interact to determine specific effects of external parasites on grazing animals.

CONCLUSION

External parasite burdens create a "cost of fitness" [98] for the grazing animal. This cost can be manifested through depressed appetite or greater investment in production and maintenance of metabolic and immune molecules, cells and tissues. Any combination of these will impact available energy resources. Behavioral manifestations of stress may also contribute to a state of protein-energy malnutrition. There are metabolic, endocrine and immune consequences with respect to parasitism which may not always be detectable by current methods used to assess animal performance measures, but which could lead to lowered subsequent performance in animals stressed by tick infestations.

REFERENCES

[1] Selye, H. (1936). A syndrome produced by diverse nocuous agents. (Lond) *Nature 138*:32.

[2] Sapolsky, R. (2003). Taming stress. *Sci. Am. 289*:86-95.

[3] Ader, R., Felton, D. L., and Cohen, N. (1991). *Physchoneuroimmunology*. Academic Press, San Diego CA. Pp. 661.

[4] Sheridan, J. F., Dobbs, C., Brown, D., and Zwilling, B. (1994). Psychoneuroimmunology: stress effects on pathogenesis and immunity during infection. *Clin. Microbiol. Rev. 7*(2):200-212.

[5] Werb, Z., Foley, R., and Munck, A. (1978). Interaction of glucocorticoids with macrophages: identification of glucocorticoid receptors in monocytes and macrophages. *J. Exp. Med. 147*:1684-1694.

[6] Zwilling, B., Brown, S. D., and Pearl, D. (1992). Induction of major histocompatibility complex class II glycoproteins by interferon-gamma: attenuation of the effects of restraint stress. *J. Neuroimmunol. 37*(1/2):115-122.

[7] Beutler, B., Krochin, N., Milsark, I. W., Luedke, C., and Cerami, A. (1986). Control of cachectin (tumor necrosis factor) synthesis: mechanisms of endotoxin resistance. *Science. 23*(4753):977-980.

[8] Brown-Borg, H. M, Klemcke, H. G., and Blecha, F. (1993). Lymphocyte proliferative responses in neonatal pigs with high or low plasma cortisol concentration after stress induced by restraint. *Am. J. Vet. Res. 54*(12):2015-2020.

[9] Merchav, S., Tatarsky, I., and Hochberg, Z. (1988). Enhancement of human granulopoiesis *in vitro* by biosynthetic insulin-like growth factor-1/somatomedin C and human growth hormone. *J. Clin. Invest. 81*(3):791-797.

[10] Berschorner, W. E., Divic, J., Pulido, H., Yao, X., Kenworthy, P., and Bruce, G. (1991). Enhancement of thymic recovery after cyclosporine recombinant human growth hormone insulin-like growth factor-I. *Transplantation. 52*:879-884.

[11] Ferrucci, L., and Guralnik, J. M. (2003). Inflammation, hormones, and body composition at a crossroad. *Am. J. Medicine. 115*(6):501-502.

[12] Wolf, M., Hansen, N., and Greten, H. (1994). Interleukin 1 beta, tumor necrosis factor-alpha and interleukin 6 decrease nuclear thyroid hormone receptor capacity in a liver cell line. *Eur. J. Endocrinol.* 131(3):307–312.

[13] Venkatraman, J. T., and Pendergast, D. R. (2002). Effect of dietary intake on immune function in athletes. *Sports Med. 32*(5): 323-337.

[14] Jasnow, A. M, Drazen, D. L., Huhman, K. L., Nelson, R. J., and Demas, G. E. (2001). Acute and chronic social defeat suppresses humoral immunity of male Syrian hamsters (Mesocricetus auratus). *Horm. Behav. 40*(3):428-433.

[15] Matalka, K. Z. (2003). Neuroendocrine and cytokines-induced responses to minutes, hours, and days of mental stress. Neuro. Endocrinol. Lett. *24*(5):283-292.

[16] Mashaly, M. M., Hendricks 3rd, G. L., Kalama, M A., Gehad, A. E., Abbas, A. O., and Patterson, P. H. (2004). Effect of heat stress on production parameters and immune responses of commercial laying hens. *Poult. Sci. 83*(6):889-894.
[17] Hangalapura, B. N., Nieuwland, M G., Buyse, J., Kemp, B., and Parmentier, H. K. (2004). Effect of duration of cold stress on plasma adrenal and thyroid hormone levels and immune responses in chicken lines divergently selected for antibody responses. *Poult.Sci. 83*(10):1644-1649.
[18] Saker, K. E., Allen, V. G., Kalnitsky, J., Thatcher, C. D., Swecker Jr,, W. S., and Fontenot, J. P. (1998). Monocyte immune cell response and copper status in beef steers that grazed endophyte-infected tall fescue. *J. Anim. Sci. 76*(10):2694-2700.
[19] Rice, R. L., Blodgett, D. J., Schurig, G. G., Swecker, W. S., Fontenot, J. P., Allen, V. G., and Akers, R. M. (1997). Evaluation of humoral immune responses in cattle grazing endophyte-infected or endophyte-free fescue. *Vet. Immunol. Immunopathol. 59*(3/4):285-291.
[20] Rook, G. A. (1999). Glucocorticoids and immune function. Baillieres Best Pract. Res. *Clin. Endocrinol. Metab. 13*(4):567-581.
[21] Pond, W. G., Church, D. C., and Pond, K. R. (1995). *Basic Animal Nutrition and Feeding*, 4th ed. John Wiley and Sons, New York, NY.
[22] Guyton, A. C., and J. E. Hall, J. E. (2000). *Textbook of Medical Physiology*, 10th ed. W.B. Saunders Company, Philadelphia, PA.
[23] Church, D. C. (1979). *Digestive Physiology and Nutrition of Ruminants,* 2nd. Ed. Oxford Press, Portland, OR.
[24] Van Soest, P. J. (1983). *Nutritional Ecology of the Ruminant*, 2nd ed. Your Town Press, Salem, OR.
[25] Nicoletti, C. (2000). Unsolved mysteries of intestinal M cells. *Gut. 47*(5):735-739.
[26] Kiyono, H., and Fukuyama, S. (2004). NALT- versus Peyer's-patch-mediated mucosal immunity. *Nature Rev. Immun. 4*(9):699-710.
[27] Kraehenbuhl, J. P., and Neutra, M. R. (2000). Epithelial M cells: differentiation and function. *Annu. Rev. Cell Dev. Biol. 16*(1):301-332.
[28] Neutra, M. R., Pringault, E., and Kraehenbuhl, J. P. (1996). Antigen sampling across epithelial barriers and induction of mucosal immune responses. *Annu. Rev. Immun. 14*(1):275-300.
[29] Alpan, O., Rudomen, G., and Matzinger, P. (2001). The role of dendritic cells, B cells, and M cells in gut-oriented immune responses. *J. Immunol. 166*(8):4843-4852.
[30] Parrott, D. M. (1976). The gut as a lymphoid organ. *Clin. Gastroenterol. 5*(2):211-228.
[31] Stagg, A. J., Hart, A. L., Knight, S. C., and Kamm, M. A. (2003). The dendritic cell: its role in intestinal inflammation and relationship with gut bacteria. *Gut. 52*(10):1522-1529.
[32] Palma, J. P., Yauch, R. L., Kang, H. K., Lee, H. G., and Kim, B. S. (2002). Preferential induction of IL-10 in APC correlates with a switch from Th1 to Th2 response following infection with a low pathogenic variant of Theiler's virus. *J. Immunology. 168*(8):4221-4230.
[33] Beutler, B. (2004). Inferences, questions and possibilities in Toll-like receptor signaling. *Nature. 430*(6996):257-263.
[34] Bilsborough, J., and Viney, J. L. (2004). Gastrointestinal dendritic cells play a role in immunity, tolerance, and disease. *Gastroenterology. 127*(1):300-309.

[35] Zhang G., and Ghosh, S. (2001). Toll-like receptor-mediated NF-kappa B activation: a phylogenetically conserved paradigm in innate immunity. *J. Clin. Invest. 107*(1):13-19.
[36] Abbas, A. K., and Lichtman, A. H. (2003). *Cellular and Molecular Immunology*, 5th ed. W.B. Saunders Company, Philadelphia.
[37] Turnbaugh, P. J., Ley, R. E., Hamady, M., Fraser-Ligget, C. M., Knight, R., and Gordon, J. I. (2007). The Human Microbiome Project. *Nature. 449*(7164):804.
[38] Savage, D. C. (1986). Gastrointestinal microflora in mammalian nutrition. *Annu. Rev.Nutr. 6*:155-178.
[39] Roediger, W. E. (1980). The colonic epithelium in ulcerative colitis: an energy-deficiency disease? *Lancet. 8197*:712-715.
[40] Cebra, J. J. (1999). Influences of microbiota on intestinal immune system development. *Am. J. Clin. Nutr. 69*(5):1046S-1051S.
[41] Henke, J. M., and Bassler, B. L. (2004). Bacterial social engagements. *Trends Cell Biol. 14*(11):648-656.
[42] Sperandio V, Torres, A. G., Jarvis, B., Nataro, J. P., and Kaper, J. B. (2003). Bacteria-host communication: the language of hormones. *Proc. Natl. Acad. Sci. USA. 100*(15):8951-8956.
[43] Burton, C. L., Chhabra, S. R., Swift, S., Baldwin, T. J., Withers, H., Hill, S. J., and Williams, P. (2002). The growth response of Escherichia coli to neurotransmitters and related catecholamine drugs requires a functional enterobactin biosynthesis and uptake system.*Infect. Immun. 70*(11):5913-5923.
[44] Blalock, J. E. (2005). The immune system as the sixth sense. *J. Intern. Med. 257*(2):126-138.
[45] Tannock, G.W. (1996). Modification of the normal microbiota by diet, stress, antimicrobial agents, and probiotics. *In: Gastrointestinal Microbiology, Vol 2: Gastrointestinal microbes and host interactions. Mackie, R.I, B.A. White, and R.E. Isaacson, (eds). Chapman and Hall Microbiology Series. International Thomson Publishing, New York.* Pp 434-465.
[46] Bailey, M.T., and Coe, C. L. (1999). Maternal separation disrupts the integrity of the intestinal microflora in infant rhesus monkeys. *Dev. Psychobiol. 35*(2):146-155.
[47] Jones, P. H., Roe, J. M., and Miller, B. G. (2001). Effects of stressors on immune parameters and on the fecal shedding of enterotoxigenic Escherichia coli in piglets following experimental inoculation. *Res. Vet. Sci. 70*(1):9-17.
[48] Bach, S. J., McAllister, T. A., Mears, G. J., and Schwartzkopf-Genswein, K. S. (2004). Long-haul transport and lack of preconditioning increases fecal shedding of Escherichia coli and Escherichia coli O157:H7 by calves. *J. Food Prot. 67*(4):672-678.
[49] Whyte, P, Collins, J. D., McGill, K., Monahan, C., and O'Mahony, H. (2001). The effect of transportation stress on excretion rates of campylobacters in market-age broilers. *Poult. Sci. 80*(6):817-820.
[50] Foley, W. J., McIlwee, A., Lawler, I. R., Aragones, L., Woolnough, A., and Berding, N. (1998). Ecological applications of near infrared spectroscopy: a tool for rapid, cost effective prediction of the composition of plant tissues and aspects of animal performance. *Oecologia. 116*(3):293 305.
[51] Stuth, J. W., Freer, M., Dove, H., and Lyons, R. K. (1999). Nutritional management for free ranging livestock. *In: H.J.G. Jung and G. C. Fahey, Jr eds. Nutritional Ecology of*

Herbivores. Proc. Vth International Symposium on Nutrition of Herbivores. ASAC, Denver, CO. Pp. 696-750.

[52] Coates, D. B. (1998). Predicting diet digestibility and crude protein content from the faeces of grazing cattle. *CSIRO, Final Report, Project CS.253.*

[53] Lyons, R. K., and Stuth, J. W. (1992). Fecal NIRS equations for predicting diet quality of free ranging cattle. *J. Range Manag. 45*(3):238 244.

[54] Purnomoadi, A., Mitsunori, K., Takehiro, N., Furminori, T., and Akira, A. (1998). Prediction of feed digestibility using differences in NIRS spectra between feeds and feces at a determined region of wavelength. *Anim. Sci. Tech. 69*(3):253 259.

[55] Ksiksi, T., Ash, A., Fry, P., and Allen, P. (2000). NIRS predicts seasonal changes in diet quality of cattle. *Trop. Grasslands. 34*:306.

[56] Leite, E. R., and Stuth, J. W. (1995). Fecal NIRS equations to assess diet quality of free-ranging goats. *Small Rumin. Res. 15*(3):223-230.

[57] Krachounov, I., Paul, C., and Kirilov, A. (2000). Application of near infrared reflectance spectroscopy (NIRS) in the analysis of faeces from sheep for estimation of foragc availability. *Zhivotnov'Dni Nauki. 37*:22 30.

[58] Li H., Tolleson, D., Stuth, J., Bai, K., Mo, F., and Kronberg, S. (2006). Fecal near infrared reflectance spectroscopy to predict dietary protein and digestibility of sheep. *Small Rumin. Res. 68*(1/2):263-268.

[59] Showers, S. E., Tolleson, D. R., Stuth, J. W., Kroll, J. C., and Koerth, B. H. (2006). Predicting diet quality of white-tailed deer via NIRS fecal profiling. *Rangeland Ecol. Manag. 59*(3):300-307.

[60] Brooks, J. M., Anderson, M., and Urness, P.J. (1984). Infrared reflectance analysis of forage quality for elk. *J. Wildlife Manag. 48*(1):254 258.

[61] Keating, M. S. (2004). Prediction of diet quality parameters of rocky mountain elk via near infrared reflectance spectroscopy fecal profiling. *Ph.D. Diss. Texas A&M University, College Station.*

[62] Tolleson, D., Stuth, J., Keating, S., and Kronberg, S. (2002). Dry matter intake of cattle, sheep, and elk determined by near infrared reflectance spectroscopy of feces. *J. Anim. Sci. 80*(Suppl. I):8 (Abstr.).

[63] Tolleson, D. R., Osborn, R. G., Stuth, J. W., Ginnett, T. F., and Applegath, M. T. (2000). Determination of dietary tannin concentration in white-tailed deer via near infrared reflectance spectroscopy of feces. *Natl. Conf. on Grazinglands. 1*:727-733.

[64] Kronberg, S. L., Walker, J. W., and Short, R. E. (1998). Predicting Ponderosa pine needle intake with NIRS and fecal samples. *Proc. West. Sec. Am. Soc. Anim. Sci. 49*:84-85.

[65] Walker, J. W., Clark, D. H., and McCoy, S. D. (1998). Fecal NIRS for predicting percent leafy spurge in diets. *J. Range Manag. 51*(4):450-455.

[66] Walker, J. W., McCoy, S. D., Launchbaugh, K. L. and Fraker, M. J. (2000). Near infrared spectroscopy of sheep feces for predicting botanical composition of diets. pp 81-94 *In: Texas Agriculture Experiment Station, Sheep and Goat, Wool and Mohair Consolidated Progress Report.*

[67] Tolleson, D.R., Randel, R.D., Stuth, J.W., and Neuendorff, D.A. (2005). Determination of sex and species in red and fallow deer by near infrared reflectance spectroscopy of the faeces. *Small Ruminant Research 57*(2/3):141-150.

[68] Tolleson, D. R., Wilson, T. W., Randel, R. D., Neuendorff, D. A., Lewis, A. W. and Stuth, J. W. (2000). Discrimination between physiologically different groups of cattle via near infrared reflectance spectroscopy of feces. *J. Anim. Sci. 78*(Suppl. I):14 (Abstr.).

[69] Tolleson, D. R., Randel, R. D., Stuth, J. W., Willard, S. T. and Gandy, B. S. (2001). Detection of pregnancy in cattle using near infrared reflectance spectroscopy of feces. *J. Anim. Sci. 79*(Suppl. I):21 (Abstr.).

[70] Tolleson, D. R., Willard, S. T., Gandy, B. S. and Stuth, J. W. (2001). Determination of reproductive status in dairy cattle using near infrared reflectance spectroscopy of feces. *J. Anim. Sci. 79*(Suppl. I):21 (Abstr.).

[71] Godfrey, R. W., Dodson, R. E., Bultman, J. K., D. R. Tolleson, D. R., Stuth, J. W., and Norman, A. J. (2001). Use of near infrared reflectance spectroscopy to differentiate pregnancy status and gender of hair sheep in the tropics. *J. Anim. Sci.79* (Suppl. I):26 (Abstr.).

[72] Tolleson, D. R., Teel, P. D., Stuth, J. W., Strey, O. F., Welsh, Jr, T. H., and Carstens, G. E. (2007). Fecal NIRS: Detection of tick infestations in cattle and horses. *Vet Parasitol. 144*(1/2):146-152.

[73] Teel, P. D., Strey, O. F., Wright, R. E., Coburn, L. A., Tolleson, D. R., and Stuth, J. W. (2004). Discrimination between two levels of indigenous external parasite burden in free-ranging beef cattle via near infrared reflectance spectroscopy of feces. *J. Anim. Sci. 82*(Suppl. II):22 (Abstr.).

[74] Soderholm, J. D., Yang, P. C., Ceponis, P., Vohra, A., Riddell, R., Sherman, P. M., and Perdue, M. H. (2002). Chronic stress induces mast cell-dependent bacterial adherence and initiates mucosal inflammation in rat intestine. *Gastroenterology 123*(4):1099-1108.

[75] Prabhu, R., Anup, R., and Balasubramanian, K. A. (2000). Surgical stress induces phospholipid degradation in the intestinal brush border membrane. *J. Surgical Res. 94*(2):178-184.

[76] Kemp, D. H., Stone, B. F., and Binnington, K. C. (1982). Tick attachment and feeding: role of the mouthparts, feeding apparatus, salivary gland secretions and the host response. I*n: Physiology of Ticks. (F.D Obenchain and R. Galun eds.) Pergamon Press, Oxford.* Pp. 119-168.

[77] Ribeiro, J. M. C. (1987). Role of saliva in blood-feeding by arthropods. *Ann. Rev. Ento. 32*:463-478.

[78] Waxman, L., and Connally, T. M. (1993). Isolation of an inhibitor selective for collagen-stimulated platelet aggregation from the soft tick Ornithodoros moubata. *J. Biol. Chem. 268*(8):5445-5449.

[79] Karczewski, J., Endris, R. G., and Connolly, T. M. (1994). Disagregin is a fibrinogen receptor antagonist lacking the Arg-Gly-Asp sequence from the tick Ornithodoros moubata. *J. Biol. Chem. 269*(9):6702-6708.

[80] Wang, X., Coons, L. B., Taylor, D. B., Stevens Jr., S. E., and Gartner, T. K. (1996). Variabilin, a novel RGD-containing antagonist of GPIIb-Iia and platelet inhibitor from the hard tick Dermacentor variabilis. *J. Biol. Chem. 271*(30)*:*17785-17790.

[81] Bowman, A. S., Dillwith, J. W., and Sauer, J. R. (1996). Tick salivary prostaglandins: presence, origin and significance. *Parasitology Today. 12*(10):388-396.

[82] Bowman, A. S., Gengler, C. L., Surdick, M. R., Zhu, K., Essenberg, R. C., Sauer, J. R., and Dillwith, J. W. (1997). Biosynthesis of salivary prostaglandins in the lone star tick Ambylomma americanum. *Insect Biochem. Mol. Biol. 25*(6):735-741.

[83] Madden, R. D., Sauer, J. R., Dillwith, J. W., and Bowman, A. S. (1996). Dietary modification of host blood lipids affect reproduction in the lone star tick Ambylomma americanum. *J. Parasitology*. *82*(2):203-209.

[84] Nuttall, P. A. (1998). Displaced tick-parasite interactions at the host interface. *Parasitology.116*(Suppl.1):S65-S72.

[85] Wang, H., and Nuttall, P. A. (1994). Excretion of host immunoglobulin in tick saliva and detection of IgG-binding proteins in tick haemolymph and salivary glands. *J. Parasitology*. *109:*525-530.

[86] Kubes, M, Fuchsberger, N., Labuda, M., Zuffova, E., and Nuttal, P. A. (1994). Salivary gland extracts of partially fed Dermacentor reticulatis ticks decrease natural killer cell activity *in vitro*. *Immunology*. *82*(1):113-116.

[87] Hajnicka, V., Fuchsberger, N., Slovac, M., Kocakova, P., Labuda, M., and Nuttal, P. A. (1998). Tick salivary gland extracts promote virus growth *in vitro*. *Parasitology*. *116*(6):533-538.

[88] Wikel, S. K., and Bergman, D. (1997). Tick-host immunology: significant advances and challenging opportunities. *Parasitol Today*. *13(10)*:383-389.

[89] Nuttall, P. A., and Jones, L. D. (1991). Non-viraemic tick-borne virus transmission: mechanism and significance. *In: Modern Acarology. (Dusdabek F. and V. Bukva eds.). Prague and the Hague: Academia and SPB Academic Publishing*. Pp. 3-6.

[90] Austyn, J. M. (1992). Antigen uptake and presentation by dendritic leukocytes. *Semin Immunol*. *4*:227-236.

[91] Tolleson, D. R., Teel, P. D., Stuth, J. W., Strey, O. F., Welsh Jr, T. H., Carstens, G. E., Banik, K. K., and Prince, S. D. (2010). Effects of a Lone Star Tick (Amblyomma americanum) burden on performance and metabolic indicators in growing beef steers. *Vet Parasitol*. *173*(1/2):99-106.

[92] Craine, J. M., Elmore, A. J., Olson, K. C., and Tolleson, D. (2010). Climate change and cattle nutritional stress. *Global Change Biol*. *16*(10): 2901.

[93] Barnard, D. R. (1985). Injury thresholds and production loss functions for the lone star tick, *Amblyomma americanum (Acari: Ixodidae)* on pastured preweaner beef cattle, *Bos taurus*. *J. Econ. Entomol*. *78*(4): 852-855.

[94] Seebeck, R. M., O'Kelley, J. C., and Springell, P. H. (1971). Alterations in host metabolism by specific and anorectic effects of cattle tick *(Boophilus microplus)*. 1. Food intake and body weight growth. *Aust. J. Biol. Sci*. *24*(2): 373-380.

[95] O' Kelley, J. C., and Kennedy, P. M. (1981). Metabolic changes in cattle due to the specific effect of the tick, *Boophilus microplus*. *Br. J. Nutr*. *45(3)*: 557-566.

[96] Jonsson, N. N., Mayer, D. G., Matschoss, A. L., Green, P. E., and Asnell, J. (1998). Production effects of cattle tick (*Boophilus microplus*) infestation of high yielding dairy cows *Vet Parasitol*. *78*(1): 65-77.

[97] D. C. [97] Willis, D. C., Knapp, F. W., Boling, J. A., and Burg, J. G. (1995). Physiological and nutritional responses of steers infested with varying densities of *Amblyomma americanum* (*Acari: Ixodidae*) *J. Econ. Entomol*. *88*(6): 1679-1683.

[98] Demas, G. E. (2004). The energetics of immunity: A neuroendocrine link between energy balance and immune function. *Horm Behav*. *43*: 75-80.

In: Ticks: Disease, Management and Control
Editor: Moges Woldemeskel
ISBN: 978-1-62081-136-8

Chapter 2

TICK SALIVA-MEDIATED IMMUNOMODULATION OF THE VERTEBRATE HOST

***Helena Langhansová*[1], *Andrezza C. Chagas*[2], *John F. Andersen*[2], *Jan Kopecký*[1] *and Michalis Kotsyfakis*[1,2]**
[1]Biology Centre of the Academy of Sciences of the Czech Republic, Institute of Parasitology, Branisovska 31, Ceske Budejovice, Czech Republic
[2]National Institutes of Health, National Institute of Allergy and Infectious Diseases, Laboratory of Malaria and Vector Research, Rockville, Maryland, US

ABSTRACT

The feeding period of hard ticks often exceeds one week and makes tick infestation a considerable intervention to host physiology. A broad range of inflammatory and immune reactions take place locally at the feeding site as well as systemically within the host body. Active modulation of the host immune response by tick saliva is thus required for the tick to complete its blood meal. Upon attachment, the tick inoculates its saliva containing a repertoire of bioactive molecules that help ticks to attach to the host, to overcome host hemostasis, and to prevent host pain and itching and the concomitant scratching. Furthermore, the tick must deal with the host innate immunity during primary infestation and with innate and adaptive immunity in subsequent infestations. The immunomodulatory properties of tick saliva and the roles of individual salivary components at the tick-host interface are described in this chapter, dealing with virtually all levels of nonspecific and specific host immune mechanisms. In addition, tick-borne pathogens, including the Lyme disease spirochete *Borrelia burgdorferi*, exploit the pharmacologic activities of tick salivary molecules to successfully maintain their life cycle. Identification of molecules responsible for this phenomenon, termed saliva-assisted transmission (SAT), can help in the design of an efficient transmission-blocking vaccine. Moreover, understanding of the immunity-based processes within the tick-host-pathogen interface opens a relatively new field for the development of drugs to treat various immunity-related diseases.

INTRODUCTION

Ticks are blood-sucking mites of high human health and veterinary importance because they are responsible for a significant percentage of all vector-borne diseases, with Lyme borreliosis being the most widespread tick-borne human infection in the northern hemisphere [1]. Concerning tick-borne diseases in humans, the hard ticks of the *Ixodes* genus are the most prominent vectors.

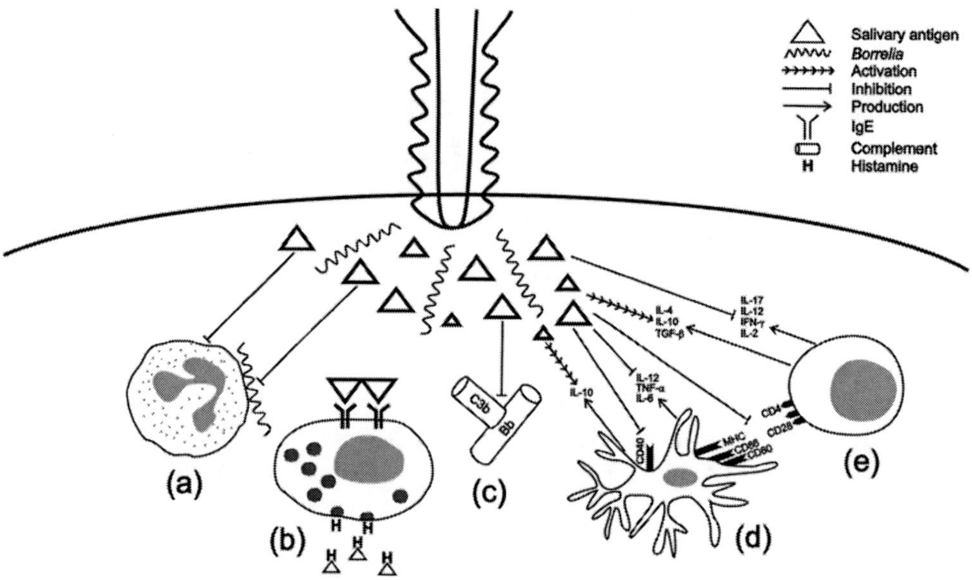

Figure 1. Schematic illustration of the effects of tick saliva on host immunity upon infection with *Borelia* spp. Salivary antigens (a) affect neutrophil influx to the tick feeding site and inhibit uptake of *Borrelia* spirochetes; (b) induce mast cell degranulation and bind the released biogenic amines such as histamine; (c) inhibit proper interactions of complement proteins; (d) down regulate co-stimulatory molecules on dendritic cell surfaces, decrease the production of proinflammatory cytokines, and increase IL-10 levels; and (e) block T cell activation and drive T lymphocytes toward Th2 immune responses.

The hard tick feeding period varies from a few days to over one week and makes tick infestation a considerable intervention to host physiology. After the tick mouthparts penetrate the host epidermis, the disruption of blood capillaries creates a feeding pool of blood. The proteinaceous content of the ingested blood is concentrated in the tick midgut, while the mostly aqueous leftover returns into the feeding pool admixed with tick saliva.

A broad range of inflammatory and immune reactions take place locally at the feeding site and also systemically within the host body, considering that the feeding period sometimes exceeds a week. Active modulation of the host immune response by tick saliva is thus required for the tick to complete its blood meal. A repertoire of bioactive molecules that help ticks to attach to the host, to overcome host hemostasis, and to prevent host pain and itching and the concomitant host scratching is present in tick salivary secretion. Thus, tick saliva contains anti-haemostatic, anti-inflammatory, and immunomodulatory molecules [2, 3]. Some representative immunomodulatory activities of tick saliva are outlined in Figure 1. Furthermore tick saliva facilitates transmission of tick-borne pathogens.

IMMUNOMODULATORY PROPERTIES OF TICK SALIVA

Several animal species including guinea pigs, rabbits, and bovines have been shown to develop resistance to tick feeding after a single or multiple tick infestations [4]. This resistance is considered to be driven by an immune reaction from the host and it is characterized by early rejection of the ticks, followed by reduced numbers and weights of engorged ticks, and finally impaired oviposition or metamorphosis to the next developmental stage. Acquired immune resistance of the host to ticks can also significantly contribute to reduced transmission of pathogens from infected ticks.

Movement of leukocytes within the body and their appropriate location in various tissues are driven by small chemotactic cytokines (chemokines). One general strategy used by ticks to evade host immune response is to sabotage the cellular communication necessary to activate immunity by neutralizing chemokines that normally recruit cells in the sites of exogenous antigens. Indeed, highly selective chemokine-binding proteins with anti-inflammatory features, termed evasins, were identified in *Rhipicephalus sanguineus* ticks [5].

Various leukocytes are involved in the tick-infestation-associated immunomodulation and the related host immune responses. Neutrophils (polymorphonuclear cells, PMN) are short-lived phagocytic granulocytes that form the major population of blood leukocytes. They are able to kill pathogens by both oxygen-dependent and oxygen-independent mechanisms, and their secretory granules contain highly cytotoxic proteins. Neutrophils are the most abundant cells in the acute inflammatory infiltrate induced upon primary tick infestation but not in subsequent infestations [6]. Saliva of ixodid ticks was previously found to inhibit neutrophil adhesion and granule release and to decrease the efficiency of PMN in uptake and killing of *Borrelia burgdorferi* [7, 8]. Two salivary proteins of *Ixodes scapularis*, ISL929 and ISL1373, have been shown to downregulate PMN β2-integrins and to inhibit production of superoxide radicals by neutrophils *in vitro*. Mice immunized with these proteins had increased numbers of PMN at the site of tick attachment and a lower spirochete burden in the skin and joints 21 days after infection compared with control-immunized animals [9] (see Table 1 for the summary of salivary immunomodulators). More recently, a novel inhibitory serpin, IRS-2 from *Ixodes ricinus* saliva, inhibited neutrophil influx to the inflamed tissue in a mouse model of acute inflammation [10]. A similar effect on the neutrophil migration has been addressed also to sialostatin L, an *I. scapularis* salivary cystatin [11]. Because neutrophils form the first line of immune defense against *Borrelia* spirochetes, failure of sufficient neutrophil recruitment and activation during the initial inflammatory response may allow spirochetes to effectively colonize the mammalian host [12].

Basophils and mast cells are key players in hypersensitivity reactions. Crosslinking of IgE molecules bound to the surface of a mast cell or a basophil via its high-affinity Fc receptors for IgE (often mediated by an allergen), triggers the rapid release of the contents of the intra-cytoplasmic granules of mast cells and basophils. THIS SHOULD BE ONE PARAGRAPH WITH THE NEXT ONEHistamine and other mediators released by this degranulation event cause the adverse symptoms of allergy but also play a role in immunity against parasites by enhancing acute inflammation. Basophils have long been documented as the predominant cell type that infiltrates the tick-bite site in the skin and plays a key role in the rejection of the tick [13]. Migration of these cells upon tick feeding and the resulting basophil-rich responses are known as cutaneous basophil hypersensitivity (CBH).

Mast cells, which are particularly concentrated beneath the epithelial surface of the skin and mucosal layers, have also been connected with the development of tick resistance. Mast cell-deficient mice failed to develop resistance to ticks, while the host anti-tick response was normalized after adoptive transfer of mast cells [14–16]. Degranulation of basophils and mast cells and the local release of mediators, including histamine and serotonin, are considered to represent an important part of the protective immune response against ticks [17]. Importantly, basophils but not mast cells, are responsible for the antibody/IgFc-mediated tick resistance, even though both types of cells are essential for the manifestation of resistance to the feeding of *Haemaphysalis longicornis* [16]. Histamine and other mediators released from mast cell granula cause pruritus and itching that can cause mechanical stress or even destruction of the tick. To prevent host scratching, histamine-binding lipocalins are present in the saliva of several soft and hard tick species [18, 19]. On the other hand, histamine-releasing factors modulate vascular permeability and increase blood flow to the feeding site, thus facilitating tick engorgement [20].

Table 1. Examples of the tick salivary molecules with immunomodulatory features

Molecule	Origin	Target cells/components	Function/mechanism	Refs
ISL929 ISL1373	*Ixodes scapularis*	Neutrophils	↓Superoxide production ↓β2-integrins	[9]
IRS-2	*Ixodes ricinus*	Neutrophils	Inhibition of neutrophil migration	[10]
HBP	*Rhipicephalus appendiculatus*	Basophils, Mast cells	Binding of histamine	[18]
HRF	*I. scapularis*	Basophils, Mast cells	Release of histamine	[20]
ISAC	*I. scapularis*	Complement	Dissociating C3 convertase	[22]
Salp20	*I. scapularis*	Complement	Dissociating C3 convertase	[23]
IRAC I/II	*I. ricinus*	Complement	Dissociating C3 convertase	[24]
IRIS	*I. ricinus*	Monocytes, Macrophages	↓ TNF-α	[29]
PGE_2	*I. scapularis* *R. sanguineus*	Dendritic cells	↓ IL-12 and TNF-α ↓ CD40, differentiation inhibitor	[34] [35]
Adenosine	*R. sanguineus*	Dendritic cells	↓ CD40	[35]
Salp15	*I. scapularis*	Dendritic cells	Raf-1/MEK activation ↓ IL-6, TNF-α, and IL-12p35	[37]
		T cells	CD4 binding ↓ T cell activation, and IL-2	[46, 47]
Sialostatin L	*I. scapularis*	Neutrophils Dendritic cells	Inhibition of neutrophil influx ↓ CD80/86 ↓ IL-12p70, TNF- α, and MHC II Ii processing	[11, 38]
IsSMase	*I. scapularis*	T cells	↑ IL-4	[48]
BIP	*I. ricinus*	B cells	Prevention of B cell activation	[49]
IgG-BP	*R. appendiculatus*	IgG	Binding of IgG	[50]

Eosinophils are specialized leukocytes whose major function is the secretion of toxic granule constituents and, thereby, extracellular killing of microorganisms. Their toxic products are also the major cause of the tissue damage seen during inflammation. Similar to mast cells and basophils, eosinophilic granulocytes are also associated with resistance to ticks, as they were detected around the tick feeding site in guinea pigs after repeated infestations [21].

Complement is a group of serum and membrane proteins involved in opsonization, chemotaxis, and osmotic lysis of a target cell. The alternative pathway of complement cascade activation is an important defense against pathogens and a key player in tick rejection reactions of the host. Several complement inhibitors have been described to be tick saliva constituents, among others Isac (*I. scapularis* anti-complement) and Salp20 [22, 23], in addition to two paralogous proteins, IRAC I and II, that protect *I. ricinus* from the complement-mediated immune response of a broad spectrum of vertebrate hosts [24].

Macrophages are important resident cells producing cytokines and chemokines that attract inflammatory cells to the tick-bite site. Activated macrophages are able to engulf bacteria including tick-borne *Borrelia* spirochetes by coiling phagocytosis and kill them by oxygen- and nitrogen-dependent mechanisms [25, 26]. Tick saliva has evolved mechanisms for regulating macrophage functions. Salivary gland extract (SGE) from *I. ricinus* impaired production of superoxide and nitric oxide, two major defense molecules of macrophages [27]. SGE reduced the proportion of active phagocytes and the number of engulfed bacteria. Treatment of macrophages with tick saliva downregulated the production of proinflammatory cytokines [28], perhaps due to the presence of the serpin Iris, which binds to monocytes/macrophages and inhibits their ability to secrete TNF-α. Notably, this action is independent of the antiprotease activity of this protein, but it is driven by exosites [29]. Salivary gland (SG) molecules of *Rhipicephalus microplus* induced differential expression of co-stimulatory molecule CD86 in murine macrophages, which may serve to promote Th2 polarization of the immune response [30].

Dendritic cells (DCs) are the most efficient cells in antigen presentation. They capture antigens in the peripheral tissue and migrate to the spleen or regional lymph nodes, where they present the processed antigen epitopes to T cells. DCs are the key initiators of T cell activation, thus connecting innate and adaptive immunity. DCs treated with ixodid saliva promote naïve $CD4^+$ T cells to Th2-type responses, and the presence of saliva impairs their maturation, migration, and overall function [31–33]. Non-protein molecules present in saliva, such as prostaglandin E2 (PGE_2) and adenosine (Ado) were reported to account for these effects [34, 35]. Ado inhibits the production of the pro-inflammatory cytokines IL-12p40 and TNF-α and stimulates the production of the anti-inflammatory cytokine IL-10 by murine DCs activated with Toll-like receptor (TLR) agonists. Additionally, PGE_2 suppresses the differentiation of DCs from blood cell precursors. Both Ado and PGE_2 exert immunomodulatory effects on cytokine production by inducing a common cAMP-PKA signaling pathway. Furthermore, both Ado and PGE2 were able to inhibit the expression of CD40 in mature DCs [35]. Tick saliva also induces regulatory DCs to secrete IL-10 and low levels of IL-12 and TNF-α when stimulated by TLR ligands. Such regulatory DCs are associated with the expression of TLR-2 and the inhibition of ERK and p38 MAP kinases, which in turn promotes the production of IL-10 and thus modulates the host immune response [36].

Salp15, a salivary protein of *I. scapularis,* interacts with the DC-SIGN lectin on DCs, activating the Raf-1/MEK pathway that leads to impaired IL-6 and TNF-α mRNA stability and nucleosome remodeling at the IL-12p35 promoter [37]. Sialostatin L, an *I. scapularis* cystatin, inhibits DC maturation by down regulating the co-stimulatory molecules CD80 and CD86 and the production of IL-12p70 and, to a lesser extent, also TNF-α, by DCs.

Furthermore, it binds to cathepsin S inside the DCs and affects processing of MHC class II invariant chain (Ii). The resulting impaired DC maturation and Ii degradation correlates with inhibition of antigen-specific $CD4^+$ T cell proliferation by a cathepsin S-dependent mechanism [38].

Interestingly, argasid salivary proteins such as the cystatin OmC2 from the soft tick *Ornithodoros moubata* can also suppress the host adaptive immune response by reducing TNF-α and IL-12 production and the proliferation of antigen-specific $CD4^+$ T cells, suggesting the presence of this mechanism in the common progenitor of soft and hard ticks [39].

Natural killer (NK) cells are lymphocytes with cytotoxic activity mainly against tumor cells and virus-infected cells. Tick salivary proteins were shown to affect the ability of NK cells to conjugate with their physiologic cell targets and, as a result, impair NK cell cytotoxicity [40, 41].

Immunomodulatory activities of tick saliva can be further demonstrated on the subsequent steps of immune activation such as the T and B cell-mediated responses. Treatment with tick SGE decreases γδ T lymphocyte load in infected skin [42]. Repeated infestations of the host with pathogen-free ticks or exposure to pure saliva itself selectively promote an anti-inflammatory Th2 cytokine profile: IL-4, IL-10, and transforming growth factor beta (TGF-β) secretions are upregulated, with a corresponding suppression of Th1 and Th17 responses (downregulation of IFN-γ, IL-2, IL-12, and IL-17) in *ex vivo* murine and human cell cultures [32, 43–45]. This polarization is partially caused by the impairment of DC function; however, direct influence of tick salivary constituents on T cells also plays some role.

Salp15 binds directly to the CD4 receptor on the surface of T cells. It inhibits early T cell activation via CD3/CD28 and MHC/CD4 by blocking T cell receptor (TCR)-mediated Lck activation and thus improper downstream signaling. The end result is a reduced production of IL-2 [46, 47]. A novel sphingomyelinase-like enzyme in *I. scapularis* saliva (IsSMase) directly programs host $CD4^+$ T cells to express IL-4, the hallmark Th2 cytokine, even if the cells are primed toward a Th1 response by viral infection. TCR transgenic $CD4^+$ T cell proliferation was also significantly increased by IsSMase [48]. Humoral immunity also can be modulated by an 18-kDa B cell inhibitory protein found in *I. ricinus* saliva [49] or by immunoglobulin G binding proteins identified in the saliva of *Rhipicephalus appendiculatus* [50].

Saliva-Assisted Transmission of Pathogens

Ixodes ricinus complex (*I. ricinus*, *I. scapularis*, *Ixodes pacificus,* and *Ixodes persulcatus*) is responsible for the transmission of at least six disease agents (*Borrelia burgdorferi* sensu lato, *Anaplasma phagocytophilum*, tick-borne encephalitis virus, Powassan virus, *Babesia microti,* and *Babesia divergens*). Tularemia and Q fever are also transmitted by *Ixodes* ticks; however, the main vectors for these pathogens are *Dermacentor* and *Amblyomma* ticks, respectively [51].

Survival and successful proliferation of tick-borne pathogens is also dependent on their ability to exploit the pharmacologic activities of tick salivary molecules. The facilitation of pathogen transmission and multiplication because of the presence of tick saliva in the transmission interface is termed saliva-assisted transmission (SAT) [52]. The SAT phenomenon has been documented for several tick-borne pathogens including Thogoto virus, tick-borne encephalitis virus, *Francisella tularensis*, and several *Borrelia* species [53–60]. Transmission of other vector-borne organisms such as *Leishmania* spp. can also be facilitated by the salivary secretion of its vector, the sand fly *Lutzomyia longipalpis* [61].

As mentioned above, salivary components of *I. ricinus* inhibit killing of spirochetes by murine macrophages *in vitro* and negatively influence their phagocytosis [27, 28]. Moreover, co-inoculation of *B. burgdorferi* in combination with SGE into naïve mice allows more efficient survival and higher infectivity of the spirochetes, as demonstrated by increased loads of spirochetes in various tissues when compared with injection of spirochetes without SGE [58, 60]. SG-derived *Borreliae* show ten times higher infectivity than those derived from the gut, indicating an important role for arthropod salivary proteins in the transmission cycle of Lyme disease [62]. Finally, the presence of saliva changes the proliferation and distribution of the spirochetes in mice, on various tissues (skin, lymph nodes, heart, and urinary bladder) and at early time points. This was demonstrated after needle inoculation of the pathogens and when comparing three different sources of salivary proteins (SGE, saliva, and co-feeding of pathogen-free nymphs in the inoculation area). Accordingly, it has been demonstrated that if spirochetes are co-inoculated with SGE/saliva, transmission of Lyme disease agents from murine host to *I. ricinus* ticks is enhanced [63].

In theory, any kind of immunomodulatory molecule present in saliva that facilitates tick feeding is a candidate SAT factor. For example, it is likely that polarization toward Th2 immune response, extensively described in the previous section of this chapter, facilitates the transmission of pathogens that would otherwise be neutralized by Th1 immunity. Not surprisingly, the suppression of IL-4 and IL-5 prior to the feeding of *B. burgdorferi*-infected ticks significantly decreased spirochetal load in the target organs of a Lyme disease-susceptible host [64].

Expression of the first molecularly characterized SAT molecule, Salp15 from *I. scapularis*, is selectively induced in *Borrelia*-infected tick SGs. Salp15 binds to the spirochetal outer surface protein C (OspC) and thus protects bacteria from antibody-mediated killing and facilitates their proliferation in target organs [65]. Moreover, the presence of Salp15 increases the levels of spirochetes in host skin, joints, and the urinary bladder. When *salp15* was silenced in ticks by RNA interference, the amounts of spirochetes transmitted to mice were significantly reduced [65]. The Salp15 homolog from *I. ricinus* protects a serum-sensitive isolate of *B. burgdorferi* against complement-mediated killing; this is because the deposition of membrane attack complex on the spirochetal surface is inhibited by Salp15 [66]. Additionally, Salp15 inhibits keratinocyte inflammation induced by *Borrelia in vitro* via down regulation of alarmins, essential mediators that mobilize and activate antigen presenting cells [67].

A cysteine protease inhibitor derived from *I. scapularis* SGs, sialostatin L2, dramatically affects the feeding success of the tick, most likely by influencing the activity of cathepsins involved in host inflammation, tissue remodeling, and angiogenesis [68]. Co-administration of *Borrelia* spirochetes with sialostatin L2 exacerbates skin infections in a murine model, as determined by spirochetal load in the skin of infected mouse 4 days after inoculation. Sialostatin L2 does not influence spirochetal growth *in vitro* and, unlike Salp15, the stimulatory effect seems to be not due to direct interaction of the protein with the spirochetes [69].

LYME DISEASE AND ANTI-TICK VACCINES

Previous efforts to develop a protective and safe vaccine against Lyme disease were based on the classical approach, where immunogenic proteins from *Borrelia* surface (OspA, OspC, DbpA) were used as the vaccination antigens [70–72]. Although these vaccines were protective under certain conditions, their general usage is not feasible due to the high heterogeneity of the targeted antigens among different *Borrelia* strains, the insufficient efficacy, and the side effects of the vaccines [73–76].

Due to immune evasion mechanisms, the spirochetal proteins, expressed on *Borrelia* surface during its life cycle inside the host, are considered as inferior vaccine candidates. Efforts have thus been turned to the development of a universal anti-tick vaccine. In contrast to acaricides, the anti-tick vaccines based on tick recombinant proteins are environmentally safe, may not be risky for human health, and the development of resistance is less likely [77]; however, the identification of suitable antigenic targets seems to be the primary limiting factor of the whole process.

Generally, there are two distinct types of vaccine antigens. The so-called "exposed tick antigens" are salivary-secreted proteins or tick cement components. They come into direct contact with host immune mechanisms; hence, immunity against an exposed antigen is enhanced by repeated tick infestations [78]. Immunization with the *I. scapularis* serpin Iris, which is declared as a component of saliva and interacts with both host immunity and hemostasis, established a significant protective immunity against tick infestation of rabbits, resulting in 30% tick mortality, diminution of weight gain, and prolongation of blood feeding time [79].

Notably, sialostatin L2, a cystatin with anti-inflammatory properties, has been termed a silent antigen. Although the protein is found in the tick–host interface, it is not recognized by humoral immunity after repeated tick infestations, possibly due to its function or the amount of its secretion; however, when sialostatin L2 is used as a vaccine antigen in guinea pigs, impaired tick feeding was observed [80].

In contrast to exposed antigens, the so-called "concealed antigens" are derived from tick tissues, most often from tick gut wall, ovaries, or tick fat body. If a concealed antigen is tightly connected with a physiologic function, its reaction with antibodies produced due to the host vaccination leads to tick poisoning, and the surviving ticks have impaired ability to molt and reproduce. However, concealed antigens do not stimulate host immunity during tick feeding, and thus host immunization has to be frequently repeated [81]. Currently, the only anti-tick vaccine commercially available is derived from Bm86, a midgut membrane-bound protein of *Boophilus* (*Rhipicephalus*) *microplus* [82].

A promising candidate for the development of an efficient anti-tick vaccine is ferritin 2, which is expressed in the tick gut tissue and it is required for iron transport to the ovaries and SGs. Immunization of rabbits and cattle with the ferritin 2 protein dramatically impaired the ability of ticks to feed and reproduce, most likely due to a resulting imbalance in iron metabolism [83].

Another tested principle for development of transmission-blocking vaccines is immunization of the host with an arthropod vector molecule that the microbe exploits during infection of the mammalian host. Passive as well as active immunization with Salp15 significantly protected mice from *B. burgdorferi* infection due to an increased clearance of Salp15-coated bacteria by phagocytes [84].

Development of vaccines against the so-called "dual antigens" is a relatively new strategy in the field, targeting both exposed antigens in the tick saliva and concealed antigenic epitopes in the tick midgut. This dual action stimulates a specific protective immune response that is boosted by natural infestations. At the same time, the host antibodies ingested with the blood meal react with antigenic epitopes on tick gut cells, leading to the rupture of the midgut [85]. A cement protein 64P derived from *R. appendiculatus* has been proposed as a candidate transmission-blocking vaccine antigen with dual function [86].

Another proposed step in anti-tick vaccine development could be the combination of different vaccine antigens composed of both *Borrelia* and tick antigens that act synergistically [87]. Using antigen orthologs for control of multiple arthropod vectors has been also proposed for the development of an effective vaccine [88]. For example, a tick-protective antigen, subolesin, has been recognized as a regulatory protein involved in the control of multiple cellular signaling pathways and innate immune response genes in ticks [89]. Targeting subolesin by RNAi or vaccination resulted in degeneration of tick gut, SGs, and reproductive tissues. Moreover, tick survival, weight, and oviposition were significantly reduced [90]. As a consequence, subolesin knock-down also affects *Anaplasma marginale* and *Anaplasma phagocytophilum* infection and acquisition by ticks [91]. Because RNAi in engorged tick females results subolesin-knockdown larvae, this technique, in combination with subolesin vaccination, is proposed as an efficient method for the control of tick infestations on cattle [92, 93]. Furthermore, subolesin is highly conserved across vector species and is thus promising for the development of a universal vaccine that will control arthropod infestations and the reduction of pathogen transmission [94].

TICK-DERIVED MOLECULES IN THE TREATMENT OF HUMAN DISEASES

The immunomodulatory properties of tick salivary molecules can be further exploited for the treatment of immunity-related diseases. Several tick-derived molecules have been successfully tested against human disorders, mainly in animal models of human diseases. A histamine-binding protein from *R. appendiculatus* was tested as a treatment of allergic conjunctivitis in mice [95]. A complement C5 inhibitor from the soft tick *O. moubata* was successful in preventing experimental autoimmune myasthenia gravis [96], and sialostatin L from the hard tick *I. scapularis* significantly prevented disease symptoms in a mouse model of multiple sclerosis, i.e., experimental autoimmune encephalomyelitis [38]. Two tick-derived proteins were successfully used in the prevention of experimental allergic asthma. These were a histamine-binding protein from *R. appendiculatus* [97] and Salp15, the immunomodulatory protein from *I. scapularis* [98]. Both proteins reduced airway hyper-responsiveness, pulmonary eosinophilia, and mucus hypersecretion.

Conclusion

Ticks are one of the most common vectors of pathogens of high importance in human and veterinary medicine, as well as in agriculture. Novel information about the tripartite interaction of ticks with hosts and tick-borne pathogens can be essential for the development of new methods to prevent the harmful consequences of tick bites. Comprehension of the events within the triangle of interactions between ticks, their vertebrate hosts, and tick-borne pathogens, explained and characterized in this chapter, combined with the description of their biochemical and physiological properties and the understanding of the mechanism of modulation of host immunity by the salivary proteins and its subsequent role in the transmission of tick-borne pathogens would help in devising a sound control strategy against tick infestations and tick-borne diseases. Because new drugs for the treatment of human diseases are constantly in demand, tick proteins can be used for the development of new treatment strategies.

Acknowledgements

This work was supported by the Intramural Research Program of the Division of Intramural Research, National Institute of Allergy and Infectious Diseases, National Institutes of Health, the Grant Agency of the Czech Republic (projects P302/11/J029 and P502/12/2409) and the Research Centre of the Ministry of Education, Youth and Sports of the Czech Republic (grant LC06009). M. K. received support from grant number Z60220518 and a Jan Evangelista Purkyne fellowship from the National Academy of Sciences of the Czech Republic, from a Marie Curie Reintegration grant of the EU-FP7(PIRG07-GA-2010-268177) and from the National Institutes of Health (R01AI093653).

We thank NIAID intramural editor Brenda Rae Marshall for assistance and Jan Langhans for his help with Figure 1.

References

[1] Dennis, D., & Hayes, E. (2002). Epidemiology of Lyme borreliosis. In: Kahl, O., Gray, J. S., Lane, R. S., & Stanek, G., editors. *Lyme borreliosis: biology, epidemiology and control.* Oxford, UK: CABI Publishing; 251–280.

[2] Ribeiro, J. M. (1995). How ticks make a living. *Parasitol. Today,* 11, 91–93.

[3] Ribeiro, J. M., & Francischetti, I. M. (2003). Role of arthropod saliva in blood feeding: sialome and post-sialome perspectives. *Annu. Rev. Entomol.*, 48, 73–88.

[4] Wikel, S. K. (1996). Host immunity to ticks. *Annu. Rev. Entomol.,* 41, 1–22.

[5] Deruaz, M., Frauenschuh, A., Alessandri, A. L., Dias, J. M., Coelho, F. M., Russo, R. C., Ferreira, B. R., Graham, G. J., Shaw, J. P., Wells, T. N., Teixeira, M. M., Power, C. A., & Proudfoot, A. E. (2008). Ticks produce highly selective chemokine binding proteins with antiinflammatory activity. *J. Exp. Med.,* 205, 2019–2031.

[6] Gill, H. S., & Walker, A. R. (1985). Differential cellular responses at *Hyalomma anatolicum anatolicum* feeding sites on susceptible and tick-resistant rabbits. *Parasitology,* 91 (Pt 3), 591–607.

[7] Ribeiro, J. M., Weis, J. J., & Telford, S. R. 3rd. (1990). Saliva of the tick *Ixodes dammini* inhibits neutrophil function. *Exp. Parasitol.,* 70, 382–388.

[8] Montgomery, R. R., Lusitani, D., De Boisfleury Chevance, A., & Malawista, S. E. (2004). Tick saliva reduces adherence and area of human neutrophils. *Infect. Immun.,* 72, 2989–2994.

[9] Guo, X., Booth, C. J., Paley, M. A., Wang, X., DePonte, K., Fikrig, E., Narasimhan, S., & Montgomery, R. R. (2009). Inhibition of neutrophil function by two tick salivary proteins. *Infect. Immun.,* 77, 2320–2329.

[10] Chmelar, J., Oliveira, C. J., Rezacova, P., Francischetti, I. M., Kovarova, Z., Pejler, G., Kopacek, P., Ribeiro, J. M., Mares, M., Kopecky, J., & Kotsyfakis, M. (2011). A tick salivary protein targets cathepsin G and chymase and inhibits host inflammation and platelet aggregation. *Blood,* 117, 736–744.

[11] Kotsyfakis, M., Sa-Nunes, A., Francischetti, I. M., Mather, T. N., Andersen, J. F., & Ribeiro, J. M. (2006). Antiinflammatory and immunosuppressive activity of sialostatin L, a salivary cystatin from the tick *Ixodes scapularis*. *J. Biol. Chem.,* 281, 26298–26307.

[12] Xu, Q., Seemanapalli, S. V., Reif, K. E., Brown, C. R., & Liang, F. T. (2007). Increasing the recruitment of neutrophils to the site of infection dramatically attenuates *Borrelia burgdorferi* infectivity. *J. Immunol.,* 178, 5109–5115.

[13] Askenase, P. W. (1977). Role of basophils, mast cells, and vasoamines in hypersensitivity reactions with a delayed time course. *Prog. Allergy,* 23, 199–320.

[14] Matsuda, H., Fukui, K., Kiso, Y., & Kitamura, Y. (1985). Inability of genetically mast cell-deficient W/Wv mice to acquire resistance against larval *Haemaphysalis longicornis* ticks. *J. Parasitol.,* 71, 443–448.

[15] Matsuda, H., Nakano, T., Kiso, Y., & Kitamura, Y. (1987). Normalization of anti-tick response of mast cell-deficient W/Wv mice by intracutaneous injection of cultured mast cells. *J. Parasitol.,* 73, 155–160.

[16] Wada, T., Ishiwata, K., Koseki, H., Ishikura, T., Ugajin, T., Ohnuma, N., Obata, K., Ishikawa, R., Yoshikawa, S., Mukai, K., Kawano, Y., Minegishi, Y., Yokozeki, H., Watanabe, N., & Karasuyama, H. (2010). Selective ablation of basophils in mice reveals their nonredundant role in acquired immunity against ticks. *J. Clin. Invest.,* 120, 2867–2875.

[17] Brown, S. J., & Askenase, P. W. (1983). Immune rejection of ectoparasites (ticks) by T cell and IgG1 antibody recruitment of basophils and eosinophils. *Fed. Proc.,* 42, 1744–1749.

[18] Paesen, G. C., Adams, P. L., Harlos, K., Nuttall, P. A., & Stuart, D. I. (1999). Tick histamine-binding proteins: isolation, cloning, and three-dimensional structure. *Mol. Cell,* 3, 661–671.

[19] Mans, B. J., Ribeiro, J. M., & Andersen, J. F. (2008). Structure, function, and evolution of biogenic amine-binding proteins in soft ticks. *J. Biol. Chem.,* 283, 18721–18733.

[20] Dai, J., Narasimhan, S., Zhang, L., Liu, L., Wang, P., & Fikrig, E. (2010). Tick histamine release factor is critical for *Ixodes scapularis* engorgement and transmission of the Lyme disease agent. *PLoS Pathog.,* 6, e1001205.

[21] McLaren, D. J., Worms, M. J., & Askenase, P. W. (1983). Cutaneous basophil associated resistance to ectoparasites (ticks). Electron microscopy of *Rhipicephalus appendiculatus* larval feeding sites in actively sensitised guinea pigs and recipients of immune serum. *J. Pathol.,* 139, 291–308.

[22] Valenzuela, J. G., Charlab, R., Mather, T. N., & Ribeiro, J. M. (2000). Purification, cloning, and expression of a novel salivary anticomplement protein from the tick, *Ixodes scapularis. J. Biol. Chem.,* 275, 18717–18723.

[23] Tyson, K., Elkins, C., Patterson, H., Fikrig, E., & de Silva, A. (2007). Biochemical and functional characterization of Salp20, an *Ixodes scapularis* tick salivary protein that inhibits the complement pathway. *Insect Mol. Biol.,* 16, 469–479.

[24] Schroeder, H., Daix, V., Gillet, L., Renauld, J. C., & Vanderplasschen, A. (2007). The paralogous salivary anti-complement proteins IRAC I and IRAC II encoded by *Ixodes ricinus* ticks have broad and complementary inhibitory activities against the complement of different host species. *Microbes Infect.,* 9, 247–250.

[25] Rittig, M. G., Krause, A., Haupl, T., Schaible, U. E., Modolell, M., Kramer, M. D., Lutjen-Drecoll, E., Simon, M. M., & Burmester, G. R. (1992). Coiling phagocytosis is the preferential phagocytic mechanism for *Borrelia burgdorferi. Infect. Immun.,* 60, 4205–4212.

[26] Modolell, M., Schaible, U. E., Rittig, M., & Simon, M. M. (1994). Killing of *Borrelia burgdorferi* by macrophages is dependent on oxygen radicals and nitric oxide and can be enhanced by antibodies to outer surface proteins of the spirochete. *Immunol. Lett.,* 40, 139–146.

[27] Kuthejlova, M., Kopecky, J., Stepanova, G., & Macela, A. (2001). Tick salivary gland extract inhibits killing of *Borrelia afzelii* spirochetes by mouse macrophages. *Infect. Immun.,* 69, 575–578.

[28] Kyckova, K., & Kopecky, J. (2006). Effect of tick saliva on mechanisms of innate immune response against *Borrelia afzelii. J. Med. Entomol.,* 43, 1208–1214.

[29] Prevot, P. P., Beschin, A., Lins, L., Beaufays, J., Grosjean, A., Bruys, L., Adam, B., Brossard, M., Brasseur, R., Zouaoui Boudjeltia, K., Vanhamme, L., & Godfroid, E. (2009). Exosites mediate the anti-inflammatory effects of a multifunctional serpin from the saliva of the tick *Ixodes ricinus. FEBS J.,* 276, 3235–3246.

[30] Brake, D. K., Wikel, S. K., Tidwell, J. P., & Perez de Leon, A. A. (2010). *Rhipicephalus microplus* salivary gland molecules induce differential CD86 expression in murine macrophages. *Parasite Vectors,* 3, 103.

[31] Mejri, N., & Brossard, M. (2007). Splenic dendritic cells pulsed with *Ixodes ricinus* tick saliva prime naive $CD4^+$ T to induce Th2 cell differentiation *in vitro* and *in vivo. Int. Immunol.,* 19, 535–543.

[32] Skallova, A., Iezzi, G., Ampenberger, F., Kopf, M., & Kopecky, J. (2008). Tick saliva inhibits dendritic cell migration, maturation, and function while promoting development of Th2 responses. *J. Immunol.,* 180, 6186–6192.

[33] Vesely, D. L., Fish, D., Shlomchik, M. J., Kaplan, D. H., & Bockenstedt, L. K. (2009). Langerhans cell deficiency impairs *Ixodes scapularis* suppression of Th1 responses in mice. *Infect. Immun.*, 77, 1881–1887.

[34] Sa-Nunes, A., Bafica, A., Lucas, D. A., Conrads, T. P., Veenstra, T. D., Andersen, J. F., Mather, T. N., Ribeiro, J. M., & Francischetti, I. M. (2007). Prostaglandin E2 is a major inhibitor of dendritic cell maturation and function in *Ixodes scapularis* saliva. *J. Immunol.*, 179, 1497–1505.

[35] Oliveira, C. J., Sa-Nunes, A., Francischetti, I. M., Carregaro, V., Anatriello, E., Silva, J. S., Santos, I. K., Ribeiro, J. M., & Ferreira, B. R. (2011). Deconstructing tick saliva: non-protein molecules with potent immunomodulatory properties. *J. Biol. Chem.*, 286, 10960–10969.

[36] Oliveira, C. J., Carvalho, W. A., Garcia, G. R., Gutierrez, F. R., de Miranda Santos, I. K., Silva, J. S., & Ferreira, B. R. (2010). Tick saliva induces regulatory dendritic cells: MAP-kinases and Toll-like receptor-2 expression as potential targets. *Vet. Parasitol.*, 167, 288–297.

[37] Hovius, J. W., de Jong, M. A., den Dunnen, J., Litjens, M., Fikrig, E., van der Poll, T., Gringhuis, S. I., & Geijtenbeek, T. B. (2008). Salp15 binding to DC-SIGN inhibits cytokine expression by impairing both nucleosome remodeling and mRNA stabilization. *PLoS Pathog.*, 4, e31.

[38] Sa-Nunes, A., Bafica, A., Antonelli, L. R., Choi, E. Y., Francischetti, I. M., Andersen, J. F., Shi, G. P., Chavakis, T., Ribeiro, J. M., & Kotsyfakis, M. (2009). The immunomodulatory action of sialostatin L on dendritic cells reveals its potential to interfere with autoimmunity. *J. Immunol.*, 182, 7422–7429.

[39] Salat, J., Paesen, G. C., Rezacova, P., Kotsyfakis, M., Kovarova, Z., Sanda, M., Majtan, J., Grunclova, L., Horka, H., Andersen, J. F., Brynda, J., Horn, M., Nunn, M. A., Kopacek, P., Kopecky, J., & Mares, M. (2010). Crystal structure and functional characterization of an immunomodulatory salivary cystatin from the soft tick *Ornithodoros moubata*. *Biochem. J.*, 429, 103–112.

[40] Kopecky, J., & Kuthcjlova, M. (1998). Suppressive effect of *Ixodes ricinus* salivary gland extract on mechanisms of natural immunity *in vitro*. *Parasite Immunol.*, 20, 169–174.

[41] Kubes, M., Kocakova, P., Slovak, M., Slavikova, M., Fuchsberger, N., & Nuttall, P. A. (2002). Heterogeneity in the effect of different ixodid tick species on human natural killer cell activity. *Parasite Immunol.*, 24, 23–28.

[42] Severinova, J., Salat, J., Krocova, Z., Reznickova, J., Demova, H., Horka, H., & Kopecky, J. (2005). Co-inoculation of *Borrelia afzelii* with tick salivary gland extract influences distribution of immunocompetent cells in the skin and lymph nodes of mice. *Folia Microbiol. (Praha)*, 50, 457–463.

[43] Ferreira, B. R., & Silva, J. S. (1999). Successive tick infestations selectively promote a T-helper 2 cytokine profile in mice. *Immunology*, 96, 434–439.

[44] Schoeler, G. B., Manweiler, S. A., & Wikel, S. K. (1999). *Ixodes scapularis*: effects of repeated infestations with pathogen-free nymphs on macrophage and T lymphocyte cytokine responses of BALB/c and C3H/HeN mice. *Exp. Parasitol.*, 92, 239–248.

[45] Kovar, L., Kopecky, J., & Rihova, B. (2001). Salivary gland extract from *Ixodes ricinus* tick polarizes the cytokine profile toward Th2 and suppresses proliferation of T lymphocytes in human PBMC culture. *J. Parasitol.*, 87, 1342–1348.

[46] Anguita, J., Ramamoorthi, N., Hovius, J. W., Das, S., Thomas, V., Persinski, R., Conze, D., Askenase, P. W., Rincon, M., Kantor, F. S., & Fikrig, E. (2002). Salp15, an *Ixodes scapularis* salivary protein, inhibits $CD4^+$ T cell activation. *Immunity,* 16, 849–859.

[47] Garg, R., Juncadella, I. J., Ramamoorthi, N., Ashish, Ananthanarayanan, S. K., Thomas, V., Rincon, M., Krueger, J. K., Fikrig, E., Yengo, C. M., & Anguita, J. (2006). CD4 is the receptor for the tick saliva immunosuppressor, Salp15. *J. Immunol.,* 177, 6579–6583.

[48] Alarcon-Chaidez, F. J., Boppana, V. D., Hagymasi, A. T., Adler, A. J., & Wikel, S. K. (2009). A novel sphingomyelinase-like enzyme in *Ixodes scapularis* tick saliva drives host CD4 T cells to express IL-4. *Parasite Immunol.,* 31, 210–219.

[49] Hannier, S., Liversidge, J., Sternberg, J. M., & Bowman, A. S. (2004). Characterization of the B cell inhibitory protein factor in *Ixodes ricinus* tick saliva: a potential role in enhanced *Borrelia burgdoferi* transmission. *Immunology,* 113, 401–408.

[50] Wang, H., & Nuttall, P. A. (1995). Immunoglobulin G binding proteins in male *Rhipicephalus appendiculatus* ticks. *Parasite Immunol.,* 17, 517–524.

[51] Gray, J. S. (2002). Biology of *Ixodes* species ticks in relation to tick-borne zoonoses. *Wien Klin. Wochenschr.,* 114, 473–478.

[52] Nuttall, P., & Jones, L. D. (1991). Non-viraemic tick-borne virus transmission: mechanism and significance. In: Dusbabek, F. & Bukva, V., editors. Modern Acarology. Prague and The Hague: Academia and SPB Academic Publishing, 3–6.

[53] Jones, L. D., Davies, C. R., Steele, G. M., & Nuttall, P. A. (1987). A novel mode of arbovirus transmission involving a nonviremic host. *Science,* 237, 775–777.

[54] Labuda, M., Jones, L. D., Williams, T., & Nuttall, P. A. (1993). Enhancement of tick-borne encephalitis virus transmission by tick salivary gland extracts. *Med. Vet. Entomol.,* 7, 193–196.

[55] Krocova, Z., Macela, A., Hernychova, L., Kroca, M., Pechova, J., & Kopecky, J. (2003). Tick salivary gland extract accelerates proliferation of *Francisella tularensis* in the host. *J. Parasitol.,* 89, 14–20.

[56] Gern, L., & Rais, O. (1996). Efficient transmission of *Borrelia burgdorferi* between cofeeding *Ixodes ricinus* ticks (Acari: Ixodidae). *J. Med. Entomol.,* 33, 189–192.

[57] Sato, Y., & Nakao, M. (1997). Transmission of the Lyme disease spirochete, *Borrelia garinii*, between infected and uninfected immature *Ixodes persulcatus* during cofeeding on mice. *J. Parasitol.,* 83, 547–550.

[58] Zeidner, N. S., Schneider, B. S., Nuncio, M. S., Gern, L., & Piesman, J. (2002). Coinoculation of *Borrelia* spp. with tick salivary gland lysate enhances spirochete load in mice and is tick species-specific. *J. Parasitol.,* 88, 1276–1278.

[59] Pechova, J., Stepanova, G., Kovar, L., & Kopecky, J. (2002). Tick salivary gland extract-activated transmission of *Borrelia afzelii* spirochaetes. *Folia Parasitol. (Praha),* 49, 153–159.

[60] Machackova, M., Obornik, M., & Kopecky, J. (2006). Effect of salivary gland extract from *Ixodes ricinus* ticks on the proliferation of *Borrelia burgdorferi* sensu stricto *in vivo*. *Folia Parasitol. (Praha),* 53, 153–158.

[61] Titus, R. G., & Ribeiro, J. M. (1988). Salivary gland lysates from the sand fly *Lutzomyia longipalpis* enhance *Leishmania* infectivity. *Science,* 239, 1306–1308.

[62] Lima, C. M., Zeidner, N. S., Beard, C. B., Soares, C. A., Dolan, M. C., Dietrich, G., & Piesman, J. (2005). Differential infectivity of the Lyme disease spirochete *Borrelia burgdorferi* derived from *Ixodes scapularis* salivary glands and midgut. *J. Med. Entomol.,* 42, 506–510.

[63] Horka, H., Cerna-Kyckova, K., Skallova, A., & Kopecky, J. (2009). Tick saliva affects both proliferation and distribution of *Borrelia burgdorferi* spirochetes in mouse organs and increases transmission of spirochetes to ticks. *Int. J. Med. Microbiol.,* 299, 373–380.

[64] Zeidner, N. S., Schneider, B. S., Rutherford, J. S., & Dolan, M. C. (2008). Suppression of Th2 cytokines reduces tick-transmitted *Borrelia burgdorferi* load in mice. *J. Parasitol.,* 94, 767–769.

[65] Ramamoorthi, N., Narasimhan, S., Pal, U., Bao, F., Yang, X. F., Fish, D., Anguita, J., Norgard, M. V., Kantor, F. S., Anderson, J. F., Koski, R. A., & Fikrig, E. (2005). The Lyme disease agent exploits a tick protein to infect the mammalian host. *Nature,* 436, 573–577.

[66] Schuijt, T. J., Hovius, J. W., van Burgel, N. D., Ramamoorthi, N., Fikrig, E., & van Dam, A. P. (2008). The tick salivary protein Salp15 inhibits the killing of serum-sensitive *Borrelia burgdorferi* sensu lato isolates. *Infect. Immun.,* 76, 2888–2894.

[67] Marchal, C., Schramm, F., Kern, A., Luft, B. J., Yang, X., Schuijt, T., Hovius, J., Jaulhac, B., & Boulanger, N. (2011). Antialarmin effect of tick saliva during the transmission of Lyme disease. *Infect. Immun.,* 79, 774–785.

[68] Kotsyfakis, M., Karim, S., Andersen, J. F., Mather, T. N., & Ribeiro, J. M. (2007). Selective cysteine protease inhibition contributes to blood-feeding success of the tick *Ixodes scapularis*. *J. Biol. Chem.,* 282, 29256–29263.

[69] Kotsyfakis, M., Horka, H., Salat, J., & Andersen, J. F. (2010). The crystal structures of two salivary cystatins from the tick *Ixodes scapularis* and the effect of these inhibitors on the establishment of *Borrelia burgdorferi* infection in a murine model. *Mol. Microbiol.,* 77, 456–470.

[70] de Silva, A. M., Telford, S. R. 3rd, Brunet, L. R., Barthold, S. W., & Fikrig, E. (1996). *Borrelia burgdorferi* OspA is an arthropod-specific transmission-blocking Lyme disease vaccine. *J. Exp. Med.,* 183, 271–275.

[71] Gilmore, R. D. Jr., Kappel, K. J., Dolan, M. C., Burkot, T. R., & Johnson, B. J. (1996). Outer surface protein C (OspC), but not P39, is a protective immunogen against a tick-transmitted *Borrelia burgdorferi* challenge: evidence for a conformational protective epitope in OspC. *Infect. Immun.,* 64, 2234–2239.

[72] Hanson, M. S., Cassatt, D. R., Guo, B. P., Patel, N. K., McCarthy, M. P., Dorward, D. W., & Hook, M. (1998). Active and passive immunity against *Borrelia burgdorferi* decorin binding protein A (DbpA) protects against infection. *Infect. Immun.,* 66, 2143–2153.

[73] Wilske, B., Preac-Mursic, V., Jauris, S., Hofmann, A., Pradel, I., Soutschek, E., Schwab, E., Will, G., & Wanner, G. (1993). Immunological and molecular polymorphisms of OspC, an immunodominant major outer surface protein of *Borrelia burgdorferi*. *Infect. Immun.,* 61, 2182–2191.

[74] Hagman, K. E., Yang, X., Wikel, S. K., Schoeler, G. B., Caimano, M. J., Radolf, J. D., & Norgard, M. V. (2000). Decorin-binding protein A (DbpA) of *Borrelia burgdorferi* is not protective when immunized mice are challenged *via* tick infestation and correlates

with the lack of DbpA expression by *B. burgdorferi* in ticks. *Infect. Immun.,* 68, 4759–4764.

[75] Rose, C. D., Fawcett, P. T., & Gibney, K. M. (2001). Arthritis following recombinant outer surface protein A vaccination for Lyme disease. *J. Rheumatol.,* 28, 2555–2557.

[76] Alexopoulou, L., Thomas, V., Schnare, M., Lobet, Y., Anguita, J., Schoen, R. T., Medzhitov, R., Fikrig, E., & Flavell, R. A. (2002). Hyporesponsiveness to vaccination with *Borrelia burgdorferi* OspA in humans and in TLR1- and TLR2-deficient mice. *Nat. Med.,* 8, 878–884.

[77] Willadsen, P. (2004). Anti-tick vaccines. *Parasitology,* 129 Suppl, S367–387.

[78] Opdebeeck, J. P. (1994). Vaccines against blood-sucking arthropods. *Vet. Parasitol.,* 54, 205–222.

[79] Prevot, P. P., Couvreur, B., Denis, V., Brossard, M., Vanhamme, L., & Godfroid, E. (2007). Protective immunity against *Ixodes ricinus* induced by a salivary serpin. *Vaccine,* 25, 3284–3292.

[80] Kotsyfakis, M., Anderson, J. M., Andersen, J. F., Calvo, E., Francischetti, I. M., Mather, T. N., Valenzuela, J. G., & Ribeiro, J. M. (2008). Immunity against a "silent" salivary antigen of the Lyme vector *Ixodes scapularis* impairs its ability to feed. *J. Immunol.,* 181, 5209–5212.

[81] Willadsen, P., Eisemann, C. H., & Tellam, R. L. (1993). "Concealed" antigens: expanding the range of immunological targets. *Parasitol. Today,* 9, 132–135.

[82] Willadsen, P., Bird, P., Cobon, G. S., & Hungerford, J. (1995). Commercialisation of a recombinant vaccine against *Boophilus microplus*. *Parasitology,* 110 Suppl, S43–50.

[83] Hajdusek, O., Almazan, C., Loosova, G., Villar, M., Canales, M., Grubhoffer, L., Kopacek, P., & de la Fuente, J. (2010). Characterization of ferritin 2 for the control of tick infestations. *Vaccine,* 28, 2993–2998.

[84] Dai, J., Wang, P., Adusumilli, S., Booth, C. J., Narasimhan, S., Anguita, J., & Fikrig, E. (2009). Antibodies against a tick protein, Salp15, protect mice from the Lyme disease agent. *Cell Host Microbe,* 6, 482–492.

[85] Trimnell, A. R., Hails, R. S., & Nuttall, P. A. (2002). Dual action ectoparasite vaccine targeting "exposed" and "concealed" antigens. *Vaccine,* 20, 3560–3568.

[86] Havlikova, S., Roller, L., Koci, J., Trimnell, A. R., Kazimirova, M., Klempa, B., & Nuttall, P. A. (2009). Functional role of 64P, the candidate transmission-blocking vaccine antigen from the tick, *Rhipicephalus appendiculatus*. *Int. J. Parasitol.,* 39, 1485–1494.

[87] Schuijt, T. J., Hovius, J. W., van der Poll, T., van Dam, A. P., & Fikrig, E. (2011). Lyme borreliosis vaccination: the facts, the challenge, the future. *Trends Parasitol.,* 27, 40–47.

[88] Canales, M., Almazan, C., Naranjo, V., Jongejan, F., & de la Fuente, J. (2009). Vaccination with recombinant *Boophilus annulatus* Bm86 ortholog protein, Ba86, protects cattle against *B. annulatus* and *B. microplus* infestations. *BMC Biotechnol.,* 9, 29.

[89] Galindo, R. C., Doncel-Perez, E., Zivkovic, Z., Naranjo, V., Gortazar, C., Mangold, A. J., Martin-Hernando, M. P., Kocan, K. M., & de la Fuente, J. (2009). Tick subolesin is an ortholog of the akirins described in insects and vertebrates. *Dev. Comp. Immunol.,* 33, 612–617.

[90] de la Fuente, J., Almazan, C., Blas-Machado, U., Naranjo, V., Mangold, A. J., Blouin, E. F., Gortazar, C., & Kocan, K. M. (2006). The tick protective antigen, 4D8, is a conserved protein involved in modulation of tick blood ingestion and reproduction. *Vaccine,* 24, 4082–4095.

[91] de la Fuente, J., Almazan, C., Blouin, E. F., Naranjo, V., & Kocan, K. M. (2006). Reduction of tick infections with *Anaplasma marginale* and *A. phagocytophilum* by targeting the tick protective antigen subolesin. *Parasitol. Res.,* 100, 85–91.

[92] Almazan, C., Lagunes, R., Villar, M., Canales, M., Rosario-Cruz, R., Jongejan, F., & de la Fuente, J. (2010). Identification and characterization of *Rhipicephalus (Boophilus) microplus* candidate protective antigens for the control of cattle tick infestations. *Parasitol. Res.,* 106, 471–479.

[93] Merino, O., Almazan, C., Canales, M., Villar, M., Moreno-Cid, J. A., Estrada-Pena, A., Kocan, K. M., & de la Fuente, J. (2011). Control of *Rhipicephalus (Boophilus) microplus* infestations by the combination of subolesin vaccination and tick autocidal control after subolesin gene knockdown in ticks fed on cattle. *Vaccine,* 29, 2248–2254.

[94] de la Fuente, J., Moreno-Cid, J. A., Canales, M., Villar, M., de la Lastra, J. M., Kocan, K. M., Galindo, R. C., Almazan, C., & Blouin, E. F. (2011). Targeting arthropod subolesin/akirin for the development of a universal vaccine for control of vector infestations and pathogen transmission. *Vet Parasitol.,* 181, 17–22.

[95] Chapin, M., Nichols, L., Ono, S., Nakamura, T., Weston-Davies, W., Paesen, G., Nuttall, P., & Abelson, M. B. (2002). Evaluation of EV131. A histamine binding protein, in a murine model of allergic conjunctivitis. *IOVS,* 43, U17–U17.

[96] Soltys, J., Kusner, L. L., Young, A., Richmonds, C., Hatala, D., Gong, B., Shanmugavel, V., & Kaminski, H. J. (2009). Novel complement inhibitor limits severity of experimentally myasthenia gravis. *Ann. Neurol.,* 65, 67–75.

[97] Couillin, I., Maillet, I., Vargaftig, B. B., Jacobs, M., Paesen, G. C., Nuttall, P. A., Lefort, J., Moser, R., Weston-Davies, W., & Ryffel, B. (2004). Arthropod-derived histamine-binding protein prevents murine allergic asthma. *J. Immunol.,* 173, 3281–3286.

[98] Paveglio, S. A., Allard, J., Mayette, J., Whittaker, L. A., Juncadella, I., Anguita, J., & Poynter, M. E. (2007). The tick salivary protein, Salp15, inhibits the development of experimental asthma. *J. Immunol.,* 178, 7064–7071.

In: Ticks: Disease, Management and Control
Editor: Moges Woldemeskel

ISBN: 978-1-62081-136-8

Chapter 3

MODELLING THE POPULATION DYNAMICS OF EUROPEAN TICK IXODES RICINUS: EXPLORING THE IMPACT OF VARIABLE ABIOTIC, BIOTIC AND CONTROL FACTORS

***Andrew D. M. Dobson*[1] *and Sarah E. Randolph*[2*]**
[1]British Trust for Ornithology, University of Stirling, Scotland
[2]Department of Zoology, University of Oxford, UK

ABSTRACT

The tick *Ixodes ricinus (I.ricinus)* is the most multi-competent vector in Europe, responsible for significant diseases of humans and livestock throughout the northern hemisphere. Quantitative analysis of tick-borne disease risk and potential tick control depends on a full understanding of the tick's complex seasonal dynamics, for which a realistic model is an invaluable tool. The effects of individual environmental factors on population dynamics can be investigated in ways not easily possible by experimental manipulation of *in situ* populations. A population model for *I. ricinus*, based on a modified stage-classified Leslie matrix, gives a good fit to field data from UK sites and, after modification of a single parameter, also replicates divergent seasonal patterns in central Spain. This modification with respect to diapause, which is required before the model is more generally applicable across wide geographic areas, highlights gaps in our knowledge of tick biology. The model is generally robust, but most sensitive to changes in parameters related to density-dependent mortality.

This model can be used to explore the relative impact of changes in climate, host density and acaricide-treated hosts on tick abundance and seasonality to identify the causes of apparent recent increases in tick populations, and to predict potential effects of future environmental change. The rise in temperature in 1989 across Europe could have increased the inter-stadial development rate of ticks, thereby reducing the cumulative effect of daily mortality rates and potentially raising population levels. An earlier onset of tick questing activity in warmer springs could increase contact between ticks and humans during recreation in the countryside in spring holidays. These tick population effects are

*E-mail address: sarah.randolph@zoo.ox.ac.uk.

more marked where initial background climate is more limiting, especially along latitudinal and altitudinal boundaries. At the same time, the observed increase in deer abundance is predicted to drive tick populations up, most markedly where deer are initially at low densities. Where host densities are already high, however, further increases may reduce the population of questing ticks as they find hosts more rapidly, even though the overall tick population (and therefore pathogen transmission potential) increases. Culling high-density deer populations could, therefore, initially cause an apparent increase in questing ticks, while adding more large hosts (e.g. sheep) would effectively reduce the number of questing ticks, and therefore the risk to humans. If such sheep were treated with acaricide, tick populations are predicted to decrease rapidly. Tick control in designated areas may be achieved by using sheep in this way as 'lethal mops', as occurred in the past.

INTRODUCTION

Model Types

Arthropod-borne diseases are among the most significant causes of human mortality worldwide [1]. An important element of disease risk is the distribution and abundance of the vector, both of which are liable to be directly affected by environmental change [2,3]. The likely impacts of future environmental change on diseases such as malaria and tick-borne encephalitis (TBE) have been investigated through statistical pattern-matching techniques that identify correlations between the recorded distribution of the vector and climatic variables, and use the resulting models to predict changes in distribution and/or abundance under projected future climate scenarios [4,5]. This 'top-down' approach has various advantages, including rapid implementation and the potential for utilising existing spatial datasets that are often widely available, and must be favoured where complete biological data for the vector are lacking. The alternative 'bottom-up' approach, whereby the directly-measured effects of environmental conditions on vector development, feeding and reproductive success are incorporated into a population model, requires a solid and comprehensive understanding of the biology of the vector, but should produce more reliable, versatile and meaningful results [6-8]. Such models may be used to produce either mapped estimates of distribution, based upon predicted population trajectories in individual grid cells (e.g. for the African tick *Rhipicephalus appendiculatus* [9]), or, more simply, simulations of population dynamics in a hypothetical environmental space under different scenarios. This latter application effectively represents a strictly controlled environment in which the effects of individual factors may be assessed in isolation.

Temperate Ticks as Subjects for Biological Models

The tick *Ixodes ricinus* is an haematophagous ectoparasite of considerable medical and veterinary importance. It occurs throughout Europe and Eurasia, from Ireland to the Urals in Russia and from northern Sweden to the Mediterranean, even into Tunisia. Its abundance, indiscriminate choice of vertebrate host species, and competence to transmit a large number of bacterial, viral and protozoan infections, make it an exceptionally effective vector. It is responsible for transmitting the agents of Lyme borreliosis (LB), human granulocytic anaplasmosis (HGA), babesiosis, tick-borne encephalitis (TBE) and Louping ill. While LB is the most common vector-borne disease of humans in the northern hemisphere, and may also affect domesticated animals, TBE is clinically more serious and untreatable, although preventable by vaccination. The other infections appear to be on the increase, for reasons that have not yet been identified.

This vector performance belies the fact that, although ticks are well equipped for blood feeding, their extreme feeding strategy makes them much less well equipped as vectors [10]. Each instar of *I. ricinus* (larva, nymph, adult female) takes a single enormous blood meal from a separate vertebrate host. Larvae and nymphs typically feed for 2-4 days, whereas adult females require about (c.)10 days, and increase their body weight 100 times. The processes of development and moulting between instars take several months, during which time ticks must survive on nutrients from this single meal. Microclimatic factors impose strong seasonal constraints on tick population dynamics, not least because most of the lifespan is spent off-host, either actively questing for hosts on vegetation, or sheltering in moist micro-habitats on the ground to avoid desiccation during development and seasonal diapause [9].

The natural life cycle of *I. ricinus* commonly includes periods of both behavioural and developmental (or morphogenetic) diapause [11]. Behavioural diapause is typically defined in the acarological literature as a pre-programmed period of functional inactivity in unfed ticks, whereas developmental diapause is a physiological state in which the ticks' bodies undergo biochemical changes, particularly related to a slowing of metabolic processes [11,12]. Both forms of diapause allow ticks to minimise energy use during times when it would be climatically unfavourable to quest for hosts. In *I. ricinus* in temperate Europe, behavioural diapause occurs to variable degrees through autumn and winter, and developmental diapause occurs as a precursor to development when ticks feed in late summer or autumn (reviewed in [9]).

A quantitative understanding of the complex life cycle of vector ticks must underpin any serious attempt to predict disease risk and investigate potential control strategies. Population models for temperate species are necessarily far more complex than for tropical species due to the marked environmental seasonality and diapause patterns. Peak abundance of *I. ricinus* nymphs and adults usually occurs during mid to late spring, followed after a variable period by the peak in larvae. All three stages may display a lesser resurgence in numbers in late summer. These basic patterns, particularly the precise timing of activity by each stage, vary geographically according to local environmental conditions, with significant consequences for the tick's vectorial capacity, especially for pathogens with short duration of infectivity in the vertebrate. For example, foci of persistent cycles of TBE virus occur where larvae are active in synchrony with nymphs [13-15]. The same is true of *I. scapularis* in North America and its ability to transmit certain strains of *Borrelia burgdorferi* [16,17].

The first biological, process-based population model for *I. ricinus* has recently been developed [18]. It incorporates all available empirical information on the biological processes that control *I. ricinus* populations, including environmental determinants of diapause initiation and cessation [9], the probability of questing [19,20], day-degree summation modelling of development [12] and estimates of density-dependent mortality [21]. The model was constructed using data from three separate field sites in the UK where detailed tick abundance and microclimatic variables were measured over a period of several years [12], and tested against data from independent sites in the UK and in central Spain. A sensitivity analysis revealed the model's considerable robustness to deviations in individual parameter values and identified factors of key sensitivity.

In contrast with the model for *I. scapularis*, that captured seasonality well and offered insights into potential range expansion of this species in Canada [22], the following model is characterized by its distinct underlying mathematical approach, protracted egg-laying phenology, use of saturation deficit (as opposed to temperature) to mediate questing probability and simulation of development periods by incremental day-degree summation.

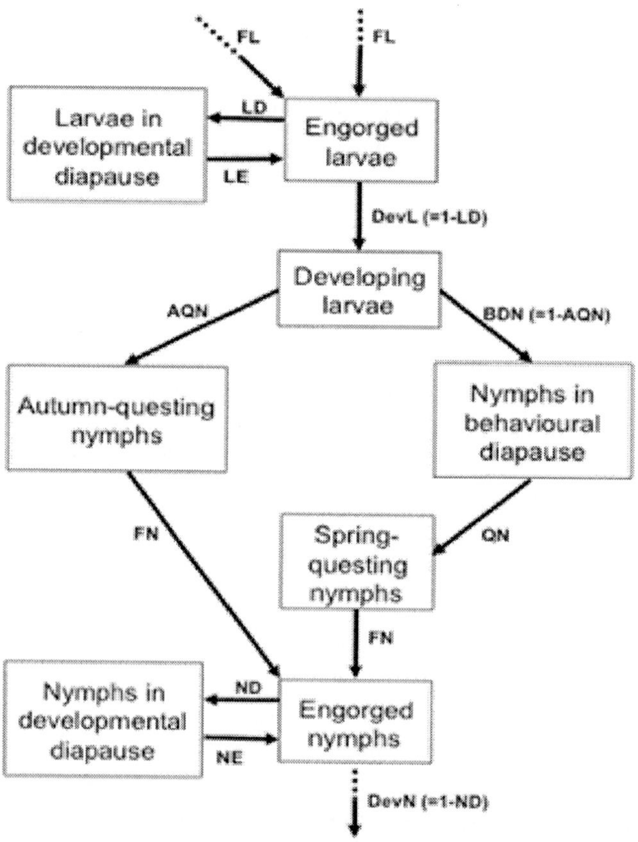

Figure 1. Schematic outline of part of the model. The full model is repeated three times, once for each instar. Autumn questing and behavioural diapause (with subsequent spring questing) are mutually exclusive alternatives. Ticks enter developmental diapause if they have fed after a certain date in late summer (see text), thus the proportions vary depending upon the development rate of the previous stage and the host contact rate. Labelled processes: FL – Larvae feeding (combines host encounter rate with probability of actively questing, mediated by saturation deficit), LD – Engorged larvae entering developmental diapauses, LE – Larvae exiting developmental diapauses, DevL – Engorged larvae beginning development, AQN – Developed nymphs questing in autumn (following a 20-day cuticular hardening period), BDN – Developed nymphs going into behavioural diapause in autumn (following a 20-day cuticular hardening period), QN – Nymphs exiting behavioural diapause to begin questing, FN – Nymph feeding (combines host encounter rate with probability of actively questing, mediated by saturation deficit), ND – Engorged nymphs entering developmental diapauses, NE – Nymphs exiting developmental diapauses. Reproduced from Ref [18], with permission.

Model Description

General

The tick model, which simulates relative numbers of ticks of each instar and physiological stage, is written in R language [23] and based around a stage-classified Leslie matrix [24,25] with some significant modifications.

It moves by daily increments; ticks in each stage experience a daily probability of moving to the next stage (Fig.1), though certain transitions incur fixed time delays to simulate such processes as feeding and post-eclosion cuticular hardening. Active processes such as temperature-dependent development and egg laying, which take more than a single day to complete, are also simulated in multi-element vectors outside the matrix.

Developmental Diapause

Developmental diapause delays the onset of development if ticks experience shortening day-length [26] by feeding after a certain date in late summer. Empirically, at three latitudinally-graded sites across Europe the observed date of diapause onset (e.g. end of July, mid-August, etc.) [9] coincided with the period at which day length shortens at a maximal rate, measured according to when the centre of the sun dips 9° beneath the horizon (dubbed 'tick twilight'). This is applied to the model. Since brief exposure to cold (at least $0^{o}C$) breaks diapause [27] and development is determined by temperature above a threshold of c.7.5 - $9^{o}C$ (depending on the stage), as long as ticks have emerged from diapause before spring warming starts, the timing of emergence (onset of development) is not critical, and is controlled by a Gaussian probability function that gradually releases ticks from 6th February onwards, a date chosen to accommodate these criteria adequately.

Development

Daily increments of development are modelled as a function of temperature with day-degree summation, using the following previously given [12] stage-specific equations derived from empirical observations [27], where T is daily mean temperature and Y is daily increments of development:

From egg to emerged larva:

$$Y = -0.0000112T^2 + 0.002305T - 0.0185 \quad (1)$$

From engorged larva to emerged nymph:

$$Y = 0.0000303T^2 + 0.000733T - 0.00706 \quad (2)$$

From engorged nymph to emerged adult:

$$Y = -0.00000796T^2 + 0.00193T - 0.0161 \quad (3)$$

Development is complete when the cumulative development reaches one, and eclosion at each instar is followed by a twenty-day period of cuticular hardening.

Questing and Behavioural Diapause

Ticks with the potential to quest (i.e. not in diapause or development) are considered to be 'questing', although, importantly, this is a nominal state in the model and does not imply that they are necessarily actively questing for hosts. Probability of active questing is incorporated into the probability of contacting a host (see below). The fat content (an indicator of physiological age) of field ticks demonstrated that, whether or not they have undergone developmental diapause, ticks emerge for questing in a largely synchronised fashion each year in the late summer/early autumn [12]. These ticks then either quest at this juncture or undergo a period of behavioural diapause. It is not known what determines which individuals start to quest during the autumn rather than entering behavioural diapause, and here the choices are modelled as mutually exclusive. The proportions that go into either autumn-questing (0.1) or behavioural diapause (0.9) were estimated for the UK by reference to the relative heights of spring and autumn peaks of abundance, and held constant for all UK field sites.

Ticks that undergo behavioural diapause over winter emerge to quest the following spring [12], thought to be triggered by increasing temperatures; the threshold for onset of questing is set at a smoothed weekly average daily maximum temperature of 7°C for nymphs and adults and 10°C for larvae, in line with field observations [9,28,29]. In years when this temperature index does not drop below 7°C, the date of activity onset is set at 22nd December, the shortest day; increasing day length is thought to stimulate questing where temperatures are sufficiently high [9]. In reality, an increasing proportion of the tick population may start to quest as more of its micro-habitat reaches the threshold temperature, so a Gaussian function is applied after the threshold date. The parameter values of this function were calibrated against field data so as to capture the timing and rate of spring appearance of questing ticks, but are nevertheless constant across the three UK field sites used for model construction.

Host Density and Feeding Probability

Effective host density is defined by the stage-specific host contact rate, represented as the probability that a tick in a questing position will make contact with a host (Table 1). This is separated into the relative contributions from large and small hosts: small hosts are programmed to feed 0% of adults, 55% of nymphs and 95% of larvae, according to limited empirical data from woodland habitats [30], but on upland moorlands in the UK large hosts (predominantly sheep) feed >90% of immature stages [31,32].* It is an individual tick's probability of finding a host, not the proportional rate of change in the questing population, that is incorporated into this model. Host contact rates were expected to vary between field sites, and values were scaled according to observed differences in tick density at the three sites used in model fitting. This is the only parameter value that is site-specific.

* New data on bloodmeal sources show that c.35-50% of larval *I. ricinus* may feed on deer in Swiss forests (Burri, C., Bastic, V., Maeder, G., Patalas, E., & Gern, L. (2011). *J. Med Entomol.*, *48*, 615-627).

Table 1. Model parameter values, specific for each field site only where shown. Values for the two validation sites are as per Exmoor

Density-dependent mortality	
Slope	
Larvae	0.219
Nymphs	0.7008
Adults	0.7008
Intercept	
Larvae	-1.45
Nymphs	-4.597
Adults	-4.597

Daily host contact rates	
Wimborne, Dorset	
Larvae	0.07325
Nymphs	0.0975
Adults	0.105
Exmoor, Somerset	
Larvae	0.0493
Nymphs	0.071
Adults	0.09
Powys, Wales	
Larvae	0.025
Nymphs	0.04
Adults	0.06

Constant daily mortality	
Questing	
Larvae	0.0068
Nymphs	0.0034
Adults	0.00136
Behavioural diapause	
Larvae	0.001428
Nymphs	0.000476
Adults	0.000408
Developmental diapause	
Larvae	0.001428
Nymphs	0.000476
Adults	0.000408
Development	
Larvae	0.001428
Nymphs	0.000476
Adults	0.000408
Eggs	0.02618

Proportion of ticks that quest in autumn	
Larvae	0.1
Nymphs	0.1
Adults	0.1

Curve controlling exit from behavioural diapause - peak height	
Larvae	0.05
Nymphs	0.125
Adults	0.2

Curve controlling exit from behavioural diapause - centre date	
Larvae	2nd August
Nymphs	4th August
Adults	4th August

Curve controlling exit from developmental diapause - peak height	
Larvae	0.2
Nymphs	0.2
Adults	0.2

Curve controlling exit from developmental diapause - centre date	
Larvae	1st April
Nymphs	1st April
Adults	1st April

A tick's feeding probability is calculated as the product of the probability of its active questing (i.e. being positioned on the vegetation ready to respond to a passing host) and the host contact rate. The first term is modelled as a function of saturation deficit (SD) (equations 4 and 5). The effects of SD on the probability of questing by nymphs (also applied to larvae in lieu of anything more suitable) and adults are based on the proportion of ticks seen questing in field arenas over several months [19], but these data do not account for tick mortality, so they were first corrected, modelling mortality as an exponential decay curve. Furthermore, relative humidity and temperature (used to calculate SD) in that study were measured at a nearby meteorological station [19] rather than by a data logger at the site, so the values had to be adjusted using the relationships between these climate variables at a relevant field site (Dorset, UK) and a nearby meteorological station. The relationship between the corrected proportions of questing ticks and adjusted SD was used to generate the functions:

$$\text{Proportion of nymphs questing} = 0.7393 \times e^{-0.1054*SD} \quad (4)$$

$$\text{Proportion of adults questing} = 0.7597 \times e^{-0.1021*SD} \quad (5)$$

The feeding probability (F) for each stage can therefore be described by the generic equation:

$$F = pq * (shc + lhc) \quad (6)$$

Where

pq = the probability of questing
shc = small host contact rate
lhc = large host contact rate

Host contact rates for each stage, combined for both host sizes, are given in Table 1.

The feeding period, three days for larvae and nymphs, ten days for adults, is incorporated into the model as a fixed delay. Density-dependent mortality is applied during this period (see below).

Egg-Laying

A fully engorged adult female *I. ricinus* is assumed to lay 2,000 eggs [33]. Egg-laying is modelled as this fixed number of eggs distributed over a 60-day period according to the egg-laying curve for *Rhipicephalus appendiculatus* [34], shown in equations 7 and 8. Adult females are removed from the model after laying, when they die.

$$\text{ELV} = k((1-(e^{-r1x(1,60)}))e^{-r2x(1,60)}) \quad (7)$$

$$k = 1000/\text{sum(ELV)} \quad (8)$$

Mortality

ELV is a 60-element vector that contains the factors by which to multiply the 60-element laying adult vector to generate the correct number of eggs laid per female per day. The term $x(1,60)$ describes the vector elements 1-60, i.e. the number of eggs laid by each female varies as she moves sequentially down the vector. The constant *k*, that depends upon the values of *r*1 and *r*2, ensures that the maximum number of eggs laid over 60 days does not exceed 2,000; the function is applied to adults of both sexes, so the maximum number of eggs is halved to account for a 50:50 sex ratio (Eqn 8). The constants *r*1 and *r*2 determine the steepness of the climb (*r*1) and decline (*r*2) of the egg-laying curve; the same values are used as previously given [34], where the values were adjusted to fit the curve given by Branagan (1973) [35].

Mortality

Daily density-independent mortality rates are applied to all stages. Background rates are based upon field observations on ticks held in permeable containers in the Czech Republic [36] and Ireland [33], but were modified *post hoc* to ensure that they were low enough to allow population equilibrium at the climatically least favourable of the field sites used for validation (Powys, Wales; see below for description of sites). To achieve this, all the published mortality values were multiplied by a factor of 0.68.

Mortality rates were not available for unfed ticks in behavioural diapause, which may suffer lower predation from arthropod predators than engorged ticks [37,38] and be less susceptible to freezing stress [39], but may be more vulnerable to desiccation (see below) due to their higher surface area to volume ratio and the relatively low levels of deposition of waxy lipids in cuticles of unfed ticks [40]. In the absence of empirical data, the most parsimonious solution is to assume equal background mortality and apply the same rates as for the engorged stages.

Mortality rates vary with the degree of desiccation stress determined by atmospheric SD, but increase only when SD rises beyond approximately 3 mm Hg [41]. This level of moisture stress was virtually never recorded in the tick habitats monitored in the UK, so this term was omitted. In drier regions of Europe, however, there may be no choice but to rely on the only available empirical parameters and relationships, even though derived for a different species (*I. scapularis*) on a different continent (N. America).

In addition to the background daily mortality rate, questing duration is limited to 60 days in larvae and 120 days in nymphs and adults, whether or not preceded by behavioural diapause, according to the approximate maximum lifespan of unfed ticks imposed by the metabolic costs of questing [31].

Besides mortality through desiccation, starvation and predation, all ticks are likely to experience density-dependent mortality, potentially mediated via host immunity [42,43], host grooming [44] and host avoidance of infested areas [45]. In East and South Africa, relatively simple demographic processes and seasonal recruitment in tropical climates allowed density-dependent effects to be estimated for *R. appendiculatus* from the relationships between interstadial mortality indices (*k*-values) and tick density [21]. For example, in South Africa the overall density-dependent relationship showed a slope of 0.387 and intercept of -0.814 during the larva-to-nymph interstadial period, and a slope of 1.24 and intercept of -2.579 during the nymph-to-adult period. These data came mainly from farms where cattle act as the dominant host. A more varied host assemblage might produce weaker density-dependent effects if some hosts are less likely to display tick avoidance, grooming and/or acquired resistance [46]. Where rodents feed a greater proportion of ticks, the rapid turnover of hosts relative to the timing of acquired immune response might be expected to impose less density-dependent tick population regulation than in a cattle-only system. Since the *I. ricinus* model was designed to simulate environments with diverse wildlife host assemblages (and because 'mortality' in the African study also included density-independent mortality, which is accounted for separately in our model), the slope of density-dependence was set below one and the intercept below zero (i.e. a positive threshold tick density). Within these constraints, the slope and intercept for the larva/nymph transition were iteratively adjusted to produce matches with the observed data; the nymph/adult values were fixed to retain the same proportional relationship to the larva/nymph values as in Randolph's (1997) study [21], with the same values applied to the adult/egg transition in the absence of any other empirical data.

Density-dependent mortality is applied during the engorgement period according to the generic equations 9 and 10 for the probability of survival (S); individual values for slope and intercept for the three stages are given in Table 1:

$$kv = (i + s(\log((qp + 1)*100))) * y \tag{9}$$

$$S = 10^{-kv} \tag{10}$$

where, kv = 'k-value'; i = intercept of line of best fit in relationship between kv and log of population size (see [21,47]); s = slope of line of best fit in relationship between kv and log of population size; qp = active questing population; y = a scaling value inversely proportional to the host contact rate. Higher host contact rates (i.e. host density) should cause concomitant reductions in density-dependent mortality, since the on-host density of a fixed number of ticks is reduced when the host density is increased.

Field Data for Validation

Tick Data

Primary data (density of each tick stage) for model construction came from blanket sampling for *I. ricinus*, carried out at approximately fortnightly intervals between 1997 and 2000 at three sites in the UK: in Powys, Wales, and in Dorset and Somerset (Exmoor), south-west England [12]. Data for independent model validation were provided by blanket sampling on Exmoor in a similar habitat to the above at three-weekly intervals from March 2008 to December 2009 [48], and for a site in central Spain where ticks were sampled monthly for nine years [49].

Climate Data

Hourly temperature (ground level and 30cm above) and relative humidity (30cm above ground level) at the three field sites were recorded using automatic weather stations (based on Squirrel data loggers, Grant Instruments, Cambridge, UK) within the same time periods (above) as the tick data collection. Hourly data were converted to daily minimum, maximum and mean figures. Saturation deficit at 30cm above ground level was calculated from temperature and relative humidity (equation in [50]). For model simulations, actual climate data for each site were preceded by a ten-year series in which each day was given a value equal to the mean for that date across the full observation period for that site; this allowed population establishment. Comparisons with observed data were based on model simulations using the actual data after these ten years of mean values.

Microclimate is rarely recorded, necessitating the use of coarser standard data from meteorological stations. In some cases, this is more appropriate as it matches our knowledge base. For example, to match the empirical basis for the observed threshold temperatures for the onset of activity in the spring (see above), triggers for the exit from behavioural diapause are set by threshold daily maximum values taken from meteorological stations closest to the field sites. These temperatures were smoothed by loess interpolation to give weekly averages. In other cases, for example for the sites used for independent tests of this model, station data are all that are available. Their use is a good test of the general robustness of the model to cruder input data. For the Exmoor test site, climate data (temperature and relative humidity at questing height) taken during the field visits were calibrated with hourly data from a nearby meteorological station for use as the input in model testing. For the field site in central Spain, recorded weekly climate data [49] were converted to daily values by simple linear interpolation.

Model Evaluation

Empirical Validation

The three field sites for which model outputs were generated differed in their input climate data and also host contact rates, which reflect estimated indices of varying host abundance. Apart from that, all other parameter values (Table 1) were held constant across all sites. The goodness of fit between the predicted seasonal patterns of numbers of ticks of each stage and those observed on each date of field sampling over three years was assessed according to the R^2 value and slope of the regression line [34]. The same procedure was used to test simulations for the two independent field sites (Exmoor and central Spain). In both cases, host contact rates were specified according to the original Exmoor site. Because interstadial development periods cannot be estimated from observations on field populations where two sorts of diapause occur, the observed and predicted relationships between interstadial mortality rates and population density cannot be compared, as they were in [34]. Simulations for the independent Exmoor site are presented for the years of data collection (2008-2009), but those for Spain (a nine-year series) are presented as mean values per sample/model day across all years, because of an observed strong upwards trend over the nine years that was demonstrably unrelated to climate [49].

Sensitivity

Following the previously described method [34], the sensitivity of model predictions to deviations in parameter values was investigated by alternately increasing and decreasing each parameter value in turn by 50%, except where parameters consisted of dates (e.g. the centre date of the curve controlling exit from behavioural diapause), which were moved forwards and backwards by fourteen days. The intercepts of the density-dependent mortality functions were not altered, since they merely pertain to the hypothetical unit of density of the model simulations, and are thus effectively arbitrary.

Models were run with each altered parameter in turn for a 13-year simulation using climate data from the Dorset data logger, with each day's values comprising the mean values from that date in each of the years for which data were recorded. Using only the final five years of simulations, the effect on R^2 values and slopes in regressions of each changed model against the final model was recorded. Changed models that returned mean population sizes lower than the final model give regression slopes between zero and one, and are presented in Figs. 7 & 8 as negative reciprocals to give equivalent magnitude of change as in models where means are higher (when slopes exceed one). A slope of 0.2 therefore becomes -5, indicating that the changed model's mean was five times lower than that of the final model.

Population growth rate calculations involving matrix eigenvectors [24] are not suitable for this model due to the number of life history events (notably density-dependent mortality and temperature-dependent development) that are modelled in vectors external to the main matrix, and which vary according to external inputs and/or population state.

Model Performance

Abundance and Seasonality at Primary Field Sites for Model Construction

The observed and predicted numbers of ticks of each stage at the three field sites (Figs. 2, 3 & 4), and correlations between them (R^2 and slope values - Table 2) indicate that, with some exceptions, the timing of appearance and decline of each questing stage was adequately predicted.

The poor performance of the model in capturing larval seasonality at Dorset and Exmoor in 1997 (Fig. 2) may reflect a lag, to some extent, between climatic causes in one year and effects on ticks the following year. Simulations in 1997 are likely to be less accurate than in subsequent years because simulated tick populations were first established by using average climate conditions in the model before actual climate data were recorded on-site from 1997. There is no obvious explanation for the apparently anomalous reversal of observed spring and autumn peaks of larvae at Exmoor in 1997 (Fig. 2), which reduced the overall correlation despite the otherwise excellent match.

Differences in goodness of fit between predictions and observations may also be due to habitat features that determine how well field samples reflect reality. For example, it was necessary to vary the multiplication factors to accommodate differing ratios of tick life stages at each site, probably due to variable sampling efficiency of each stage in different vegetation or host composition. Both the primary and validation sites at Exmoor were open deciduous woodlands with an understorey of *Vaccinium* spp., which is ideal for blanket dragging, being pliable enough for the blanket to make contact equally with all parts of the vegetation throughout the year. In Dorset, the mixed deciduous/coniferous woodland, with ground cover of *Hyacinthoides non-scripta* and *Mercurialis canis*, allowed equally good sampling efficiency, despite some sparse *Pteridium aquilinum*. The Welsh site, in contrast, was an area of unimproved upland sheep grazing, comprising grasses of predominantly *Festuca*, *Agrostis* spp. and also much widespread *Molinia* spp., a substrate that strongly impedes the blanket's contact with lower parts of the vegetation where larval ticks quest. Additionally, the exposed nature of this latter site would render the tops of the vegetation unfavourable for larvae. Sample sizes were therefore very small. These factors may explain why larvae in Wales are one exception to the good model fit, suffering from significant over-prediction.

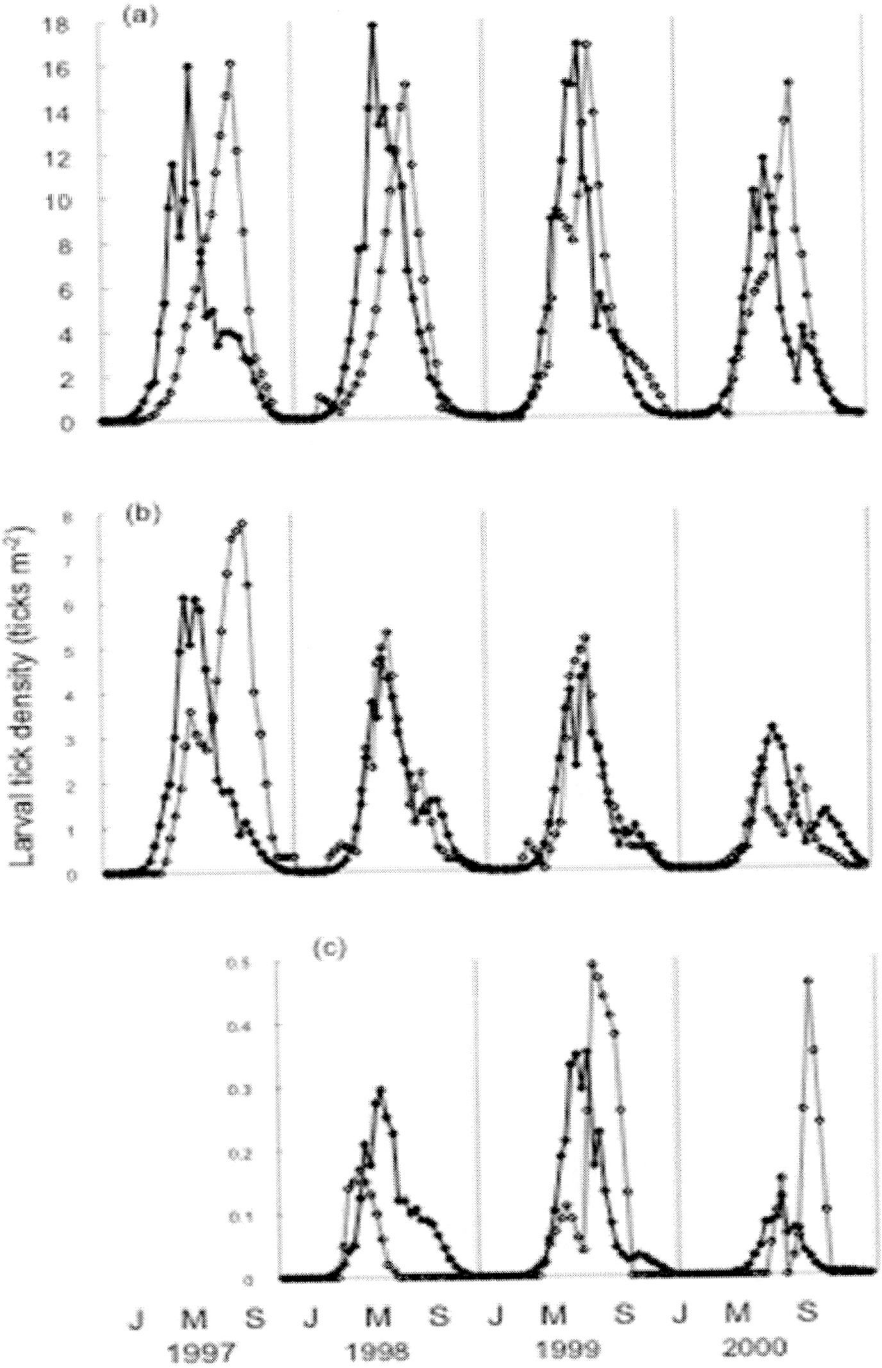

Figure 2. Observed (O) and model simulations (●) of seasonal abundance of larval ticks at (a) Wimborne, Dorset, (b) Exmoor, Somerset, (c) Powys, Wales. Field data were smoothed to dekad intervals, and simulation data taken on concurrent dates. The predicted output is re-scaled by the following factors: (a) x 1.16, (b) x 1.4, (c) x 0.16 (see text). Reproduced from Ref [18], with permission.

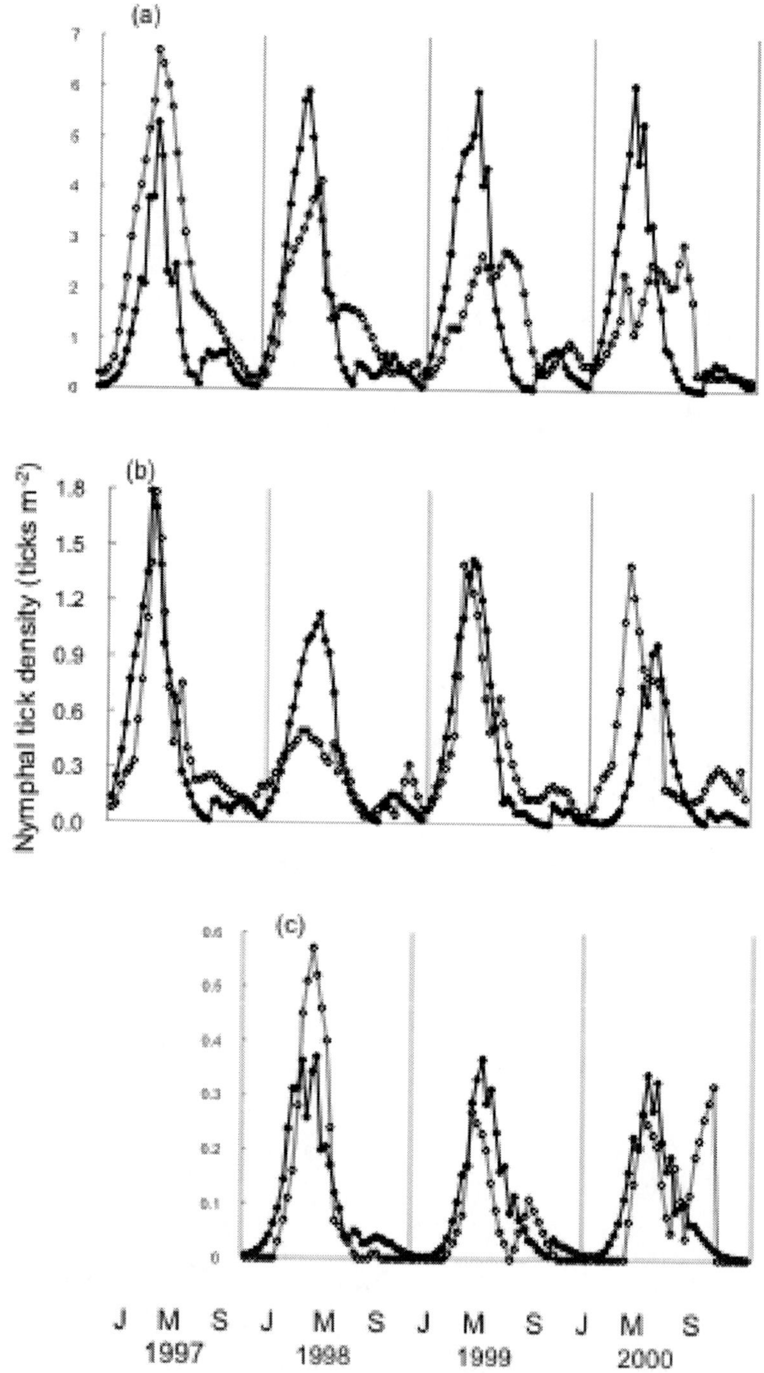

Figure 3. Observed (O) and model simulations (●) of seasonal abundance of nymphal ticks at (a) Wimborne, Dorset, (b) Exmoor, Somerset, (c) Powys, Wales. Field data were smoothed to dekad intervals, and simulation data taken on concurrent dates. None of the predicted outputs was re-scaled. Reproduced from Ref [18], with permission.

Table 2. R^2 values for correlations between predicted and observed tick abundances at each field site

		n	R^2	Slope
Wimborne, Dorset	Larvae	144	0.353	0.507
	Nymphs	144	0.303	0.656
	Adults	144	0.198	0.909
Exmoor, Somerset	Larvae	144	0.370	0.357
	Nymphs	144	0.573	0.901
	Adults	144	0.005	0.076
Powys, Wales	Larvae	108	0.120	0.619
	Nymphs	108	0.549	0.611
	Adults	108	0.454	0.788
Exmoor, validation site	Larvae	30	0.401	0.555
	Nymphs	30	0.782	0.216
	Adults	30	0.502	0.195
Spain, validation site*	Larvae	108	0.294	0.0035
	Nymphs	108	0.305	0.0023
	Adults	108	0.044	-0.0022

* For the validation site in Spain, figures refer to the unmodified model, with the same proportions of ticks questing in the autumn as in the UK.

Despite the potentially greater stochastic effects of small numbers in each sample of adults (see *y*-axes on Fig. 4), the model fit for this stage was generally acceptable. The adults in both Dorset and Wales were slightly over-predicted. This may reflect partly their relative rarity, and hence the greater element of chance in their observed abundance, and partly the greater likelihood of their being brushed off the blanket before being counted, due to their greater size. The exception to the good model fit for adults, at Exmoor, was due entirely to the anomalous peak of this stage observed at the end of 2000. This remains unexplained (and the Foot and Mouth Disease outbreak in 2001 prevented continuing fieldwork to explore any possible impact on future generations), but is not thought to indicate a significant change of conditions for questing adults.

The absolute heights of the spring and autumn peaks were less precisely captured, probably because they are so sensitive to host abundance. Seasonal variations in host abundance affect the rates of both population growth and removal of questing stages from the vegetation (see below).

Abundance and Seasonality at Independent Field Sites for Model Validation

Predictions of tick seasonal abundance at the two independent field sites threw up some instructive results. Those for a different site on Exmoor at a later period (2008-09) gave a good match to the seasonal pattern, but predicted higher absolute numbers of nymphs and adults (Fig. 5), as would be expected if host abundance were greater at this site than at the primary site for which the host contact rates were calibrated (based on local reports of deer abundance, this is likely but cannot be tested empirically).

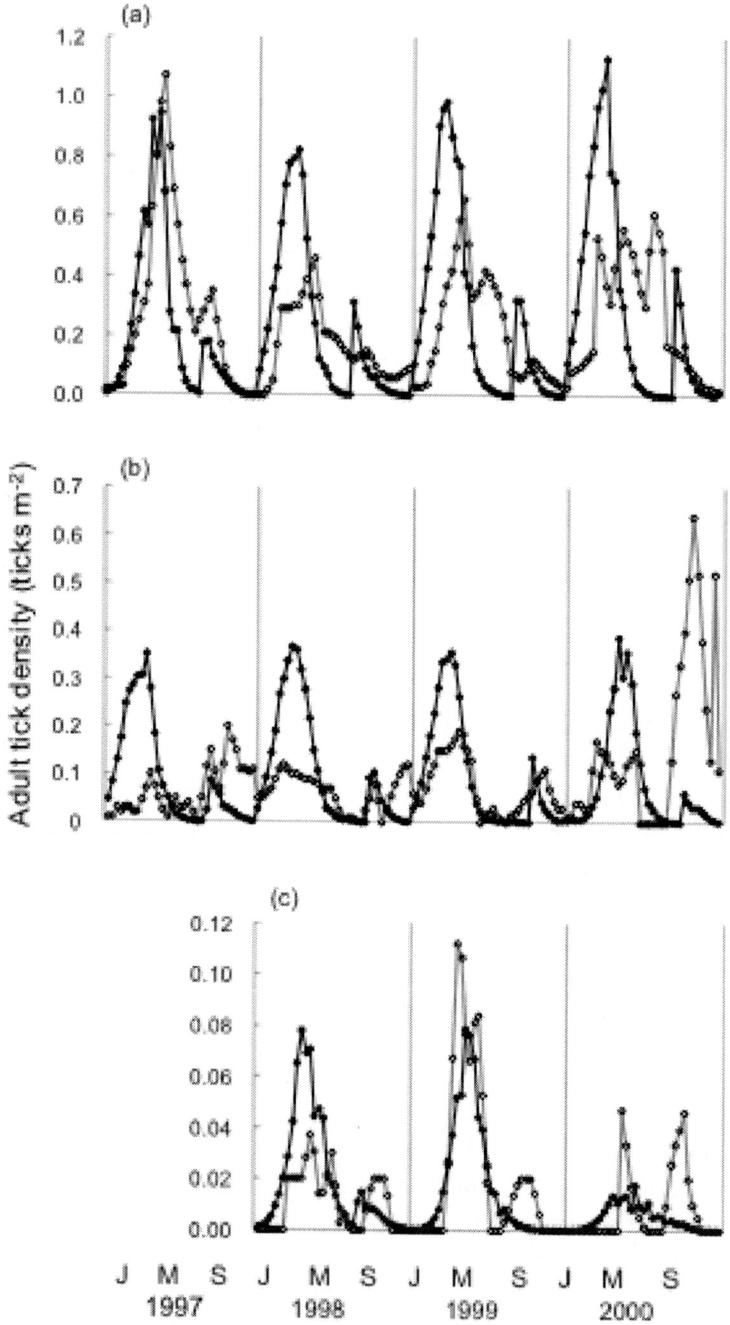

Figure 4. Observed (O) and model simulations (●) of seasonal abundance of adult ticks at (a) Wimborne, Dorset, (b) Exmoor, Somerset, (c) Powys, Wales. Field data were smoothed to dekad intervals, and simulation data taken on concurrent dates. Observed adult numbers = 2 x females. The predicted output is re-scaled by the following factors: (a) x 0.7, (b) not re-scaled, (c) x 0.75 (see text). Reproduced from Ref [18], with permission.

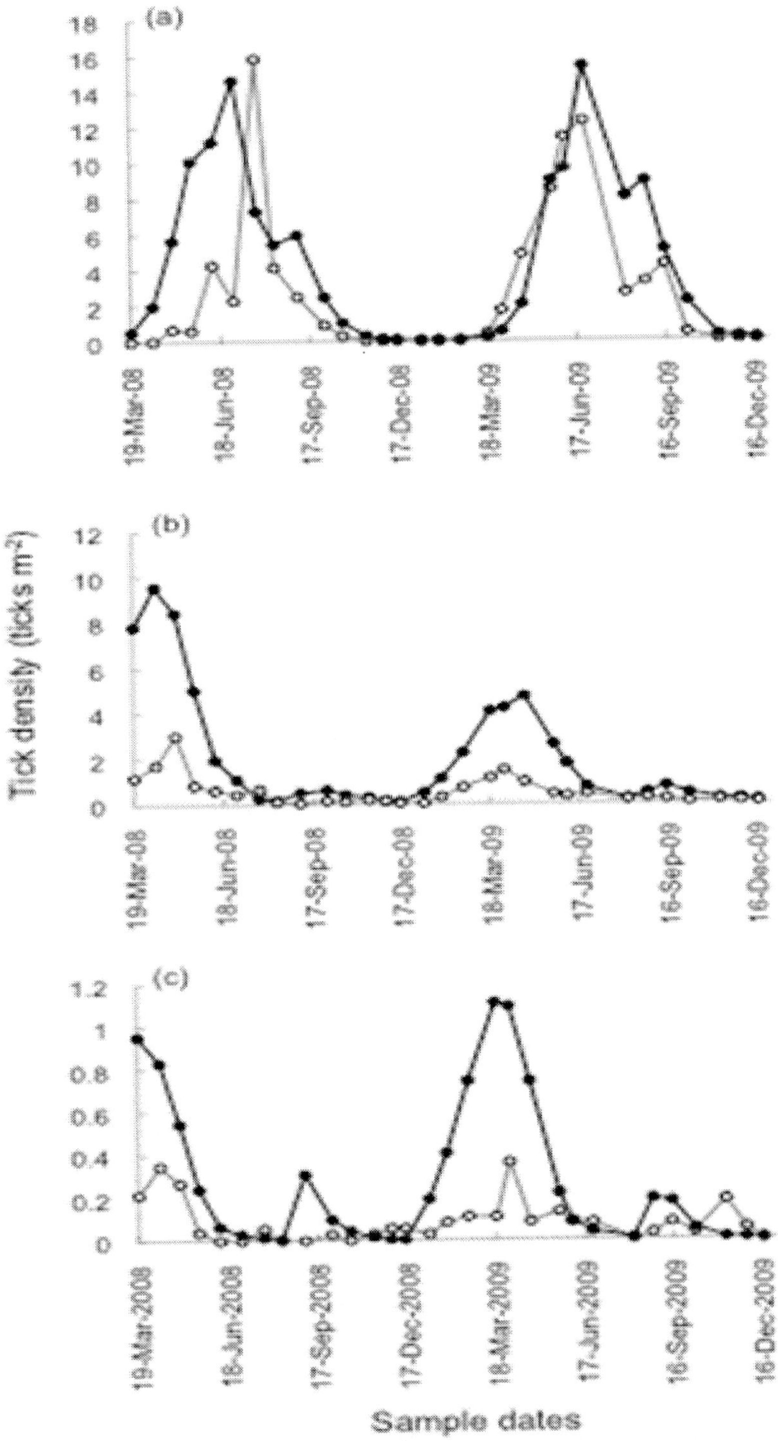

Figure 5. Model simulations (●) and observed (O) numbers of larval (a), nymphal (b) and adult (c) ticks from an independent test site on Exmoor. Reproduced from Ref [18], with permission.

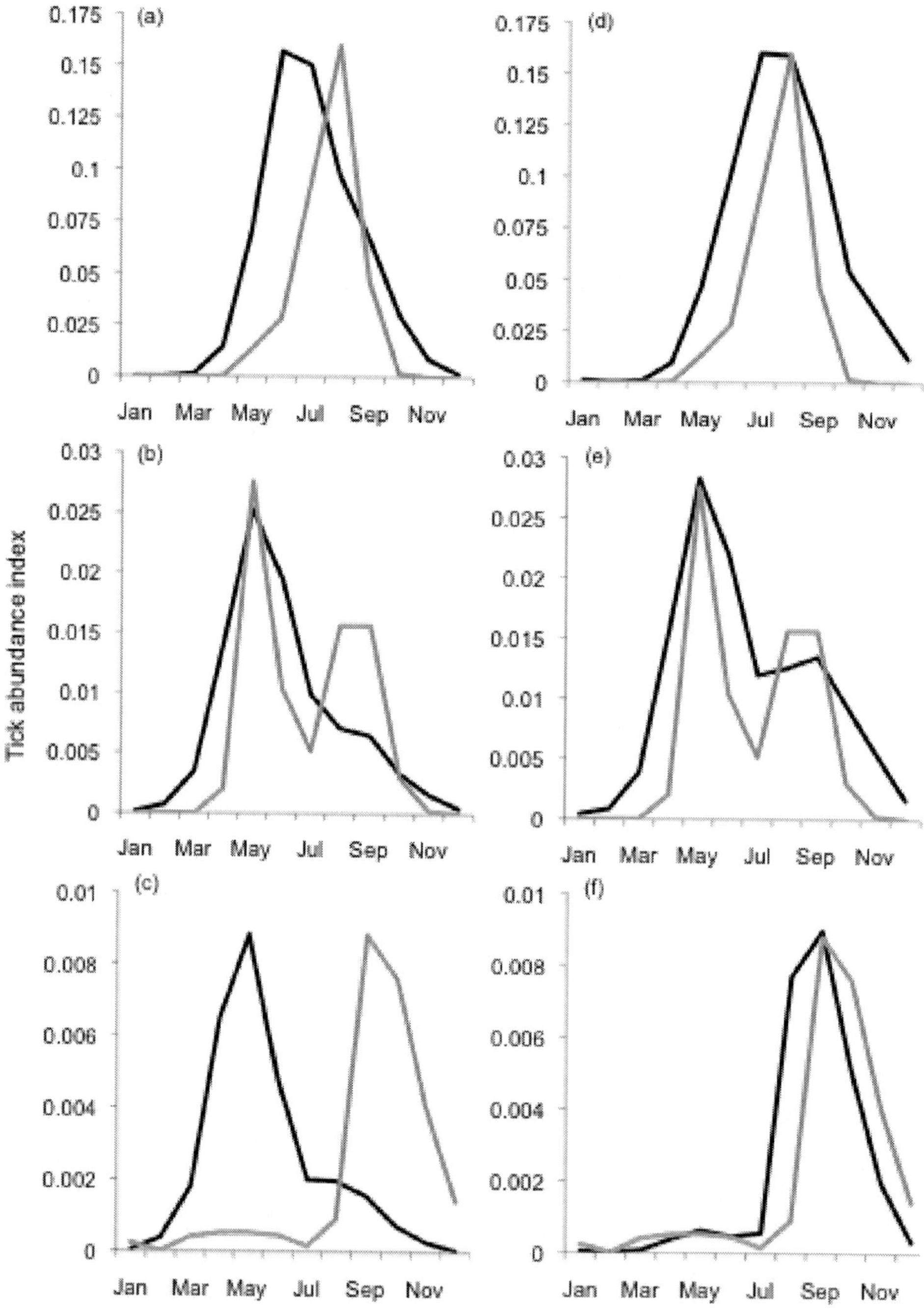

Figure 6. Model simulations (black) and observed (grey) tick density/m^2 from central Spain recorded over 1994-2002 by Estrada-Peña *et al.* (2004) [49]. Models were run for all nine years but the figure shows mean values per sampling/modelling day across all years (see text). Model simulations were scaled to fit observations to aid visual interpretation. Left side (a), (b) and (c), model predictions for larvae, nymphs and adults, respectively, where host contact rates are set as per the Exmoor, UK, site, with no adjustment to the proportions of ticks showing autumn questing (10%) or overwinter behavioural diapause (90%). Right side, simulations with questing proportions altered to: (d) larvae – 40%, (e) nymphs – 25%, (f) adults – 95%. Reproduced from Ref [18], with permission.

In central Spain, however, the same parameter values failed to capture the seasonal timing of activity of larvae and adults (Fig. 6, left). This was rectified when the percentage of ticks that quest in autumn, rather than entering behavioural diapause, was changed from 10% for all stages as in the model for the UK, to 40% for larvae, 25% for nymphs, and 95% for adults. The model then captured much more closely the starkly different seasonal patterns seen in central Spain (Fig. 6, right).

This parameter, the fraction of moulted unfed ticks that, upon recruitment in the early autumn, quest immediately rather than enter behavioural diapause overwinter and re-emerge the following spring, is an important function in the model. In north-western Europe, the majority do the latter [12], so the ratio of questing to diapausing ticks is set at 1:9 for all stages in the UK model. Although population seasonality was relatively insensitive to 50% changes in this parameter (see below), larger changes generated markedly different patterns and were indeed necessary to capture the pattern recorded for *I. ricinus* in central Spain [49], with other parameter values unchanged. Crude analyses suggest that in northern Europe this behavioural diapause may be triggered by responses to shortening day-length [9], but the degree of variable genetic determination or phenotypic plasticity to allow ticks in geographically distinct environments to respond appropriately is entirely unknown. Ignorance of the degree of geographical phenotypic variation, as seen in *I. scapularis* [41,51], limits the operation of any population model across the very extensive full range of *I. ricinus* without significant *post hoc* alterations to certain parameters. Observations on the behaviour of *I. ricinus* translocated to opposite extremes of its range would indicate whether geographical races act as discrete entities when considering population dynamics.

Parameter Sensitivity

Sensitivity results for simulations of nymph numbers (Figs. 7 & 8), and extremely similar results for adults and larvae (not shown), indicate a robustness of the final model to perturbations, as was seen for the African tick *R. appendiculatus* [34]. Changes in host contact rates and parameters associated with density-dependent mortality had greater effects than changes in background daily mortality rates. Model simulations were, however, particularly sensitive to egg survival. In addition, changes in the curve dictating the spring-time onset of questing activity after behavioural diapause caused considerable deviation from observed seasonal patterns, but the reduced R^2 values were always >0.85 of the original values (Fig. 7). Comparisons within groups of parameters (mortalities, autumn questing fractions, etc.) are more meaningful than between-group comparisons, given, for example, that a 50% increase in the fraction of ticks that quest in the autumn is only arbitrarily equivalent to a 50% increase in egg survival.

Perturbations to the slope of predicted/observed regressions (Fig. 8) caused by changes in density-independent mortality values were generally small, and were in proportion to the initial values; the largest perturbation thus came from changes in egg survival (EgS), which had the largest starting value (0.02618).

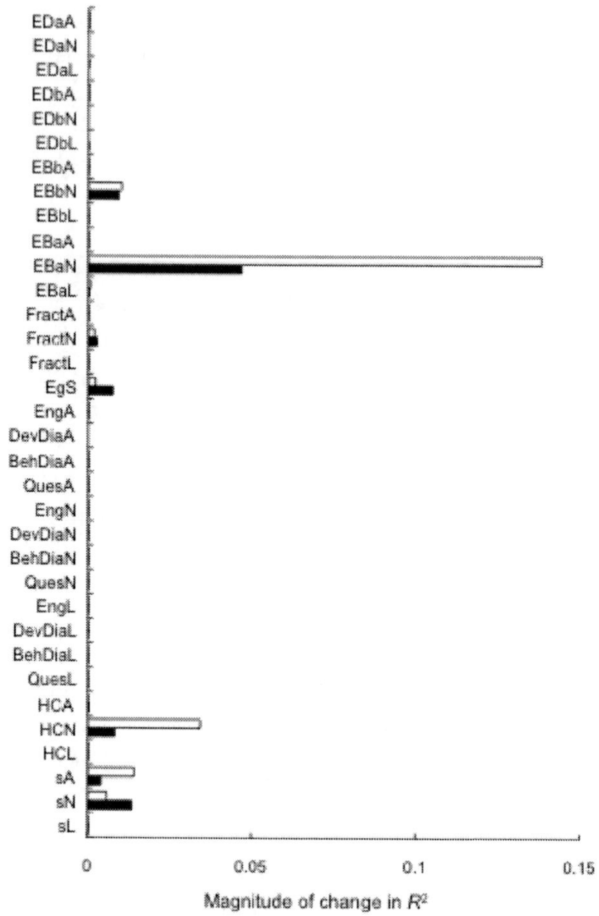

Figure 7. Sensitivity analysis: reduction in regression R^2 value relative to final model after 50% increase (filled bars) or decrease (open bars) in parameter values. Parameter definitions: Controls on Gaussian function controlling exit from developmental diapause: EDbA - Adult curve centre date; EDaA - Adult curve height; EDbN - Nymph curve centre date; EDaN - Nymph curve height; EDbL - Larva curve centre date; EDaL - Larva curve height. Controls on Gaussian function controlling exit from behavioural diapause: EBbA - Adult curve centre point; EBaA - Adult curve height; EBbN - Nymph curve centre point; EBaN - Nymph curve height; EBbL - Larva curve centre point; EBaL - Larva curve height. Fraction questing in autumn: FractA – Adults; FractN – Nymphs; FractL – Larvae. Mortality values: EgS – Eggs; EngA - Engorged adults; DevDiaA - Adults in developmental diapause; BehDiaA - Adults in behavioural diapause; QuesA - Questing adults; EngN - Engorged nymphs; DevDiaN - Nymphs in developmental diapause; BehDiaN - Nymphs in behavioural diapause; QuesN - Questing nymphs; EngL - Engorged larvae; DevDiaL - Larvae in developmental diapause; BehDiaL - Larvae in behavioural diapause; QuesL - Questing larvae. Host contact rates: HCA – Adults; HCN – Nymphs; HCL – Larvae. Density-dependent mortality regression values: sA - Adult slope; sN - Nymph slope; sL - Larva slope. Reproduced from Ref [18], with permission.

Changes in density-dependent mortality parameters caused considerably more perturbations in the slope than in the R^2 values, indicating that such changes cause significant changes in equilibrium density but not seasonal patterns. These large changes occurred when density-dependence was made more severe (up to 30-fold decrease in nymphal equilibrium population density by increasing the slope of the adult density-dependent mortality regression by 50% - almost certainly an unrealistic perturbation). These results match the assumptions made when choosing the density-dependence parameters; the population is modelled as experiencing only moderate density-dependence, such that any increase in its effect will immediately reduce population density, whereas decreases will have more negligible effects.

Overall, the sensitivity analysis indicates that the predicted seasonal patterns are robust to realistic deviations in almost all parameters, so that knowledge of only approximate values within biological plausibility does not hinder the model's utility. The predicted population density, however, may be highly sensitive to exact values of egg survival, necessitating further laboratory experiments and field observations on this key life cycle stage.

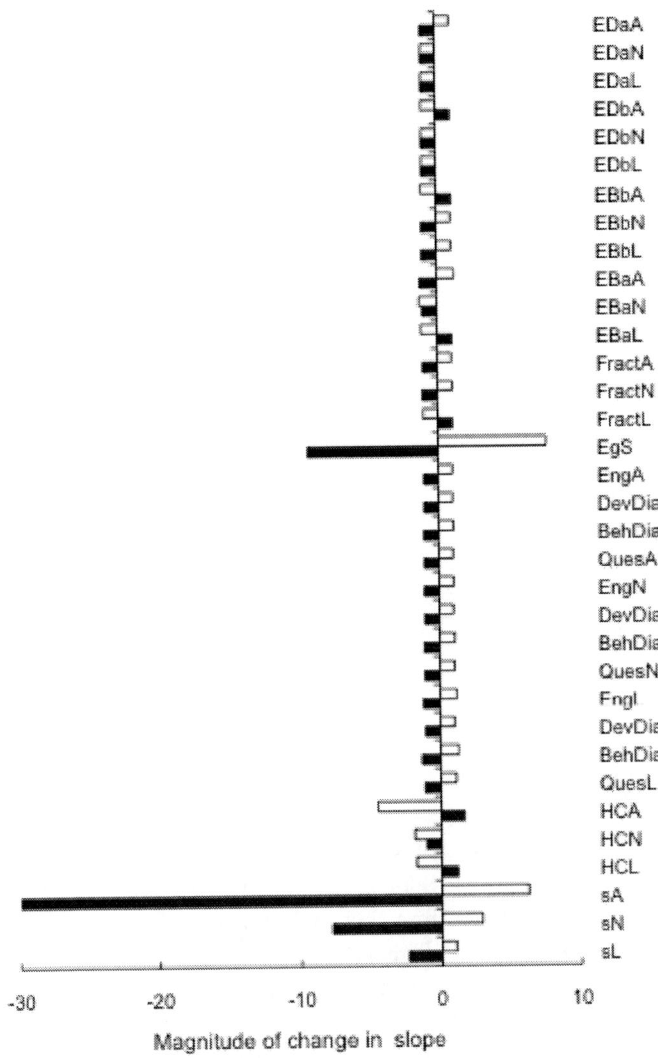

Figure 8. Sensitivity analysis: change in regression slope relative to final model after 50% increase (filled bars) or decrease (open bars) in parameter values. Positive numbers are equal to the slope; negative numbers are equal to the negative reciprocal of the slope, thus the axis reflects the factor by which slope is either multiplied or reduced. Parameter labels as in Fig. 7 legend. Reproduced from Ref [18], with permission.

Model Strengths and Weaknesses

This first process-based population model for *I. ricinus* shows considerable promise in its predictive efficacy. At this early stage in its development, its function is less to predict seasonal dynamics and abundance across the tick's extensive range, and more to test our understanding of the component population processes. In capturing the seasonal dynamics of *I. ricinus* in a north-western European climate, the model supports current ideas on the fundamental patterns of diapause, development and recruitment [12], even though extrinsic determinants of the first of these processes require much more detailed investigation.

The model shows that a small number of climatic variables may satisfactorily explain spatial variation in seasonal patterns of tick numbers within the UK. It also shows that relatively crude meteorological records from nearby stations, in the absence of data from within the tick's habitat, may offer satisfactory input variables. Given the importance of these seasonal patterns to the transmission potential of some tick-borne pathogens [13,16,17], this model should be extended to include elements in the transmission cycle of one of the many pathogens vectored by *I. ricinus*. Such a model exists for Lyme borreliosis in North America in a simplified system involving a single host species [16] and in a more complex host system [41].

To achieve the same predictability of overall abundance, however, host contact rates had to be varied within biologically reasonable limits, thereby indicating the importance of host abundance in determining tick abundance [52-54] (and see below). Unfortunately, this crucial biotic variable is highly variable in time and space, and rarely quantified beyond crude relative indices. Indeed, a number of assumptions have been made that warn against using the model to make specific, quantitative predictions about absolute tick population size at equilibrium in different parts of the UK. The model is, however, perfectly adequate and very informative when used for investigating the effects of specific environmental change on tick density or for providing practical management recommendations where ticks pose an epidemiological and/or economic threat [54], as follows.

Model Applications

In the absence of good quantitative data from long-term systematic monitoring of tick populations (but see [55] for data on Scottish grouse moors, 1985-2004), qualitative reports from many diverse sources (farmers, foresters, wildlife biologists, gamekeepers, the general public, etc.) consistently indicate an increase in tick distribution and abundance in the UK over the past one or two decades, either by anecdotal verbal reports or from a postal survey carried out in 2006 [56]. In 73% of the locations covered by this survey, ticks were perceived to have increased in abundance over the previous 5-10 years. Possible causes for such tick population increases include climate change, increased abundance of large animal hosts for ticks (deer), changes in farming and livestock husbandry practices, and changes in land cover. The tick population model has been used to investigate the relative impacts on tick density of actual recent changes in some of these factors [54]. Uncertainty of some parameter values in the model prevents reliable estimates of absolute tick abundance, but changes in indices of tick abundance may be estimated. The focus here is on the numbers of questing ticks, as an index of active ticks that are a subset of the full population. This modelling based on changing conditions in the recent past provides a basis for developing scenarios for the future under anticipated environmental changes, clarifying the relative contributions of abiotic and biotic factors. The model is also used to assess variable practices of acaricide application to hosts as an effective control strategy, to allow land managers and livestock owners to weigh up the costs and benefits of manipulating host densities in attempts to reduce tick abundance.

The Impact of Recent Climate change on Tick Populations

The above analyses of demographic processes show how climatic variables can influence tick populations in non-synergistic ways: rates of development are increased by rising temperature, but survival and questing opportunities are adversely affected by increased saturation deficit. Recent reports of extensions in the ranges of *I. ricinus* in continental Europe, putatively attributable to temperature increases having made new areas permissive for tick survival [57-59], have fuelled speculation over the effects that climate change may have on tick distribution and hence future pathogen transmission [60-62], especially along latitudinal and altitudinal boundaries where temperature is typically limiting.

Actual Patterns of Climate Change

Meteorological records show a consistent pattern of temperature change over the past 40 years throughout Europe north of c.42°N: against a background of quasi-decadel fluctuations [63], a step increase in annual mean temperatures of up to c.1-2°C in 1989, with consistently warm conditions but no upward trend since 2000 [64-66]. This contrasts with the smoothed continuous pattern commonly depicted by 10-year running means as presented by the Intergovernmental Panel on Climate Change (IPCC). This real pattern of change and its misleading depiction are also seen in government reports for the UK (e.g. the Central England temperature series http://ukclimateprojections.defra.gov.uk /content/view/751/9). The cause of the sudden warming is attributed, at least partly, to increased solar irradiance due to reduced aerosol concentrations [67].

Long-term temperature and relative humidity data for the period 1970-2008 were downloaded from http://badc.nerc.ac.uk/data for meteorological stations at Hurn, Dorset (-1.84°E, 50.78°N), close to the tick population field site at Wimborne St. Giles, and Bala, Gwynedd (-3.58°E, 52.91°N), north of the field site in Powys, Wales [12]. In common with the above patterns seen across Europe, these 39-year time series reveal that mean annual temperatures at these two UK sites showed quasi-decadel fluctuations, and also a marked step-increase around 1989, followed by consistently warm conditions but no upward trend since 1997 (Fig. 9). Seasonally, most of this warming occurred from January to early August, which includes early spring when temperatures cross the threshold for tick questing activity, and also the period from April to July that is critical for the interstadial development of the majority of ticks and has shown a slight warming trend since 1997. During the present century, autumns have also been warmer. At the same time, there has been no consistent decrease in rainfall or relative humidity. An increase of 1-2°C would not therefore increase maximum (summer time) saturation deficit as experienced by ticks within their woodland habitat to a level (>4 mm Hg) likely to cause increased mortality or decreased questing activity [19,68].

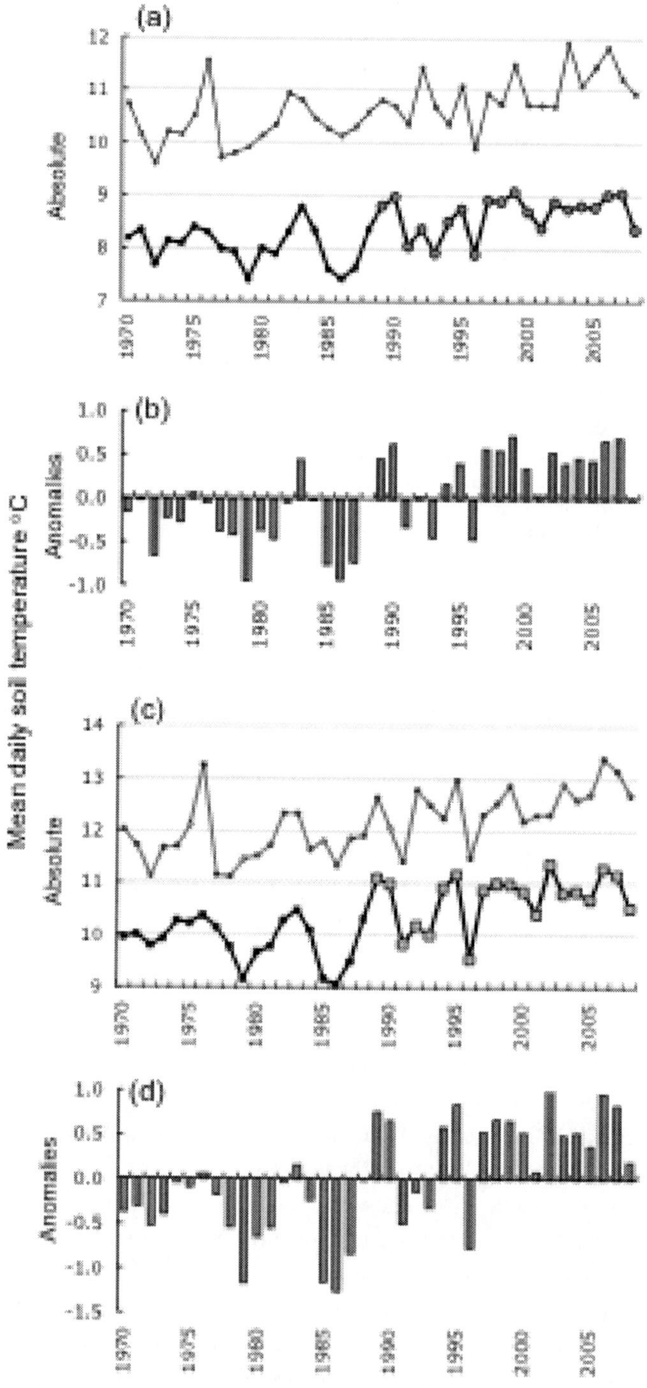

Figure 9. Mean annual (thick line) and April-July (thin line) temperatures recorded at Bala, Wales (a) and Hurst, Dorset (c). Conditions before and after the step increase in temperature in 1989 are distinguished. (b) and (d) anomalies from the 1970-2008 mean for the above mean annual temperature series. Reproduced from Ref [54], with permission.

Calibrations between simultaneous mean daily temperatures from each meteorological station and its corresponding data logger at the field sites for tick observations (1996-2001) were used to convert station readings to microclimate over the full 39-year period required, as follows. For the years where station and logger over-lapped, daily average series were produced where each day was given a value equal to the mean for that date across the full time period. Sine curves were fitted to these averaged series to produce idealised curves. The daily scale factors between the sine curves from the meteorological station and those from the data logger were calculated and used to produce values for mean daily soil temperature (T_s - for tick development), and temperature 30 cm above ground level (T_q – where ticks quest) from meteorological station mean air temperature (T_a) figures for 1970-2008. Relative humidity data from stations were similarly calibrated. Saturation deficit was calculated from the calibrated mean temperature (T_q) and relative humidity 30 cm above ground level according to the equation in [50].

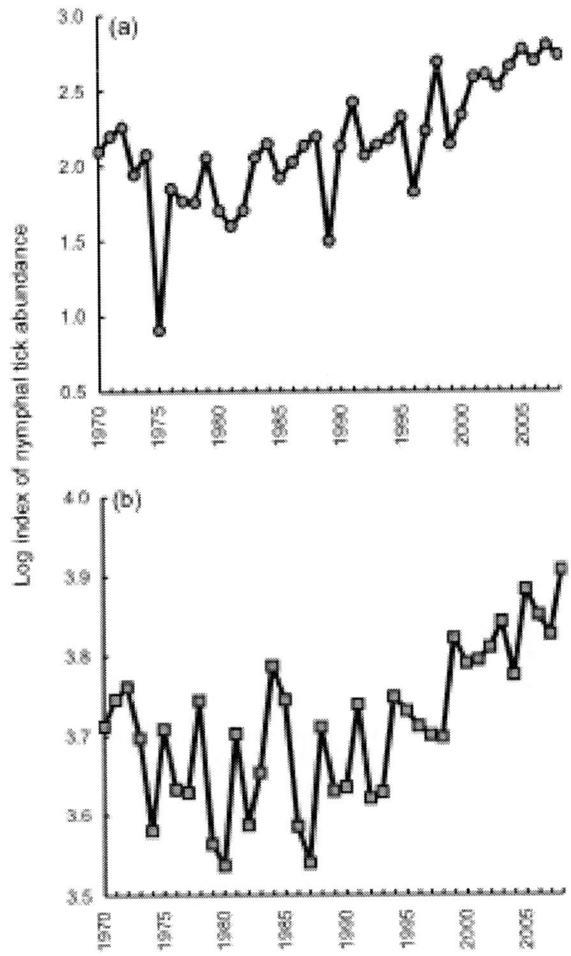

Figure 10. Simulation of mean annual questing nymphal tick populations in Wales (a) and Dorset (b) under actual climate for 1970-2008. (NB different *y*-axis scales.) Reproduced from Ref [54], with permission.

Simulated Tick Populations with Climate Change

Simulated tick populations based on daily climate data for 1970-2008 revealed the effects of the above changes in climate. To allow tick population establishment, each model run was preceded by a ten-year period in which daily temperature and SD were averaged across 1980-89 to provide a single year of 'mean' conditions, which was looped ten times. Relative to the tick abundance predicted under mean 1980-89 conditions, tick populations were predicted to vary in response to the real fluctuations in temperature from 1970 to 2008 at both sites (Fig. 10). Within each site, predicted annual total indices of nymphs were positively correlated with soil temperature of the previous year, with annual means generally better than selected seasonal mean temperatures (e.g. April-July) as significant predictors (Fig. 11). There were, however, considerable differences between the sites in the effect of warming: although in Dorset warmer initial conditions were followed by greater temperature increases (annual means by 0.69°C from 1970-88 to 1989-99 and a further 0.30°C to 2000-08) than in Wales (0.51 and 0.19°C, respectively), nymphal tick numbers increased by 7 and 36% over the same periods in Dorset, but by 77 and 159% in Wales (0.03 and 0.13 compared with 0.23 and 0.48 on the log scale in Fig. 10). Tick abundance was more variable from year to year in Wales, presumably reflecting temperatures that fluctuate closer to limiting levels. As expected from such a model with no inherent stochasticity, the abundance of each stage was strongly correlated with that of the previous developmental stage one year before, but again the tick population in Wales showed more variability in this respect (R^2 values 0.75 and 0.82 for nymphs-adults and larvae-nymphs in Wales compared with 0.99 and 0.92, respectively, in Dorset).

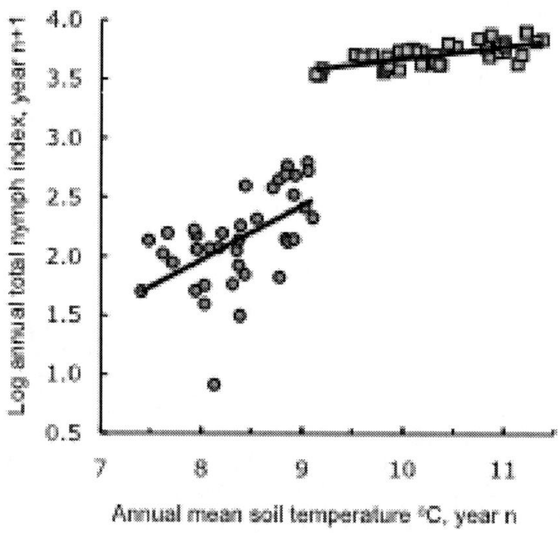

Figure 11. Correlation of predicted log total annual numbers of nymphal ticks with mean annual soil temperatures of the previous year for Wales (o) and Dorset (□). Reproduced from Ref [54], with permission.

On average, the tick phenology was predicted to change after 1988, but again with site differences (Fig. 12). Under the relatively cool conditions of Wales, warmer temperatures from 1989 allowed nymphs to start questing earlier, to reach their spring peak about a fortnight sooner, and to show a more distinct second rise in the autumn; these changes continued with further warming from 2000. In Dorset, neither the timing nor the shape of either the spring or autumn nymphal peak changed significantly. For larvae, the timing of the spring peak barely shifted in Wales, but occurred about a fortnight earlier in Dorset. The degree of synchrony between larvae and nymphs, essential for the transmission of certain pathogens with short-lived infections in vertebrates, is very low at both sites and decreased further by 20% in Wales from 1970-88 to 2000-08, but increased by 36% in Dorset.

The model allows the mechanistic basis for these responses of tick populations to climate change to be identified. The step-wise increase in temperature around 1989 will have been generally favourable to tick populations as warmer temperature will speed the development of ticks from one stage to the next, thereby reducing the cumulative effect of constant daily mortality rates. The predicted population patterns suggest that more consistent temperatures from year to year, as seen post-1997, allow tick populations to build up. This is indicated by a) the initial decline in tick numbers from 1970, especially in Wales, once real temperature fluctuations drive the model after the initial uniform 10-year bedding-in period (Fig. 10), and b) the continuing increases in ticks post-2000 when mean annual temperatures did not increase, although mean April-July temperatures did show a slight upward trend.

There are two possible reasons why the predicted questing nymph populations responded more markedly to raised temperatures, and were much more variable from year to year, under the more marginal conditions in Wales than in Dorset. First, development rates of ticks increase non-linearly with temperature, with lesser effects at higher temperatures [9]. Secondly, density-dependent mortality (a key feature of tick populations, see above) will have a greater stabilising effect at the higher population levels and less harsh climatic conditions seen in Dorset. Most of the current range of *I. ricinus* in mainland Europe experiences warmer conditions during April-July than even the mildest parts of the UK, suggesting that the impact of climate warming on tick populations will be less marked there, but more significant at the latitudinal and altitudinal peripheries. At the same time, any more marked increase in summertime saturation deficit closer to critical levels at higher mainland temperatures could increase tick mortality associated with moisture stress, thereby dampening the predicted tick population increases as European climate changed over the past 20 years.

Any effect of warming conditions on reducing the average time taken to complete a full life cycle (thought to be at least 3 years in the UK [12]) will depend on whether ticks develop fast enough to escape from the constraints of daylength-induced overwinter diapause (both developmental and behavioural [11]). Conversely, occasional unusually low spring or autumn temperatures can prevent ticks from completing their normal patterns of development and/or activity within the year. As seen in the simulations during the 1970s in Wales, this causes a sharp decline in numbers and a rebound the following year as delayed ticks emerge (as apparent from dissecting the detailed outputs of the model). The prediction of altered phenology, such that the spring peak of questing ticks occurs earlier in the year (Fig. 12), arises from the empirically-supported assumption in the model that emergence from behavioural diapause in the spring is triggered by the weekly mean of the daily maximum temperature (about 7°C for adults and nymphs, and 10°C for larvae – see above). A change in the timing of spring emergence has significant implications for pathogen transmission to humans when the vector is one frequently encountered on recreational visits to the countryside, whose timing will be partly constrained by public holidays. Any management activities designed to reduce the risk of infection, such as the dissemination of information for awareness and self-protection, or cutting the vegetation back from the edge of maintained paths, may need to start earlier than in the past, before the spring holidays.

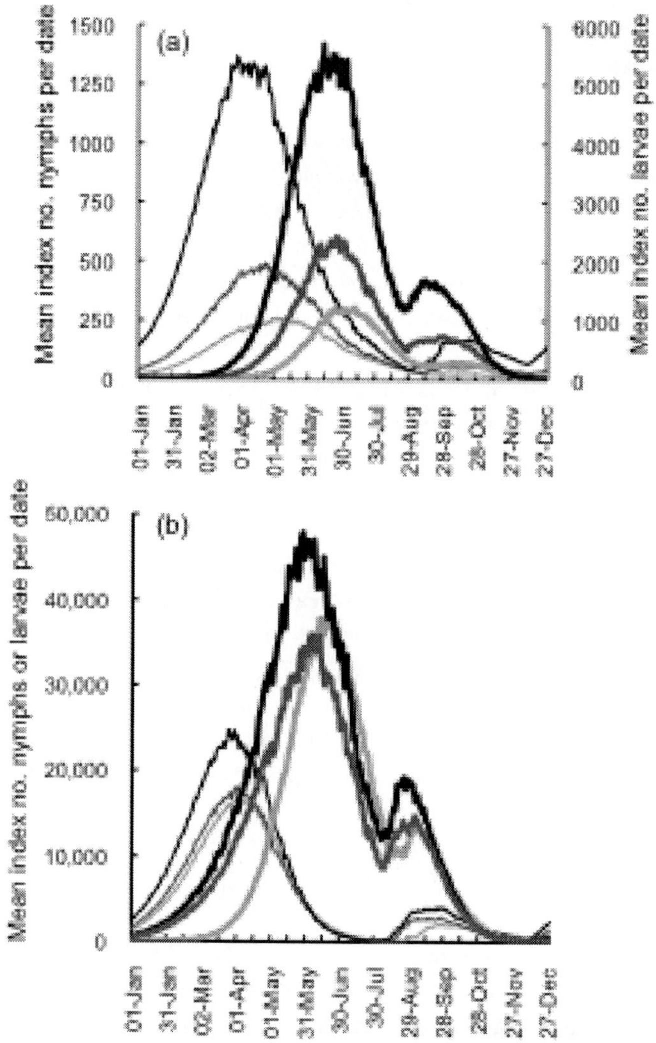

Figure 12. Simulation of seasonal patterns of the mean daily abundance of questing nymphal ticks (thin line) and larvae (thick line) for 1970-88 (pale), 1989-99 (medium) and 2000-08 (dark) for Wales (a) and Dorset (b). Reproduced from Ref [54], with permission.

The Impact of Recent Increases in Deer Abundance on Tick Populations

The major biotic determinant of tick population dynamics is the abundance of vertebrate hosts, of which deer are quantitatively the most important for all tick stages including the reproductive adults. Although the oft-observed correlation between numbers of *Ixodes* species ticks and deer densities in Europe and North America confirms that deer are instrumental in maintaining robust populations of ticks [52-54,69-71], detailed mechanistic analyses are lacking. When sampling vegetation for questing ticks, it is impossible to distinguish between those that have been removed from the population by mortality or quiescence/diapause, and those removed by attachment to a host [72]; field data relating tick abundance to host abundance are therefore difficult to interpret [73].

Host Density Trends

In the UK, the perceived increases in tick abundance in most areas over the past one or two decades largely coincide spatially and temporally with perceived increases in the density of deer, although the nature of the data collection prohibits firm conclusions [56]. Compared with temperatures, which have changed very little over the past 10 years, deer populations have expanded rapidly. Data on geographical range expansion of UK deer species between 1972 and 2002 show that roe deer (*Capreolus capreolus)* increased by 2.3% per annum, fallow deer (*Dama dama)* by 1.8%, red deer (*Cervus elaphus)* by 0.3%, Reeves' muntjac (*Muntiacus reevesi)* by 8.2%, sika deer (*Cervus nippon)* by 5.3%, and Chinese water deer (*Hydropotes inermis)* by 2.0% [74]. Recent figures suggest an accelerated rate of increase between 2003 and 2007 in all species ranging from 5.2% per year (p.a.) for roe to 22.2% p.a. for Chinese water deer [75]. Thus the first three native species occupied 2-4 times as many 10 km^2 cells in 2007 as in 1972, while the three introduced species spread into 6-20 times as many cells [75]. Range expansions are likely to follow increases in density in core areas, but unfortunately there exist few data on the real nationwide changes in numbers of these key tick hosts. The British Trust for Ornithology Breeding Bird Survey at 3139 1 km grid squares recorded an increase in their index of deer abundance of 55% for roe deer and 46% for muntjac from 1995 to 2002 [76]. Although these figures are insufficient to give an indication of full changes in deer density, either locally or across the UK, it is clear that numbers of deer in the UK have increased significantly in recent decades in line with estimated densities across most of Europe. Tick population responses to arbitrary changes in tick host availability, in line with these crude indices of deer population growth, were therefore modelled.

Effective host density (as defined above in the model description, with the daily probabilities as shown in Table 1) is divided between large and small hosts (see above). This sets the basic levels, which are much higher for Dorset to reflect the exceptionally high density of hosts, particularly deer and pheasants (*Phasianus colchicus)*, known to occur in the woodland site in Dorset, compared with the sparse host availability, mostly hill-grazing sheep, in Wales. Simulations were run under mean 1980-89 climatic conditions and the following variable host conditions: large host contact rates incrementally increased such that they doubled over 20 years; large host contact rates incrementally decreased such that they halved over 20 years. In all cases, the final conditions after 20 years were left constant for a further 10 years.

Simulated Tick Populations with Increased Deer Density

The effect of increased host density differed depending on the initial levels. In Wales (Fig. 13), an increase from the original low level caused little predicted change in nymph numbers over the first 10 years, but as host levels reached 1.4-2 times the initial level (years 10-20) the predicted questing nymph populations started to rise, reaching 47% increase by year 20.

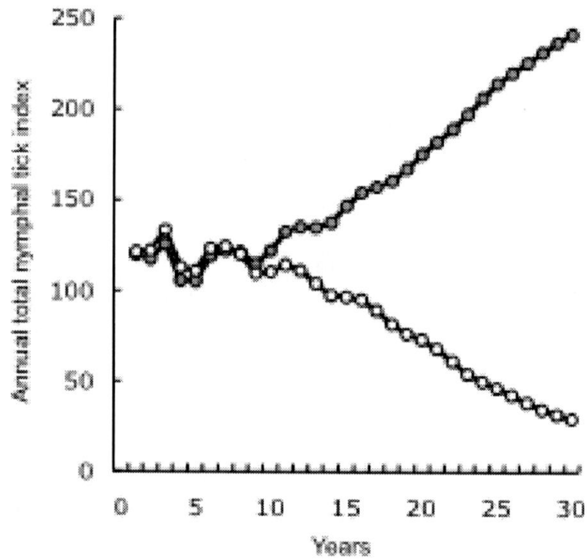

Figure 13. Simulation of effects of doubling or halving the density of large hosts (deer) on the mean annual questing nymphal tick stock with initially low host contact rates (as in Wales). Reproduced from Ref [54], with permission.

This increase continued over years 20-30, to reach a doubling of the original level and no levelling off, despite no further change in host densities. The impact of decreased host density also became apparent only after the first decade, and halving the abundance of large hosts reduced the questing nymphal population by 40% after 20 years and 76% after 30 years (Fig. 13). For the Dorset site, however, where initial host density values were considerably higher (see above), doubling the host density caused numbers of questing nymphs to decline by 16% after 30 years (Fig. 14a). Decreasing the host density also reduced the abundance of this tick stage (Fig. 14a), but only by 9%.

The model shows, however, that, whereas in Wales engorged adult densities (i.e. overall population fecundity) mirrored questing nymph densities, in Dorset they changed independently of questing nymphs, responding more intuitively to host densities and reaching asymptotes at 30% above or below the original level when host densities were doubled or halved, respectively (Fig. 14b). Evidently questing ticks were simply removed at such a high rate when host contact rates were enhanced that their numbers were quickly exhausted each year. There was no apparent change in tick phenology.

Our model suggests that under typical circumstances greater host abundance can drive tick population increases, which might account for the perceived recent changes in tick abundance, including their appearance in new places [56]. In areas that deer have only recently colonised, a doubling of large hosts for ticks is likely to be a very conservative estimate, and our predictions indicate that tick numbers could have risen dramatically in such places.

The finding that increasing host density from an initially high level may cause a decline in the questing tick population might appear non-intuitive, but nevertheless conforms to theoretical expectations. Finite fat reserves mean that ticks can quest for a finite time period; typically, a certain proportion of the questing tick population will not come into contact with a host, and will therefore perish [12,72]. Under these circumstances, a greater host contact rate will ensure that a higher proportion of the original cohort of ticks is fed, which will increase population fecundity [72]. At some point, however, the host contact rate will become so high that all questing ticks can be expected to find a host before exhausting their fat reserves (i.e. the proportion of the cohort that finds a host will reach 1 and so be unable to go higher). Increasing the host density thereafter will hasten the removal of the questing tick population whilst only slightly increasing population fecundity (by shortening the average period over which ticks suffer mortality through moisture stress or predation). Thus the evident reduction in this one phase - questing ticks (Fig. 14a) - masks an increased total population comprising all physiological phases, as demonstrated by the growing number of engorged adults (Fig. 14b), which will augment the transmission potential for tick-borne pathogens. In a tick population experiencing such high host density, a deer cull (for example) would initially cause an apparent increase [77], since the questing stages would not be removed so quickly from the vegetation. The long-term population trajectory would depend on the severity of the cull, but land managers wishing to instigate such a measure to control ticks should be advised that benefits may only be seen in the longer term, after an initial period during which results may appear to be counter-productive. Given approximate tick densities and host densities, our model would be able to predict the likely tick population trajectory under a given culling regime. Conversely, in a tick habitat with such high initial host densities, the further addition of large hosts (such as sheep) would effectively reduce the number of questing ticks, and therefore the risk to humans.

Relative Effects of Abiotic and Biotic Environmental Changes on Tick Populations

It is apparent from this modelling exercise that both climate warming and increased host densities can result in greater tick populations. As with all complex systems, however, the responses of tick populations to simple changes in abiotic and/or biotic conditions vary depending on the initial conditions. The risk, to humans and livestock, of infection with tick-borne pathogens may change more markedly where the background environment is less favourable (lower temperatures and host densities). This is important along the environmental limits of tick populations, potentially allowing ticks and associated diseases to move up latitudinal and altitudinal gradients, as has been recorded recently in Sweden, the Czech Republic, Slovakia and Austria [57-59,78]. Within the core endemic regions of Europe, however, the changes to date in the typical continental climates, even the marked step increase of 1-2°C in annual mean temperature in 1989, may have had less impact on tick population levels than the changes in biotic factors. This model is not designed to take account of changes in land cover and land use that have accompanied the socio-economic revolution or evolution in eastern and western Europe, respectively, over the past few decades. However, where these changes have involved reductions in agricultural pressure and consequent natural succession of vegetation, re-colonisation by wildlife hosts for ticks is likely to have been more important than climate change in enhancing tick populations.

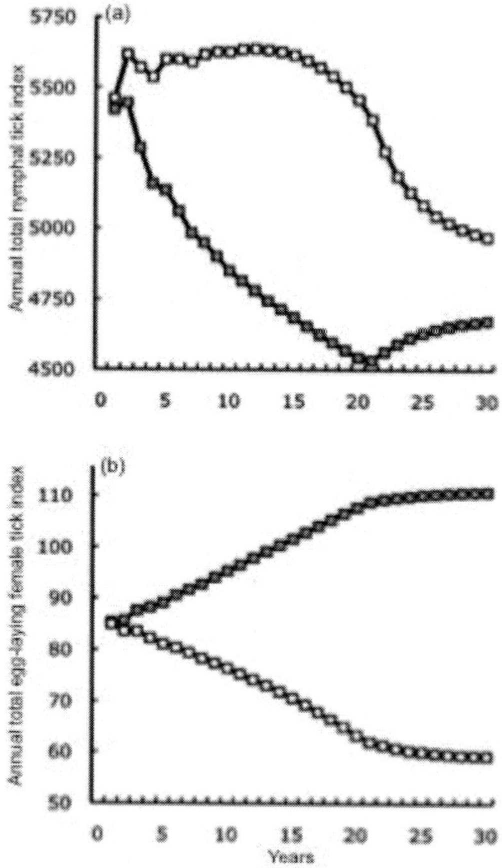

Figure 14. Simulation of effects of doubling or halving the density of large hosts (deer) on the mean annual questing nymphal tick stock (a) and mean annual stock of post-engorgement adult ticks in a tick population (b) with initially high host contact rates (as in Dorset). Reproduced from Ref [54], with permission.

In principle, this population model, developed here for two UK scenarios, could be rolled out across the broader range of conditions throughout the European distribution of *I. ricinus*, determining not only the relative change in abundance of ticks but also their potential to transmit various pathogens each with their own biological, and therefore environmental, prerequisites. In practice, however, this exercise awaits better data on microclimatic conditions experienced by ticks and better knowledge of the geographical heterogeneity of tick phenotype in relation to environmental factors.

The Impact of Variable Acaricide Treatment on Tick Populations

In the past, sheep were compulsorily treated against ectoparasites (scab mites, blowflies and ticks) by being dipped in organochlorines, replaced in the 1980s by organophosphates and later supplemented with pyrethroids, that gave up to 6-8 weeks protection [79]. Sheep therefore acted as 'lethal mops', sweeping questing ticks from the vegetation and killing them. Since the early 1990s, these practices were phased out when the law was changed, reinforced by concern over environmental contamination and operator health. Instead, a growing proportion of farmers now use spray or pour-on insect growth regulators or high-cis cypermethrin [79], whose efficacy in providing long-lasting protection against ticks appears to be less certain.

Despite this uncertainty in the effectiveness of acaricidal activity in current products, especially when combined with normal practices of treating hill sheep only at times related to husbandry (e.g. lambing, shearing, mating), the effect of acaricidal treatment was modelled under the ideal prior scenario of 100% tick mortality on contact with treated sheep [31]. The real efficacy of acaricide treatment can be scaled accordingly in future simulations. Here, two scenarios of acaricide application are considered. In the first scenario, a given proportion of the existing large host population is treated, simulated in the model by reducing on-host survival appropriately. In the second scenario, additional large hosts are added to the system, all of which are treated. In both scenarios, the effects of treating 10% or 50% of the original number of hosts were simulated.

Acaricide application at both sites (Dorset and Wales) caused a predicted drop in questing tick numbers (Fig. 15). Applying acaricide to existing hosts was slightly more efficient in reducing tick populations than when acaricide-treated hosts were added to the existing host assemblage, presumably because there were no counter-active positive effects of increased host density (above). Treating 10% of the host population was sufficient to reduce tick abundance within four years by about 50%, while treating 50% rapidly drove the tick population more or less to extinction. The efficacy of acaricide-treatment, in terms of the proportional reduction in numbers of questing ticks after 30 years, was predicted to be similar at the two sites.

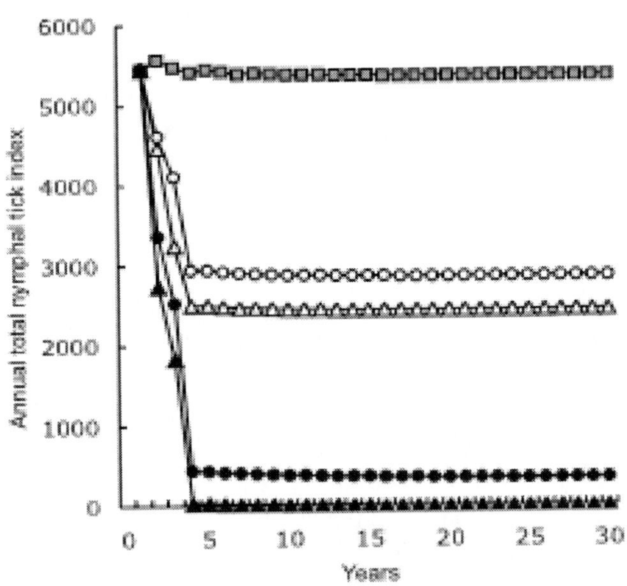

Figure 15. Simulation of the effects of acaricide application on a tick population in Dorset, UK. Control (■); scenario 1 (△), acaricide is applied to 10% (open symbol) or 50% (filled symbol) of the existing large host community; scenario 2 (O), acaricide is applied to 10% (open symbol) or 50% (filled symbol) of newly-added large hosts, quantified as the percentage of the pre-existing large host community. Reproduced from Ref [54], with permission.

If adding hosts to tick habitats might reduce the abundance of questing ticks (above), treating any additional hosts with acaricide would be even more effective. The predicted efficacy depended more on the proportion of the host population that was treated, and less on whether those animals were already part of the host community or were simply added to it. This has important management implications, since the treatment of wild hosts is clearly logistically highly demanding [80] (although less so when deer anoint themselves with acaricides via the corn-baited 4-poster devices used successfully, but at considerable cost, in the US [81]), whereas simply releasing treated (domestic) hosts into the area may be almost as effective.

These predicted results are in accordance with the past effectiveness of treated sheep as 'lethal mops', especially in habitats where alternative hosts are scarce. Washing sheep at 2-3 monthly intervals with the acaricide Top Clip Gold Shield (Ciba Geigy) containing 60% w/v diazinon proved sufficient to reduce ticks by c.90% within one year in an enclosure on a Welsh hillside with low diversity of alternative vertebrate hosts [72]. Treating domestic hosts has the additional advantage that the areas of ground over which they roam can be designated by fencing, such that the control measure is strictly focussed in the right place. Sheep are indeed being used in this way to protect grouse on Scottish moors where the value of shooting rights now far exceeds the value of sheep products, with the further advantage that such sheep may be treated at times independent of usual husbandry practices, if necessary.

Conclusion

Models have in the past been dismissed as the product of 'armchair ecologists'. Happily, they are now seen more correctly as the essential framework within which to define questions more precisely and to focus empirical work on the key factors within complex systems. With a biological system as opaque as that of tick population ecology, in which field data cannot illuminate all the covert demographic elements (e.g. due to the multiple causes for ticks to leave the countable questing population, and the wide range of hosts upon which many tick species feed), a model is essential to disentangle some of the internal dynamics. At the same time, one cannot produce a realistic yet tractable model without considerable experience and understanding gained from struggling to collect and make sense of field data, thus there is still some cause to mistrust an ecologist with clean boots!

While this new model for *I. ricinus* is not yet fully applicable across the species' wide geographical range, it nevertheless captures some key elements that underpin the tick's performance as a very effective vector of many different microbes. Of particular importance are the seasonal dynamics not only in the numbers of questing ticks, but also in the inter-stadial and transovarial routes of transmission. Furthermore, the model suggests that, in addition to local climate conditions, a single parameter (propensity to enter behavioural diapause after development, rather than quest, in autumn – also likely to be an adaptation to local climate) may be responsible for strongly divergent phenological patterns seen at the latitudinal extremes of the tick's range. It has also proved useful in assessing the relative contributions of past changes in abiotic and biotic environmental factors in driving apparent increases in tick populations, and conversely, in predicting possible success in new control efforts.

Future work must focus upon the environmental triggers underlying the afore-mentioned diapause/quest 'decision' (if indeed a deterministic relationship exists, as opposed to an abrupt phenotypic change), density-independent mortality rates of eggs in field conditions, and the effects of saturation deficit across the full range of values experienced in Europe. Coupling a full tick model with variable assemblages of hosts would allow the simulation of complex pathogen transmission systems in some considerable detail.

ACKNOWLEDGMENTS

We acknowledge the primary tick fieldwork carried out by Mick Peacey and early work on the population model by Tom Finnie. This work was funded by the UK Department of the Environment and Rural Affairs (Defra) (project SE4105), the Research Councils' Rural Economy and Land Use (RELU) Programme (funded by the Economic and Social Research Council, Biotechnology and Biological Sciences Research Council and Natural Environment Research Council, with additional funding from Defra and the Scottish Executive Environment and Rural Affairs Department) (project RES-229-25-0007), and EU grant GOCE-2003-010284 EDEN. The contents are the sole responsibility of the authors and do not necessarily reflect the views of the European Commission. This book chapter is based on two papers published in the Journal of Applied Ecology 2011, volume 48 (Wiley-Blackwell, Oxford, UK), available online at http://dx.doi.org/10.1111/j.1365-2664.2011.02003.x and http://dx.doi.org/10.1111/j.1365-2664.2011.02004.x reproduced with modification under the normal conditions.

REFERENCES

[1] Hill, C. A., Kafatos, F. C., Stansfield, S. K., & Collins, F. H. (2005). Arthropod-borne diseases: vector control in the genomics era. *Nature Rev. Microbiol., 3*, 262-268.

[2] Dobson, A. (2009). Climate variability, global change, immunity, and the dynamics of infectious diseases. *Ecology, 90*, 920-927.

[3] Pascual, M., & Bouma, M. J. (2009). Do rising temperatures matter? *Ecology, 90,* 906-912.

[4] Rogers, D. J., & Randolph, S. E. (2000). The global spread of malaria in a future, warmer world. *Science, 289*, 1763-1766.

[5] Randolph, S. E., & Rogers, D. J. (2000). Fragile transmission cycles of tick-borne encephalitis virus may be disrupted by predicted climate change. *Proc. Roy. Soc. Lond. B, 267*, 1741-1744.

[6] Ogden, N. H., Bigras-Poulin, M., O'Callaghan, C. J., Barker, I. K., Lindsay, L. R., Maarouf, A., Smoyer-Tomic, K. E., Waltner-Toews, D., & Charron, D. (2005). A dynamic population model to investigate effects of climate on geographic range and seasonality of the tick *Ixodes scapularis*. *Int. J. Parasitol., 35*, 375-389.

[7] Rogers, D. J. (2006). Models for vectors and vector-borne diseases. *Adv. Parasitol., 62*, 1-37.

[8] Hay, S. I., Tatem, A. J., Guerra, C. A., & Snow, R. W. (2006). Population at malaria risk in Africa: 2005, 2015 and 2030. In *UK Government Foresight project, Infectious Diseases: preparing for the future.* London, UK: Government Office of Science and Innovation.

[9] Randolph, S. E. (2004). Tick ecology: processes and patterns behind the epidemiological risk posed by ixodid ticks as vectors. *Parasitology, 129,* S37-66.

[10] Randolph, S. E. (1998). Ticks are not insects: consequences of contrasting vector biology for transmission potential. *Parasitol. Today, 14*, 186-192.

[11] Belozerov, V. N. (1982). Diapause and biological rhythms in ticks. In F. D. Obenchain, & R. Galun (Eds.), *Physiology of Ticks* (pp. 469-500). Oxford, UK: Pergamon Press.

[12] Randolph, S. E., Green, R. M., Hoodless, A. N., & Peacey, M. F. (2002). An empirical quantitative framework for the seasonal population dynamics of the tick *Ixodes ricinus. Int. J. Parasitol., 32*, 979-989.

[13] Randolph, S. E., Green, R. M., Peacey, M. F., & Rogers, D. J. (2000). Seasonal synchrony: the key to tick-borne encephalitis foci identified by satellite data. *Parasitology, 121*, 15-23.

[14] Hudson, P. J., Rizzoli, A., Rosa, R., Chemini, C., Jones, L. D., & Gould, E. A. (2001). Tick-borne encephalitis virus in northern Italy: molecular analysis, relationships with density and seasonal dynamics of *Ixodes ricinus. Med. Vet. Entomol., 15*, 304-313.

[15] Carpi, G., Cagnacci, F., Neteler, M., & Rizzoli, A. (2008). Tick infestation on roe deer in relation to geographic and remotely-sensed climatic variables in a tick-borne encephalitis endemic area. *Epidem. Infect., 136*, 1416-1424.

[16] Ogden, N. H., Bigras-Poulin, M., O'Callaghan, C. J., Barker, I. K., Kurtenbach, K., Lindsay, L. R., Maarouf, A., & Charron, D. F. (2007). Vector seasonality, host infection dynamics and fitness of pathogens transmitted by the tick *Ixodes scapularis. Parasitology, 134*, 209-227.

[17] Gatewood, A. G., Liebman, K. A., Vour'ch, G., Bunikis, J., Hamer, S. A., Cortinas, R., Melton, F., Cislo, P., Kitron, U., Tsao, J. I., Barbour, A. G., Fish, D., & Diuk-Wasser, M. (2009). Climate and tick seasonality are predictors of *Borrelia burgdorferi* genotype distribution. *Appl. Environ. Microbiol., 75*, 2476-2483.

[18] Dobson, A. D. M., Finnie, T. J. R., & Randolph, S. E. (2011). A modified matrix model to descrobe the seasonal population ecology of the European tick *Ixodes ricinus. J. Appl. Ecol., 48,* 1017-1028.

[19] Perret, J.-L., Rais, O., & Gern, L. (2004). Influence of climate on the proportion of *Ixodes ricinus* nymphs and adults questing in a tick population. *J. Med. Entomol., 41*, 361-365.

[20] Alekseev, A. N., Jensen, P. M., Dubinina, H. V., Smirnova, L. A., Makrouchina, N. A., & Zharkov, S. D. (2000). Peculiarities of behaviour of taiga (*Ixodes persulcatus*) and sheep (*Ixodes ricinus*) ticks (Acari: Ixodidae) determined by different mehods. *Folia Parasitol., 47*, 147-153.

[21] Randolph, S. E. (1997). Abiotic and biotic determinants of the seasonal dynamics of the tick *Rhipicephalus appendiculatus* in South Africa. *Med. Vet. Entomol., 11*, 25-37.

[22] Ogden, N. H., Maarouf, A., Barker, I. K., Bigras-Poulin, M., Lindsay, L. R., Morshed, M. G., O'Callaghan, C. J., Ramay, F., Waltner-Toews, D., & Charron, D. F. (2006).

Climate change and the potential for range expansion of the Lyme disease vector *Ixodes scapularis* in Canada. *Int. J. Parasitol., 36*, 63-70.

[23] R Core Development Team (2005). In *Foundation for Statistical Computing,* Vienna, Austria.

[24] Caswell, H. (2001). *Matrix population models: Construction, analysis and interpretation.* Sunderland, USA: Sinauer Associates Inc.

[25] Leslie, P. H. (1945). On the use of matrices in certain population mathematics. *Biometrika, 33*, 183-212.

[26] Belozerov, V. N. (1998). Participation of two-step photoperiodic reaction in control of development and diapause in nymphs of *Ixodes persulcatus* (in Russian). *Zool. Zhurnal, 77*, 885-890.

[27] Campbell, J. A., (1948). The life history and development of the sheep tick *Ixodes ricinus* Linaeus in Scotland, under natural and controlled conditions. PhD thesis, University of Edinburgh, UK.

[28] Macleod, J. (1936). *Ixodes ricinus* in relation to its physical environment. IV. An analysis of the ecological complexes controlling distribution and activities. *Parasitology, 28*, 295-319.

[29] Perret, J.-L., Guigoz, E., Rais, O., & Gern, L. (2000). Influence of saturation deficit and temperature on *Ixodes ricinus* tick questing activity in a Lyme borrelosis-endemic area (Switzerland). *Parasitol. Res., 86*, 554-557.

[30] Talleklint, L., & Jaenson, T. G. T. (1994). Transmission of *Borrelia burgdorferi* s.l. from mammal reservoirs to the primary vector of Lyme borreliosis, *Ixodes ricinus* (Acari: Ixodidae), in Sweden. *J. Med. Entomol., 31*, 880-886.

[31] Steele, G. M., & Randolph, S. E. (1985). An experimental evaluation of conventional control measures against the sheep tick *Ixodes ricinus* (L) (Acari: Ixodidae). I. A unimodal seasonal activity pattern. *Bull. Entomol. Res., 75*, 489-499.

[32] Ogden, N. H., Nuttall, P. A., & Randolph, S. E. (1997). Natural Lyme disease cycles maintained *via* sheep by co-feeding ticks. *Parasitology, 115*, 591-599.

[33] Gray, J. S. (1981). The fecundity of *Ixodes ricinus* (L.) (Acarina: Ixodidae) and the mortality of its developmental stages under field conditions. *Bull. Entomol. Res., 71*, 533-542.

[34] Randolph, S. E., & Rogers, D. J. (1997). A generic population model for the African tick *Rhipicephalus appendiculatus*. *Parasitology, 115*, 265-279.

[35] Branagan, D. (1973). The developmental periods of the Ixodid tick *Rhipicephalus appendiculatus* Neum. under laboratory conditions. *Bull. Entomol. Res., 63*, 155-168.

[36] Daniel, M., Cerny, V., Dusbabek, F., Honzakova, E., & Olejnicek, J. (1976). Influence of microclimate on the life cycle of the common tick *Ixodes ricinus* (L.) in thermophilic oak forest. *Folia Parasitol., 23*, 327-342.

[37] Gigon, F., (1985). Biologie d'*Ixodes ricinus* L. sur le Plateau Suisse - une contribution à l'écologie de ce vecteur. PhD thesis, Université de Neuchâtel, Switzerland.

[38] Samish, M., & Alekseev, E. (2001). Arthropods as predators of ticks (Ixodidae). *J. Med. Entomol., 38*, 1-11.

[39] Dautel, H., & Knulle, W. (1997). Cold hardiness, supercooling ability and causes of low-temperature mortality in the soft tick, *Argas reflexus*, and the hard tick, *Ixodes ricinus* (Acari: Ixodidae) from central Europe. *J. Insect Physiol., 43*, 843-854.

[40] Yoder, J. A., Selim, M. E., & Needham, G. R. (1997). Impact of feeding, molting and relative humidity on cuticular wax deposition and water loss in the Lone Star tick, *Amblyomma americanum*. *J. Insect Physiol., 43*, 547-551.

[41] Mount, G. A., Haile, D. G., & Daniels, E. (1997). Simulation of blacklegged tick (Acari: Ixodidae) population dynamics and transmission of *Borrelia burgdorferi*. *J. Med. Entomol., 34*, 461-484.

[42] Randolph, S. E. (1994). Density-dependent acquired resistance to ticks in natural hosts, independent of concurrent infection with *Babesia microti*. *Parasitology, 108*, 413-419.

[43] Ogden, N. H., Casey, A. N. J., French, N. P., Adams, J. D. W., & Woldehiwet, Z. (2002). Field evidence for density-dependent facilitation amongst *Ixodes ricinus* ticks feeding on sheep. *Parasitology, 124*, 117-125.

[44] Levin, M. L., & Fish, D. (1998). Density-dependent factors regulating feeding success of *Ixodes scapularis* larvae (Acari: Ixodidae). *J. Parasitol., 84*, 36-43.

[45] Sutherst, R. W., Floyd, R. B., Bourne, A. S., & Dallwitz, M. J. (1986). Cattle grazing behavior regulates tick populations. *Experientia, 42*, 194-196.

[46] Randolph, S. E. (1979). Population regulation in ticks: the role of acquired resistance in natural and unnatural hosts. *Parasitology, 79*, 141-156.

[47] Randolph, S. E. (1994). Population dynamics and density-dependent seasonal mortality indices of the tick *Rhipicephalus appendiculatus* in east and southern Africa. *Med. Vet. Entomol., 8*, 351-368.

[48] Dobson, A. D. M., Taylor, J. L., & Randolph, S. E. (2011). Tick (*Ixodes ricinus*) abundance and seasonality at recreational sites in the UK: hazards in relation to fine-scale habitat types revealed by complementary sampling methods. *Ticks & Tick-borne Dis., 2*, 67-74.

[49] Estrada-Peña, A., Martinez, J. M., Sanchez Acedo, C., Quilez, J., & Del Cacho, E. (2004). Phenology of the tick, *Ixodes ricinus*, in its southern distribution range (central Spain). *Med. Vet. Entomol., 18*, 387-397.

[50] Randolph, S. E., & Storey, K. (1999). Impact of microclimate on immature tick-rodent interactions (Acari: Ixodidae): implications for parasite transmission. *J. Med. Entomol., 36*, 741-748.

[51] Keirans, J. E., Hutcheson, H. J., Durden, L. A., & Klompen, J. S. H. (1996). *Ixodes (Ixodes) scapularis* (Acari: Ixodidae): redescription of all active stages, distribution, hosts, geographical variation, and medical and veterinary importance. *J. Med. Entomol., 33*, 297-318.

[52] Wilson, M. L., Adler, G. H., & Spielman, A. (1985). Correlation between abundance of deer and that of the deer tick, *Ixodes dammini* (Acari: Ixodidae). *Anns. Entomol. Soc. Am., 78*, 172-176.

[53] Rand, P. W., Lubelczyk, C., Lavigne, G. R., Elias, S., Holman, M. S., Lacombe, E. H., & Smith, R. P. J. (2003). Deer density and the abundance of *Ixodes scapularis* (Acari: Ixodidae). *J. Med. Entomol., 40*, 179-184.

[54] Dobson, A. D. M., & Randolph, S. E. (2011). Modelling the effects of recent changes in climate, host density and acaricide treatments on population dynamics of *Ixodes ricinus* in the UK. *J. Appl. Ecol., 48,* 1029-1037.

[55] Kirby, A. D., Smith, A. A., Benton, T. G., & Hudson, P. J. (2004). Rising burden of immature sheep ticks (*Ixodes ricinus*) on red grouse (*Lagopus lagopus scoticus*) chicks in the Scottish uplands. *Med. Vet. Entomol., 18*, 67-70.

[56] Scharlemann, J. P. W., Johnson, P. J., Smith, A. A., Macdonald, D. W., & Randolph, S. E. (2008). Trends in ixodid tick abundance and distribution in Great Britain. *Med. Vet. Entomol., 22*, 238-247.

[57] Lindgren, E., Talleklint, L., & Polfeldt, T. (2000). Impact of climatic change on the northern latitude limit and population density of the disease-transmitting European tick *Ixodes ricinus*. *Environ. Hlth. Perspect., 108*, 119 123.

[58] Daniel, M., Danielová, V., Kriz, B., Jirsa, A., & Nozicka, J. (2003). Shift of the tick *Ixodes ricinus* and tick-borne encephalitis to higher altitudes in central Europe. *Europ. J. Clin. Microbiol. Infect. Dis., 22*, 327-328.

[59] Holzmann, H., Aberle, S. W., Stiasny, K., Werner, P. C., Mischak, A., Zainer, B., Netzer, M., Koppi, S., Bechter, E., & Heinz, F. X. (2009). Tick-borne encephalitis from eating goat cheese in a mountain region of Austria. *Emerg. Infect. Dis., 15*, 1671-1673.

[60] Gray, J. S. (2008). *Ixodes ricinus* seasonal activity: implications of global warming indicated by revisiting tick and weather data. *Int. J. Med. Microbiol., 298(S1)*, 19-24.

[61] Gray, J. S., Dautel, H., Estrada-Pena, A., Kahl, O., & Lindgren, E. (2009). Effects of climate change on ticks and tick-borne diseases in Europe. *Interdisc. Perspect. Infect. Dis.,* ID 593232 (12pp).

[62] Danielova, V., Schwarzova, L., Materna, J., Daniel, M., Metelka, L., Holubova, J., & Kriz, B. (2008). Tick-borne encephalitis virus expansion to higher altitudes correlated with climate warming. *Int. J. Med. Microbiol., 298 (S1)*, 68-72.

[63] Pisoft, P., Kalvova, J., & Brazdil, R. (2004). Cycles and trends in the Czech temperature series using wavelet transforms. *Int. J. Climatol., 24*, 1661-1670.

[64] Marty, C. (2008). Regime shift of snow days in Switzerland. *Geophys. Res. Lett., 35*, L12501.

[65] Šumilo, D., Asokliene, L., Bormane, A., Vasilenko, V., Golovljova, I., & Randolph, S. E. (2007). Climate change cannot explain the upsurge of tick-borne encephalitis in the Baltics. *PLoS ONE, 2*, e500.

[66] Ruckstuhl, C., Philipona, R., Behrans, K., Coen, M. C., Durr, B., Heimo, A., Matzler, C., Nyeki, S., Ohmura, A., Vuilleumier, L., Weller, M., Wehrli, C., & Zelenka, A. (2008). Aerosol and cloud effects on solar brightening and the recent rapid warming. *Geophys. Res. Lett., 35*, L12708.

[67] Wild, M., Gilgen, H., Roesch, A., Ohmura, A., Long, C. N., Dutton, E. G., Forgan, B., Kallis, A., Russak, V., & Tsvetkov, A. (2005). From dimming to brightening: decadal changes in solar radiation at Earth's surface. *Science, 308*, 847-850.

[68] Macleod, J. (1935). *Ixodes ricinus* in relation to its physical environment. II. The factors governing survival and activity. *Parasitology, 27*, 123-144.

[69] Gray, J. S., Kahl, O., Janetzki, C., & Stein, J. (1992). Studies on the ecology of Lyme disease in a deer forest in County Galway, Ireland. *J. Med. Entomol., 29*, 915-920.

[70] Deblinger, R. D., Wilson, M. L., Rimmer, D. W., & Spielman, A. (1993). Reduced abundance of immature *Ixodes dammini* (Acari: Ixodidae) following incremental removal of deer. *J. Med. Entomol., 30*, 144-150.

[71] Jensen, P. M., Hansen, H., & Frandsen, F. (2000). Spatial risk assessment for Lyme borreliosis in Denmark. *Scand. J. Infect. Dis., 35*, 545-550.

[72] Randolph, S. E., & Steele, G. M. (1985). An experimental evaluation of conventional control measures against the sheep tick *Ixodes ricinus* (L) (Acari: Ixodidae). II. The dynamics of the tick-host interaction. *Bull. Entomol. Res., 75*, 501-518.

[73] Eisen, L., Eisen, R. J., & Lane, R. S. (2002). Seasonal activity patterns of *Ixodes pacificus* nymphs in relation to climatic conditions. *Med. Vet. Entomol., 16*, 235-244.
[74] Ward, A. I. (2005). Expanding ranges of wild and feral deer in Great Britain. *Mammal Rev., 35*, 165-173.
[75] Ward, A., Etherington, T., & Ewald, J. (2008). Five years of change. *Deer, 14*, 17-20.
[76] Battersby J. (Ed.) and Tracking Mammal Partnership (2005). *UK Mammals: Species status and population trends. First report by the Tracking Mammals Partnership.* Peterborough, UK: JNCC/Tracking Mammals Partnership.
[77] Rand, P. W., Lubelczyk, C., Holman, M. S., Lacombe, E. H., & Smith, R. P. Jr. (2004). Abundance of *Ixodes scapularis* (Acari: Ixodidae) after the complete removal of deer from an isolated offshore island, endemic for Lyme disease. *J. Med. Entomol., 41*, 779-784.
[78] Lukan, M., Bullova, E., & Petko, B. (2010). Climate warming and tick-borne encephalitis, Slovakia. *Emerg. Infect. Dis., 16*, 524-526.
[79] Bisdorff, B., & Wall, R. (2008). Sheep blowfly strike risk and management in Great Britain: a survey of current practice. *Med. Vet. Entomol., 22*, 303-308.
[80] Laurenson, M. K., Hudson, P. J., McGuire, K., Thirgood, S. J., & Reid, H. W. (1997). Efficacy of acaricidal tags and pour-on as prophylaxis against ticks and louping-ill in red grouse. *Med. Vet. Entomol., 11*, 389-393.
[81] Pound, J. M., Miller, J. A., & George, J. E. (2000). Efficacy of amitraz applied to white-tailed deer by the '4-poster' topical treatment device in controlling free-living lone star ticks (Acari: Ixodidae). *J. Med. Entomol., 37*, 878-884.

In: Ticks: Disease, Management and Control
Editor: Moges Woldemeskel
ISBN: 978-1-62081-136-8

Chapter 4

TICKS INFESTING DOMESTIC AND WILD ANIMALS IN THE TROPICS (ETHIOPIA) AND THEIR POSSIBLE ROLES IN DISEASE TRANSMISSION

Sileshi Mekonnen[1,*], Moges Woldemeskel[2] and Solomon Gebre[3]

[1]Food and Agriculture Organization (FAO) Office, Addis Ababa, Ethiopia

[2]The University of Georgia, College of Vet. Med., Dept. of Pathology, Tifton Vet. Diagnostic & Investigational Lab, Tifton, GA, US

[3]National Animal Health & Disease Investigation Center, Sebeta, Ethiopia

ABSTRACT

Ethiopia has the largest livestock population in Africa, and lies in the tropics encompassing wide ranges of agro-climatic zones including the cool central highlands, wet tropical forest in the west and southwest and dry arid and semi-arid areas in the east, south and southeast. Therefore it would represent various climatic zones and livestock production systems in tropical Africa. Ticks are the main ectoparasites infesting domestic and wild animals and are among the major stumbling blocks in livestock production in Ethiopia and tropical Africa. Ixodid and argasid ticks collected from cattle, sheep, goats, camels, equines and a few from wild animals in Ethiopia are described. In this chapter, forty-four tick species in 9 genera recorded in different agro-ecological zones of the country and their possible role in disease transmission are briefly described. The identified ticks infesting domestic and wild animals belonged to the genus *Rhipicephalus* (15 species), *Hyalomma* (9 species), *Amblyomma* (8 species), *Haemaphysalis* (4 species), *Aponomma* (2 species), *Boophilus* (2 species), *Ornithodorous* (2 species), *Ixodes* (1 species), and *Argas* (1 species). Understanding the distribution, seasonal dynamics and the host range on various tick species infesting domestic and wild animals is very important to envisage and implement sound and effective tick- and tick-borne disease control strategies.

[*] Corresponding author. Sileshimekonnen99@yahoo.com; Sileshi.Mekonnen@fao.org.

1. Introduction

According to Central Statistical Authority of Ethiopia (2006) Ethiopia's livestock population is one of the largest in Africa and the estimated number were 47.5 million cattle, 26.1 million sheep, 21.7 million goats, 2.1 million horses and mules, 5.6 million donkeys and 1 million camels.

Ixodid ticks are responsible for a great variety of livestock health problems [1]. Apart from transmitting protozoal, rickettsial and viral diseases, ticks downgrade the quality of hides and skins, reduce meat, milk and wool production, and increase susceptibility of the infested host to other diseases [2]. The introduction of exotic cattle breeds with high productivity into the tropics, including Ethiopia, is largely hampered by ticks and tick-borne diseases since ticks are important vectors of several fatal diseases of animals. Ticks are common in tropical Africa and in all agro-ecological zones of Ethiopa [3] and the major tick genera reported were *Amblyomma, Boophilus, Haemaphysalis, Hyalomma* and *Rhipicephalus* [4]. Tick-borne diseases (TBDs) of livestock include anaplasmosis (*A. marginale*), babesiosis (*B. bigemina*), Ehrlichiosis (*E. ruminantium*) and theileriosis (*T. mutans*). Although there are no reports of the presence of either bovine tropical theileriosis (*T. annulata*) or East Coast fever (*T. parva*), the relatively uncontrolled cattle movement from south Sudan, where both these diseases and their vectors are found, indicates that there is a considerable risk of these diseases being introduced to Ethiopia. This chapter describes the major tick species affecting animals in various climatic zones in Ethiopia, indicates their possible role in disease transmission and would highlight on the damage they cause to the livestock industry in the tropics.

Most of the information included in this review is results of tick surveys reported by the Acarology and Entomology Laboratory of the Ethiopian National Animal Health Research Centre, as well as results of tick surveys by experts of the national Institutions. Experts from internationally funded projects [3, 5, 6] have contributed greatly in recording various tick species. The ticks were identified according to the descriptions given by various authors [7-11]. The species of ticks belonging to the various genera collected and identified from different domestic and wild animals are described below.

2. Genus Amblyomma (Koch, 1844)

Eight species of *Amblyomma* are identified in Ethiopia and all require three hosts to complete their lifecycles. From these, *A. variegatum* is the most important tick of livestock mainly because it is the principal vector of economically important tick-borne diseases (TBDs).

Amblyomma variegatum (Fabricius, 1794)*: A. variegatum* known by its common name “the Tropical bont tick” is widely distributed in Ethiopia (Table 1). The preferred hosts of the adult tick are cattle, but it also infests sheep and goats. This tick requires moistures and warmth for its survival and is not found in open grassland. Activity of the adult ticks commences when spring rains begin between March and June [3]. *A. variegatum* is a potential vector of *E. ruminantium*, *T. mutans* and *T. velifera* (benign bovine theileriosis) and the virus of Nairobi Sheep Disease. Infestations with adult *A. variegatum* were frequently associated with damages to the udder and teat in cows and the scrotum in bulls. It also facilitates the establishment of clinical bovine dermatophilosis (*D. congolensis*), particularly in exotic and cross-bred cattle breeds.

Amblyomma cohaerens (Donitz, 1809): A. cohaerens is a specific parasite of the buffalo [5]. Pegram [12] observed large and small varieties of this tick species in Ethiopia. The larger type is believed to be associated with African buffalo (*Syncerus caffer*) and the smaller one with cattle. It is the most abundant tick infesting cattle in southwest Ethiopia [6] recorded between 1750-2500 m above sea level and 750-1500 mm rainfall [5]. The tick prefers medium to high altitudes, moist climates and is absent in the drier regions. All developmental stages feed on the same host. It was collected from the west, south-west and central Ethiopia with moderate to ample amount of rain, including Illubabor, Keffa, East and West Wollega, East, West and North Shoa, Gurage, Hadiya, Kembata and Arsi administrative zones (Fig. 1). *A. cohaerens* is believed to be a vector of *E. ruminantium,* the causative agent of ehrlichiosis.

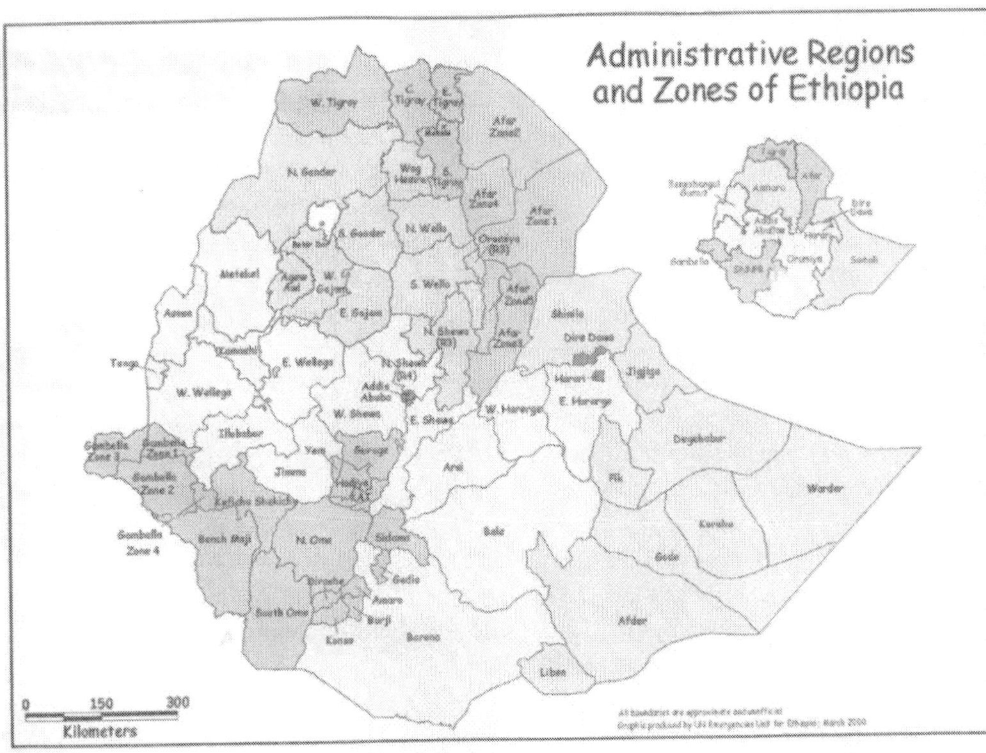

Figure 1. Administrative map of Ethiopia.

Amblyomma gemma (Donitz, 1809): Cattle are the preferred hosts of the adult *A. gemma*. The site of attachment in cattle is the ventral surface of the body from the dewlap to the escutcheon, including the axillae and groin. The tick prefers woodland, bush-land, and wooded and bushed grassland in the arid and semi-arid areas. It was mainly abundant between 500 -1750 m above sea level and 350 -750 mm rainfall [5]. *A. gemma* was mainly collected from cattle and camels in semi-arid and arid areas of eastern and southeastern parts of Ethiopia, particularly in Afar, Somali, Harari and eastern part of Tigray, Amhara and Southern Nations Nationalities and Peoples (SNNP) regional states (Table 1). Pegram et al.

[3] also reported this tick species in the Rift Valley and eastward restricted to the semi-arid plains and bush land receiving 100 - 800 mm rainfall annually. *A. gemma* also transmits *E. ruminantium, the* causative agent of ehrlichiosis.

Amblyomma lepidum (Donitz, 1809): *A. lepidum* is also known as the "East African bont-tick". The adults prefer cattle, although camels harbor quite large infestations. Adults are predominantly found attached to the ventral surface of the host from the lower dewlap to the escutcheon, including the axillae, and on the heels. The tick prefers arid and semi-arid areas, and also occurs in woodland and bush-land as well as in grassland with either trees or bushes present [13]. It is irregularly dispersed through most of the country and was collected from various semi-arid and arid areas of Tigray, Afar, Amhara, Oromia, Somali, SNNP and Harari Regional states (Table 1). There were few collections of this tick species in Gambella regional state in the west with wet tropical forest. Similar to *A. gemma* and *A. cohaerens, A. lepidum* also transmits *E. ruminantium*, the causative agent of ehrlichiosis.

Amblyomma marmoreum (Koch, 1844): *A. marmoreum* (tortoise tick) feeds mainly on the tortoise and rarely on lizards and large snakes [14]. The site of attachment is the soft parts of tortoise around the base of the legs and tails. Few collections in small number of ticks were made from tortoises in the lowlands with hot climate including Diredawa, Awash National Park, Zeway, Metehara, Harar [5], Turmi (South Omo zone) and recently also from tortoises in Mordillow district of the Somali region.

Amblyomma nuttalli (Doenitz, 1909)*: A. nuttalli* is mainly a parasite of tortoise. The tick was collected from tortoise in various regions of the country including Diredawa in the east, Arsi and Sidama zones in the south, Itang in the west (Gambella) [3, 6, 15 & 16] and also recently collected from tortoise in Mordillow district (Somali region) in the east and Illubabor zone in the west.

Amblyomma sparsum (Neumann, 1899): Rare collections were performed by Pegram *et al.* [3] in Ethiopia. The commonest mammalian host of adult *A. sparsum* is black rhinoceros (*Diceros bicornis*) and the buffalo (*Syncerus caffer*). Amongst reptiles, tortoises are most commonly parasitized [14]. *A. sparsum* was recorded from tortoises in hot and semi-arid areas in Babile (East Hararge), Awash National Park in central and Negelle Borana in southern [5] and from rhinoceros (*D. bicornis*) in Diredawa in eastern Ethiopia [15].

Amblyomma tholloni (Neumann, 1899): A single collection on tortoise from Bedelle (Illubabor zone) was recorded by [6]. There is no further information on this tick species and on its role as a vector in Ethiopia.

3. GENUS APONOMMA (NEUMANN, 1899)

There are two species of the Genus *Aponomma* recorded in Ethiopia with no reported roles in disease transmission.

Aponomma flavomaculatum (Lucas, 1846): Ap. flavomaculatum is the usual parasite of varanus (V. *niloticus*) and is collected in Debre Zeit, central Ethiopia, on the scarpment of the Ethiopian Rift Valley, and under tree bark at Arba Minch, in southwest Ethiopia [5].

Aponomma latum (Koch, 1844): *Ap. latum* was collected from snakes in Assosa, in the west, Awash National Park in central and Omo valley in southwest Ethiopia (Morel and Rodhain, 1972). Further information on this tick species in Ethiopia is not available.

Table 1. Distribution of Amblyomma, Aponomma and Boophilus ticks in different administrative regions of Ethiopia

		Administrative regions										
Tick species		Tigray	Afar	Amhara	Oromia	Somali	Benshangul & Gumuz	SNNP	Gambella	Harari	Addis Ababa	Diredawa
A. cohaerens	3-host	-	-	++	+++	-	+	++	+	-	+	-
A. gemma	3-host	+	++	+	+++	++	-	++	-	+	-	+
A. lepidum	3-host	+	+	+	++	+	-	+	+	-	-	+
A. marmoreum	3-host	-	-	-	(+)	(+)	-	(+)	(+)	-	-	(+)
A. nuttalli	3-host	-	-	-	(+)	(+)	-	(+)	(+)	-	-	(+)
A. sparsum	3-host	-	(+)	-	(+)	-	-	-	-	(+)	-	(+)
A. tholloni	3-host	-	-	-	(+)	-	-	-	-	-	-	-
A. variegatum	3-host	+	+	+++	+++	+	+	++	+	+	+	+
Ap. flavomaculatum		-	-	-	(+)	-	-	(+)	-	-	-	-
Ap. latum	3-host	-	-	-	(+)	-	(+)	(+)	-	-	-	-
B. annulatus	1-host	-	-	-	-	-	-	-	(+)	-	-	-
B. decoloratus	1-host	+	-	++	+++	+	+	+	+	+	+	+

Key
+++ Very common.
++ Common.
+ Rare.
(+) Very rare.

Table 2. Distribution of Haemaphysalis, Hyalomma and Ixodes ticks in different administrative regions of Ethiopia

		Administrative regions										
Tick species		Tigray	Afar	Amhara	Oromia	Somali	Benshangul & Gumuz	SNNP	Gambella	Harari	Addis Ababa	Diredawa
Haem. aciculifer	3-host	-	-	-	+	-	-	+	-	-	-	-
Haem. leachi	3-host	-	-	-	+	-	-	+	-	-	+	-
Haem. parmata	3-host	-	-	-	+	-	-	+	-	-	+	-
Haem. spinulosa		-	-	-	+	-	-	+	-	-	-	-
H. a. anatolicum	2 or 3-host	-	(+)	-	-	-	-	-	-	-	-	-
H. a. excavatum	2 or 3-host	-	(+)	-	-	-	-	-	-	-	-	-
H. dromedarii	2 or 3-host	+	++	+	++	++	-	-	-	+	-	+
H. erythraeum	2 or 3-host	-	(+)	-	-	-	-	-	-	-	-	-
H. impeltatum	3-host	(+)	(+)	-	(+)	-	-	(+)	-	-	-	(+)
H. m. marginatum	2-host	-	(+)	-	-	-	-	-	-	-	-	-
H. m. rufipes	2-host	+	+	+	++	+	+	+	-	+	+	+
H. punt		-	(+)	-	-	-	-	-	-	-	-	-
H. truncatum	2-host	++	+	++	++	+	-	++	-	-	-	+
Ixodes rasus	3-host	-	-	-	-	-	-	-	(+)	-	-	-

Key
+++ Very common.
++ Common.
+ Rare.
(+) Very rare.

Table 3. Distribution of Rhipicephalus, Argas and Ornithodorous ticks in different administrative regions of Ethiopia

		Administrative regions										
Tick species		Tigray	Afar	Amhara	Oromia	Somali	Benshangul & Gumuz	SNNP	Gambella	Harari	Addis Ababa	Diredawa
R. bergeoni	3-host	-	-	-	++	-	-	+	-	-	-	-
R. camicasi	3-host	(+)	-	-	+	-	-	-	-	-	-	-
R. e. evertsi	2-host	+	+	++	++	+	+	++	(+)	+	+	+
R. guilhoni	3-host	-	-	-	(+)	-	-	-	-	-	-	-
R. longicoxatus		-	(+)	-	-	-	-	-	-	-	-	-
R. longus	3-host	-	-	-	(+)	-	-	(+)	-	-	-	-
R. lunulatus	3-host	-	+	-	++	-	-	++	+	-	-	-
R. muhsamae	3-host	-	-	-	+	-	-	+	+	-	-	-
R. praetextatus	3-host	+	+	++	++	-	+	++	+	-	-	+
R. pravus	3-host	+	+	-	++	++	+	+	-	-	-	+
R. pulchellus	3-host	+	++	+	++	++	-	+	+	+	-	+
R. sanguineus	3-host	-	+	-	+	-	-	+	+	-	-	+
R. senegalensis	3-host	+	-	-	+	-	-	-	+	-	+	+
R. sulcatus		-	+	-	+	+	-	-	-	-	-	+
R. turanicus	3-host	+	-	-	+	-	-	-	+	-	-	-
Ar. persicus		+	+	+	+	-	+	–	+	+	+	+
Or. moubata		-	-	+	-	-	-	-	-	+	-	+
Or. savignyi		+	-	-	+	+	-	-	-	-	-	-

Key
+++ Very common.
++ Common.
+ Rare.
(+) Very rare.

4. GENUS BOOPHILUS (CURTICE, 1891)

Two species of *Boophilus, B. decoloratus* and *B. annulatus* are recorded in animals from Ethiopia. *B. decoloratus* is economically more important than *B. annulatus* due to its occurrence in almost all regions in the country, its impact on the quality of hides and skins and disease transmission.

Boophilus decoloratus (Koch, 1844): *B. decoloratus* is indigenous to Africa and most likely evolved as a parasite of ungulates in East Africa [14]. It is the commonest and most widespread tick species in Ethiopia and is collected from animals in all administrative regions except Afar, a semi-arid to arid region in the northeast (Table 1). It was mainly found in wetter highlands and sub-highlands receiving more than 800 mm of rain annually. The distribution pattern is similar to that of *A. variegatum* [3]. Cattle are the main hosts of *B. decoloratus* [17], but it also feeds on sheep, goats and equines. In central Ethiopia, *B. decoloratus* was more abundant on cross and exotic cattle breeds possibly due to developed resistance to the organophosphate group of acaricides in common use in these breeds [4]. The attachment sites are the dewlap, neck, shoulders and the sides of the body of the infested animals. The immature stages are found on the tips and edges of the ears and on the legs. Apart from downgrading the quality of hides and skins *B. decoloratus* also transmits *B. bigemina* and *A. marginale*.

Boophilus annulatus (Say, 1821): *B. annulatus* is restricted to Gambella regional state in western Ethiopia with wet tropical forest and was recorded in Itang and Jikao zones and Gambella town (Table 1). It was found between 400-600 m above sea level receiving between 800-1200 mm annual rainfalls in riverine forest [6]. The preferred host of this tick species is cattle and the predilection sites are the neck, dewlap, belly and legs. *B. annulatus* is a potential vector of *Babesia bovis* and *B. bigemina.* The tick also causes damages to hides and skins of infested animals.

5. GENUS HAEMAPHYSALIS (KOCH, 1844)

There are 4 species of ticks belonging to the genus *Haemaphysalis* so far recorded in Ethiopia and only *Haem. leachi* is known as a vector of TBD.

*Haemaphysalis aciculifer (*Warburton, 1913*): Haem. aciculifer* is found in the highland or sub-highland areas in Ethiopia [3]. It was collected from cattle, sheep, goats and equines at an altitude of 1400–2600 m receiving between 1400-2600 mm annual rain where it rains all over the year [6]. It was collected from animals in different areas in the country with the favorable climate indicated and was mainly recorded in East and West Wollega, Illubabor, Kaffa, Wolaita Sodo and Hagere Selam in the west and southwest and Tulu Bolo, and East Shoa in central Ethiopia [4, 5, 6 & 12]. It has adapted to a variety of habitats, but never present in large numbers. There is no report on its role in disease transmission in Ethiopian livestock.

Haemaphysalis leachi (Audouin, 1826): Adult *Haem. leachi* "the Yellow dog tick" infests mainly domestic dogs. The immature stages prefer rodents. Adults attach to the head, neck and shoulders of dogs. It has been recorded in urban dogs in Addis Ababa, Debre Zeit, Awassa and Shashemene. It was also collected from cattle and other livestock suggesting the close association between domestic dogs and the livestock. Morel [5] recorded the tick from spotted Hyaena (*Crocuta crocuta*) in Awash National Park. Adult *Haem. leachi* is the major vector of *B. canis* in dogs from Addis Ababa [18].

Haemaphysalis parmata (Neumann, 1905): *Haem. parmata* is found in humid forests and sub-highland woodlands in Ethiopia [3]. It has been collected from southwest Ethiopia at an altitude of 1600-2600 m receiving between 1200-2600 mm rainfall [6]. Morel [5] also reported the tick in cattle from Debre Zeit on the escarpment of the Ethiopian Rift Valley and in horses from Arsi, in central Ethiopia. The tick was also recorded in western Ethiopia (Kaffa & Sheka and Dawro zones) regional state (Table 2). The role of this tick in disease transmission is not well known.

Haemaphysalis spinulosa (Neumann, 1906): *Haem. spinulosa* is mainly a parasite of the cat. Morel & Rodhain [15] collected this tick species from the African wild cat (*Felis libyca)*, caracal cats (*Felis caracal)*, the bat-eared fox (*Otocyon megalotis*) and black-backed jackal (*Canis mesomelas*) in Gamogofa zone of southern Ethiopia (Shangura Omo camp). Recorded. The tick was also recorded in cats from Debre Zeit [16] and in spotted Hyaena (*C. crocuta*) from Awash National Park [5].

6. Genus Hyalomma (Koch, 1844)

There are 9 species of ticks belonging to the genus *Hyalomma* "Bont legged ticks" recorded in Ethiopia. *H. marginatum rufipes* and *H. truncatum* are important tick species causing damages to livestock in Ethiopia.

Hyalomma marginatum rufipes (Koch, 1844): *H. m. rufipes* "The hairy *Hyalomma*" is widespread in Ethiopia (Table 2). It is recorded between 500 and 3500 m above sea level [3]. The tick is found in the hairless areas around the anus and on the genitalia on camels and cattle and around the hooves on sheep. The immature stages infest the necks of the scrub hares, the heads and necks of birds and burrowing rodents. It is the most important vector of the virus Crimean-Congo hemorrhagic fever and the bacteria *Rickettsia Conorii,* incriminated for tick typhus in humans. Its long mouthparts cause tissue damage to its host and secondary bacterial infection that may lead to abscess formation.

Hyalomma truncatum (Koch, 1844): *H. truncatum* occur throughout Ethiopia in different ecological zones ranging from 300 to 3500 m above sea level, which receive between 250-1500 mm annual rainfall. The tick is adapted to dry habitat and is common in deserts, steppes and savanna climatic regions. The adults prefer cattle followed by sheep, goats and horses. The predilection sites of this tick are the tail switch, areas around the anus, the lower perineum and the legs [9].

H. truncatum is recorded from various regions including south (Bale, Yabello, Moyalle and Dollo) central (East Shoa, West Shoa, North Shoa, Gurage, Kembata-Alaba-Timbaro, mid-west (Anger Gutin in East Wollega) and eastern (Afar) Ethiopia in cattle [4, 5, & 6] and from camels and goats in eastern Ethiopia (Diredawa) (Table 2). Seasonal activities reach the peak in August and September [3]. The toxin contained in the saliva of *H. truncatum* causes skin diseases known as sweating sickness in cattle, particularly calves. The long mouthparts cause tissue damage in cattle and sheep with a consequent secondary bacterial infection that may lead to abscess formation and lameness in lambs. It also transmits *B. caballi* to horses causing equine piroplasmosis and *R. conorii* causing tick typhus in humans [9].

Hyalomma anatolicum anatolicum (Koch, 1844): *H. a. anatolicum* has a mixed two-host and three-host feeding patterns when feeding on different hosts (Table 2). The immature stages behave as two-host tick when feeding on hares and rodents, and as three-host tick when feeding on cattle [9]. Cattle are the main host of adult *H. a. anatolicum*, but sheep, goats, camels, horses and donkeys are also infested. The predilection sites of adult *H. a. anatolicum* on cattle are the axillae, groin, genital areas, perineum and udder. It is adapted to dry areas. Pegram et al. [3] collected a very few specimens of this tick species from camels and rarely from cattle in arid and semi-arid lowlands below 1500 m above sea level in the Danakil, around Worenzo (Afar, northeast Ethiopia). The tick is found in the neighboring Northern Sudan and in Somalia. Further information concerning the presence of this tick species in Ethiopia is not available. *H. a. anatolicum* is a potential vector of *T. annulata, T. equi, B. caballi* and Crimean-Congo hemorrhagic fever virus [9].

Hyalomma anatolicum excavatum (Koch, 1944): *H. a. excavatum* requires two or three hosts to complete its life cycle. Cattle, sheep, goats, horses and donkeys are the hosts of adult *H. a. excavatum,* while larvae and nymphs feed on hares and rodents. The adults attach to the hindquarters of their hosts (perineum, udder) [9]. The tick is found in arid and semi-arid areas between 100-500 m above sea level receiving 100-350 mm annual rainfall [5]. Pegram *et al.* [3], reported rare cases of *H. a. excavatum* in camels in Afar region. *H. a. excavatum* does not play a role in the natural transmission of *T. annulata*, since transovarian passage is not carried out with this tick [19]. There is no report on the presence of this tick species in other parts of the country.

Hyalomma dromedarii (Koch, 1844): *H. dromedarii*, also known as "The camel *Hyalomma*" needs two or three hosts to complete its life cycle. It is well adapted to extremely dry habitat and to its camel hosts. This tick was found in arid ecological zones of eastern and southern Ethiopia. Its distribution coincided with that of pastoralists and their camels [3]. *H. dromedarii* was collected from camels in Diredawa, Harrar, Jijiga and Negelle Borana, Tigray and Afar areas (Table 2). The sites of attachment of the adults are the medial aspect of the thighs, the udder (causing teat losses and reduced milk production) and the scrotum (causing infections leading to infertility in male camels). The role of *H. dromedarii* as a vector of TBDs is not well known.

Hyalomma impeltatum (Schulze & Schlottke, 1930): There are few reports on the life cycle of *H. impeltatum*. In the laboratory, the tick behaved as a three-host-tick to complete its life cycle [9]. *H. impeltatum* occurs in drier areas infesting all large domestic animals, particularly camels. The immature stages feed on small animals such as rodents, hares and ground birds. Morel [5] collected the tick in semi-arid regions from camels in Diredawa, Bale and Gamogofa, cattle in Negelle Borana and eland (*Taurotragus Oryx*) in Awash National Park. It has also been collected from Tsenkanit and Hawzein districts, in northern Ethiopia. *H. impeltatum* and *H. dromedarii* share similar morphological features, hosts and geographical areas [9]. It has been reported that the tick can transmit the virus which causes the Crimean-Congo hemorrhagic fever to man [9].

Hyalomma marginatum marginatum (Koch, 1844): *H. m. marginatum* mainly infests cattle. The adults attach only to the hindquarters of animals (perineum, genital areas and udder) while the immature stages prefer to feed on the ears of hares and rabbits. Pegram et al. [3] reported this tick species in Ethiopia with no further details on its distribution. *H. m.*

marginatum transmits the virus of Crimean-Congo hemorrhagic fever to man and *B. caballi,* the cause of equine babesiosis.

Hyalomma erythraeum (Tonelli-Rondelli, 1932): Pegram et al. [3] reported this tick species from eastern low lands of Ethiopia. No further information is available on the occurrence of *H. erythraeum* elsewhere in the country.

Hyalomma punt (Hoogstraal, Kaiser & Pedersen, 1969): *Hyalomma punt* is reported as rarely occurring tick from eastern Ethiopia [3]. There is no record of this tick species from the other parts of the country. Its role in disease transmission is also not documented

7. Genus Rhipicephalus (Koch, 1844)

There are 15 species of ticks belonging to the genus *Rhipicephalus* recorded in Ethiopia and out of these *R. e. evertsi, R. praetextatus* and *R. pulchellus* are the most economically important ticks in the country.

Rhipicephalus evertsi evertsi (Neumann, 1897): All stages of *R. e. evertsi* ("the red-legged tick") frequently feed on horses, donkeys and cattle. The adults are found on the hairless areas around the anus as well as in the groin region of the host while the immature stages attach on the deeper parts of the inner surface of the ear and in the outer ear canal. *R. e. evertsi* is widely distributed and common on livestock in Ethiopia (Table 3). The tick shows no apparent preference for any particular altitude, rainfall or season [3].

R. e. evertsi transmits *B. caballi* and *T. equi* to horses both causing equine piroplasmosis and *A. marginale,* which causes anaplasmosis in cattle. The saliva of female ticks contains a toxin that causes paralysis, particularly in lambs. The toxin may also affect calves and adult sheep [9].

Rhipicephalus praetextatus (Gerstäcker, 1873): *R. praetextatus* is a member of the *simus* group. The adult ticks are most abundant on cattle attaching to the tail brush and around the feet. The immature stages prefer to feed on rodents. The tick is more abundant during the rainy season [9]. *R. praetextatus* is found in a wide range of ecological habitats, from semiarid areas having 250 mm annual rainfall, through tropical and subtropical savanna, to wooded highlands with 1500 mm annual rainfall. It is rare in the more arid areas of southwestern Ethiopia but recorded in almost all-administrative regions (Table 3). *R. praetextatus* can transmit the virus of Nairobi sheep disease to sheep and the bacterium *R. conorii* which causes tick typhus in humans. The tick can cause toxicosis in cattle, resulting in paralysis [9].

Previously, there appeared to be difficulty in morphologically differentiating *R. simus* from *R. praetextatus*. However, recent findings indicated that *R. simus* occur across central and southern Africa around 6° south [8]. This suggests that all ticks previously recorded as *R. simus* should be replaced by *R. praetextatus* [11]. Based on the above clarifications, it can be suggested that *R. simus* does not occur in Ethiopia.

Rhipicephalus pulchellus (Gerstäcker, 1873): *R. pulchellus* ("the zebra tick") prefers to feed on cattle but also infests camels, sheep and goats. The tick attaches in large numbers to the ears of cattle, and also is found on the underside of the body. It survives the savanna, steppes and desert climates [9].

R. pulchellus was mainly collected from livestock in the eastern part of the country predominantly eastern Oromia, eastern Amhara, eastern Tigray, eastern and southern SNNP, Afar, Harar, Somali and Diredawa (Table 3). Few collections were also made from Dimma (Gambella) on Sudanese refugee cattle in the west. It transmits the virus causing Nairobi sheep disease to sheep and *R. conorii*, and the virus causing Crimean-Congo hemorrhagic fever to humans.

Rhipicephalus bergeoni (Morel & Balis, 1976): *R. bergeoni* is largely confined to Ethiopia except for a single collection from the Sudan [8]. It occurs predominantly on cattle in the wetter highlands and sub-highlands of Ethiopia [10]. *R. bergeoni* was collected from cattle in Bale zone (Dodolla at 2500 m above sea level, Goba at 2700 m above sea level, Adola at 1900 m above sea level) and in Gamogofa zone around Chencha at 2900 m above sea level [20] (Table 3). Its predilection site is in the ears of cattle. Its role in disease transmission is not clearly known.

Rhipicephalus camicasi (Morel, Mouchet & Rodhain, 1976): *R. camicasi* belongs to *R. sanguineus* group occurring mainly in the arid and semi-arid habitats receiving less than 250 mm annual rainfall. It appears to be most active in dry season [10]. Recently, this tick has been collected from cattle in Tigray and Oromia regions. *R. camicasi* was previously collected from *Taurotragus Oryx*, *Phacochoerus aethiopicus*, *Otocyon megalotis* in Awash National Park and *Acinonyx jubatus* in Elbah, from goats in Hararge, from *Lepus capensis*, *Equus grevyi* and *Litocranius walleri* in Kersa [5]. The association of this tick species with disease transmission is not well known.

Rhipicephalus guilhoni (Morel & Vassiliades, 1963): *R. guilhoni* is a member of the *R. sanguineus* group that occurs in habitats receiving 250-750 mm annual rainfall. The tick also extends into more arid habitats and is found in steppe and savanna climatic regions (Walker, *et al*, 2003). Few collections of this tick species were made in central [10] and western Ethiopia. The preferred hosts of adult *R. guilhoni* are cattle, sheep, horses and dogs. The immature stages feed on rodents.

Rhipicephalus longicoxatus (Neumann, 1905): *R. longicoxatus* was collected from cattle in Harar (Elbah) and Borana (Wuchale) [16]. Pegram et al. [3] also indicated the possible existence of this tick species in Ethiopia.

Rhipicephalus longus (Neumann, 1907): *R. longus* was collected from various places in Ethiopia including East Wollega [12], Illubabor, Bale, Jimma, Southwest Shoa and Kaffa & Sheka administrative zones (Table 3). The tick survives well in evergreen forest, woodlands and thickets with at least 1200 mm of annual rainfall [3]. There is no report of disease transmission by this tick species.

Rhipicephalus lunulatus (Neumann, 1907): Adults of *R. lunulatus* prefer to infest cattle, but also can be found on sheep, pigs and dogs. On cattle, the adult ticks attach to the legs and in the tail brush, while the immature stages infest rodents and hares.

In Ethiopia, it was mainly collected from cattle, goats, equine and sheep in western (Gambella/Godere, Wollega, Jimma, Illubabor), southern (Borana and Bale zones) south west (Kaffa & Sheka and Dawro zones), central (East Shoa) and eastern (Gewane, Afar) parts of Ethiopia (Table 3). Adult tick predominates in areas between 1200-2600 m above sea level receiving 1200-2400 mm annual rainfall [6]. There is no report of disease transmission by this tick species.

Rhipicephalus muhsamae (Morel & Vassiliades, 1965): The preferred hosts of the adult *R. muhsamae* are cattle, dogs and wild carnivores [9]. In cattle, adults attach to the ears, neck, dewlap, genitalia and hooves. The immature ticks prefer to infest rodents. In Ethiopia, the tick was collected from cattle, sheep and goats in the western (Illubabor, Gambella), southwestern (Kaffa & Sheka and Dawro zones), southern (Borana zones), and eastern (Hararge) parts of Ethiopia. The tick was mainly recorded in areas between 1200-2600 m above sea level receiving1600-2600 mm annual rainfall [6]. Its role in disease transmission is not known.

Rhipicephalus pravus (Dönitz, 1910): Adults of *R. pravus* prefer to feed on cattle, sheep, goats, camels and dogs. However, they can also be found on equines. They attach to the ears, head, neck and dewlap of cattle. The immature stages prefer to feed on elephant shrews and hares. It is found only in east African countries such as Ethiopia, Somalia, Kenya and Tanzania [9]. The tick was collected from livestock in the lowlands of Ethiopia (Diredawa, Tigray, Afar/Gewane, Borana, Bale, South Omo, and Somali Regional State/Filtu).

Rhipicephalus sanguineus (Latreille, 1806): *R. sanguineus* "The kennel tick" prefers dogs as a host, but can also be found on cattle. Adults attach to the ears, neck and shoulder of the host. It occurs in warm and moist climates [9]. *R. sanguineus* was collected from cattle and dogs in various places in central, west and eastern parts of Ethiopia (Diredawa Afar, Oromia, Addis Ababa and Gambella). It is a potential vector of *B. canis* in dogs.

Rhipicephalus senegalensis (Koch, 1844): *R. senegalensis* is present on cattle, dogs, warthog and buffaloes. The site of attachment of adult *R. senegalensis* is the ears in cattle, but also can attach to the head, shoulders, flanks, tail and around the feet of other hosts. The immature stages feed on rodents. In Ethiopia, it was collected from cattle in west (Anger Gutin, East Wollega) and central (Metehara, East Shoa) Ethiopia. Recently there was a collection of this tick from cattle in Southwest Shoa, central Ethiopia (Abebe Mekonnen, 2006, personal communication).

Rhipicephalus sulcatus (Neumann, 1908): *R. sulcatus* was found in woodland and tropical or sub-tropical savanna receiving more than 500 mm rainfall. Morel [5] collected *R. sulcatus* from goats (Negelle Borana) and cattle (Jijiga and Diredawa areas) in the lowland areas. However, these records may have been confused with *R. turanicus* [10]. The tick was also recovered from cattle in Afar (Sheket, Abala), Sidama and Oromia (East Shoa, East Wollega, Jimma, Borana and Southwest Shoa). Morel and Rodhain [15] also collected the tick from *Otocyon megalotis*, *Felis libyca*, *Sagittarius serpentarius* and *Ciconia nigra* in Shangura camp in Gamogofa, south Ethiopia.

Rhipicephalus turanicus (Pomerantsev, 1936): The adults of *R. turanicus* infest cattle, sheep and dogs and occasionally also horses, camels and goats. Immature stages prefer rodents and hares [13]. In Ethiopia, it has been collected from livestock in Oromia (Southwest Shoa), Tigray and Gambella regional states (Table 3).

8. Genus Ornithodoros

Ornithodoros moubata (Murray, 1877): *Or. moubata* is a soft tick also known by its common name as "the eyeless" or "warthog tampan". It is similar in appearance to *Or. savignyi* but has no eyes. The hosts of this tick are humans, poultry, domestic pigs and warthogs [13]. The habitat of the tick is defined by the housing and nests of its hosts. It is an important vector of the virus of African swine fever to domestic pigs and *Borrelia duttoni*, the cause of relapsing fever in humans. Warthogs serve as carriers of African swine fever.

Transovarian transmission of the virus occurs in *Or. moubata*. African swine fever virus may also pass between male and female ticks during mating [9]. In Ethiopia, *Or. moubata* was collected from warthog and hut ground in northwest (Gondar), and eastern (Harar and Diredawa) Ethiopia [5].

Ornithodoros savignyi (Audouin, 1827)*: Or. savignyi* belongs to the family Argasidae, and also known as "the sand tampan" or "the legged tampan". Camels and cattle are the preferred hosts of this tick species. It feeds on the lower legs of its hosts causing severe biting stress. Toxins in tick saliva can cause paralysis and toxaemia, which is sometimes fatal. *Or. savignyi* feeds at night and is more active during the dry season than the rainy season [5]. In Ethiopia, it has been collected from the sand in Tigray, Somali (Dagabur), Oromia (Bale), Harawa (El Bahe) and Meto areas.

9. Genus Argas

The genus belongs to the family Argasidae, also known as "the soft ticks". There is only one species of argasid tick recorded in Ethiopia.

Argas persicus (Oken, 1881): *Ar. persicus* "the fowl tick" is an argasid tick that has been recorded and widely spread in the warmer parts of the country because of its close association with poultry. It is a parasite of domestic fowls, turkeys, ducks and geese.

Ar. persicus transmits the bacterium *Borrelia anserine,* the causative agent of avian spirochaetosis, and the protozoan *Aegyptianella pullorum,* which causes avian piroplasmosis. Heavy infestation can cause anemia. The larval ticks produce a toxin that causes paralysis in chickens [9].

10. Genus Ixodes (Latreille, 1796)

Ixodes rasus: I. rasus was the only species of the Genus Ixodes recorded in Ethiopia. Adult *Ixodes rasus* infests cattle, dogs, leopards, small and large antelopes, mice and elephant shrews. The tick was collected from bushbuck in southwest part of the country [6]. To our knowledge, there is no other published record of this tick species in Ethiopia.

Conclusion

Ethiopia has the largest livestock population in Africa. The diverse climate ranging from wet tropical forest through cool highland to arid and semi-arid areas makes it a good representative of various tropical agro-climatic zones. The knowledge on the distribution of ticks in these diverse climatic zones and in various animal species similarly would also represent many areas in the tropics.

The most common ixodid ticks infesting domestic animals in Ethiopia are *A. cohaerens*, *A. variegatum*, *B. decoloratus*, *R. pulchellus*, *R. e. evertsi*, *R. praetextatus* and *H. m. rufipes*. The remaining ticks were present only in small numbers. *A. variegatum, H. m. rufipes* and *R. e. evertsi* were widely distributed in the country. *B. annulatus* was restricted to Gambella

Region with wet tropical forest in the west, and *A. cohaerens* and *R. bergeoni* were mostly collected from the southwest part of the country with similar climatic condition. *R. pulchellus* and *A. gemma* were predominantly collected from the semi-arid and arid regions of east and southeast Ethiopia. Most ticks of the genus *Hyalomma* ticks and *R. camicasi* were collected from the lowlands while *Haem. aciculifer* and *R. bergeoni* were predominantly found in the highland areas. There were few tick collections of the genera *Argas* and *Ornithodorous*.

Knowledge on tick species and their negative impact on the development of livestock industry is very important tool when developing cost effective, sustainable and economically viable tick and TBD control strategies.

ACKNOWLEDGMENTS

The authors are most grateful to Ejigu Zeben, Tadewos Kassa, Tucho Enkosa, Kahsay Negussie and Dagne Guta for their valuable assistance with the field and laboratory work.

REFERENCES

[1] Jongejan, F. & Uilenberg, G. (2004). The global importance of ticks. *Parasitology*, 129: S1-S12.

[2] De Castro, J.J. (1997). Sustainable tick & tick-borne disease control in livestock improvement in developing countries. *Vet. Parasitol.* 71, 77-97.

[3] Pegram, R.G., Hoogstraal, H., & Wassef, H.Y. (1981). Ticks (Acari: Ixodoidea) of Ethiopia. I. Distribution, ecology and host relationship of species infesting livestock. *Bull. Ent. Res.,* 71, 339-359.

[4] Mekonnen, S., Hussein, I. & Bedane, B. (2001). The distribution of ixodid ticks (Acari: Ixodidae) in central Ethiopia. *Onderstepoort J. Vet. Res.*, 68, 243-251.

[5] Morel, P. C. (1980). Study on Ethiopian ticks (Acarida, Ixodida). Institut d'elevage et de Me'dicineVeterinaire des Pays tropicaux. Pp. 332.

[6] De Castro, J.J. (1994). Tick survey. A survey of the tick species in Western Ethiopia; AG: DP/ETH/83/023. Technical report. FAO. Rome. Pp. 83.

[7] Hoogstraal, H. (1956). African Ixodoidea. I. Ticks of the Sudan (with special reference to Equatoria province and with preliminary reviews of the genera *Boophilus, Margaropus* and *Hyalomma*) Pp.1101. Washington DC., Department of Navy, Bureau of Medicine and Surgery.

[8] Walker, J.B., Keirans, J.E. & Horak, I.G. (2000). The Genus Rhipicephalus (Acari, Ixodidae): a guide to the brown ticks of the world. Cambridge, UK. Cambridge University Press.

[9] Walker, A.R., Bouattour, A., Camicas, J.L., Estrada-Penã, A., Horak, I.G., Latif, A.A., Pegram, R.G. & Preston, P.M. (2003). Ticks of domestic animals in Africa: a guide to identification of species. Pp. 1-219.

[10] Pegram, R.G., Clifford, C.M., Walker, J.B. & Keirans, J.E. (1987a). Clarification of the *Rhipicephalus sanguineus* group (Acari, Ixodoidea, Ixodidae). I. *R. sulcatus* Neumann 1980 and *R. turanicus* Pomarantsev, 1936. *Syst. Parasitol.* 10, 3-26.

[11] Pegram, R.G., Walker, J.B., Clifford, C.M. & Keirans, J.E. (1987b). Comparison of populations of the *Rhipicephalus simus* group: *R. simus*, *R. praetextatus,* and *R. muhsamae* (Acari: Ixodidae). *J. Med. Entomol.* 24, 666-682.

[12] Pegram, R.G. (1979). Ticks (Ixodoidea) of Ethiopia with special reference to cattle and a critical review of the taxonomic status of species with in the *Rhipicephalus sanguineus* group. Thesis (M Phil., Brunel University); Uxbridge, England. Pp. 169.

[13] Horak, I.G., Stoltzs, W.H. & Heyne, H. (2003). Short course in the identification of southern and northeast African ticks. Department of Veterinary Tropical Diseases, Faculty of Veterinary Science, University of Pretoria, South Africa. Pp. 1-110.

[14] Walker, J.B. (1991). A review of the ixodid ticks (Acari, Ixodidae) occurring in Southern Africa. *Onderstepoort J. Vet. Res.*, 58, 81-105.

[15] Morel, P. C. & Rodhain, F. (1972). Contribution à la connaissance des tiques (Ixodina) du sud de I' Ethiope. Premiére partie. *Bull. Soc. Path. Exot.,* 65 (5): 725-732.

[16] Bergeon, P. & Balis, J. (1974). Contribution à l'étude de la répartition des tiques en Ethiopie (enquéte effectués de 1965 à 1969). *Rev. Elev. Méd. vét. Pays trop.,* 27(3): 285-299.

[17] Gebre, S. & Kaaya, G.P. (1997). Development, reproductive capacity and survival of *Amblyomma variegatum* and *Boophilus decoloratus* in relation to host resistance and climatic factors under field conditions. *Vet. Parasitol.* 75, 241-253.

[18] Mekonnen, S., Tsigie, Z., Abditcho, S., Semegn, Y. & Alemayehu, T. (1999). Canine babesiosis and its vector tick in Addis Ababa region. *Ethiop. Vet. Assoc. J.* 3 (1), 34-39.

[19] Markov, A.A., Gildenblat, A.A., Kurchatov, V.I. & Petunin, F.A. (1948). A new vector of casual agent of theileriasis of cattle, the tick Hyalomma scupense P. Schulze. *Veterinarija*. 25 (9):13.

[20] Morel, P. C. & Balis, J. (1976). Description de *Rhipicephalus bergeoni* n. sp. (acariens, Ixodida) des montagnes d'Ethiopie. *Revue Élev. Méd. Vét. Pays trop.* 29, 141-148.

In: Ticks: Disease, Management and Control
Editor: Moges Woldemeskel
ISBN: 978-1-62081-136-8

Chapter 5

MANAGEMENT OF TICK-BORNE CANINE BABESIOSIS

Hui-Pi Huang**, Yueh-Lun Hsu, Tzu-Chi Tai and Hung-Yin Chen
Institute of Veterinary Clinical Science, Veterinary School,
National Taiwan University, Taipei, Taiwan

ABSTRACT

Canine babesiosis is an important worldwide, tick-borne disease caused by hemoprotozoan parasites of the genus *Babesia*. It is a small pleomorphic intraerythrocytic parasite that can cause erythrocyte destruction and hemolytic anemia. The parasite is transmitted through the bites of ixodid ticks, such as *Rhipicephalus sanguineus*. Definitive diagnosis of canine babesiosis is based on medical history and clinical signs together with the identification of *Babesia* spp. within infected erythrocytes, positive serologic results, or detection of amplification of nuclcic acid extracted from blood or tissues. Subjects with high mortality can be identified using several prognostic factors. Combination of babesicidal agents and antibiotics, and supportive cares are integrated as treatment program for canine babesiosis. Tick (vector) control is the key to prevent canine babesiosis. This chapter describes the current information regarding epidemiology, pathogenesis, and pathophysiology of the clinical manifestations associated with tick-borne canine babesiosis, and the methods of diagnosis, prognostic factors, treatments and prevention of canine babesiosis.

INTRODUCTION

Canine babesiosis is caused by hemoprotozoan parasites of the genus *Babesia* [1]. *Babesia* spp. are small pleomorphic intraerythrocytic parasites that cause erythrocyte destruction leading to hemolytic anemia as a major clinical presentation [1, 2]. The clinical presentation of babesiosis can vary, ranging from mild to life-threatening and from peracute

*Correspondence: E-mail: hphuang@ntu.edu.tw.

to chronic or the infection may be subclinical [1]. To date, babesiosis may be the most prevalent tick-borne disease affecting the canine population worldwide.

Etiology

Babesia was first identified at the end of the 19th century [1]. Numerous species associated with various domestic and wild animals have been reported. *Babesia canis* and *Babesia gibsoni* are the two species that predominately affect dogs and other canine species [3-6]. Other *Babesia* spp. such as *Babesia conradae* are isolated and reported to possibly affect canine species, [1, 7-9].

Classification

The initial description and classification of *Babesia* spp. was based on morphological characteristics, of which the pattern of merozoite assembly in red blood cells was the main defining characteristic [10]. *Babesia* spp. multiply in the erythrocytes of the vertebrate host and are transovarially transmitted in the vector host. Thus, the *Babesia* spp. clearly differs from *Theileria* spp., which multiply in the lymphocytes before the erythrocytes and are not transovarially transmitted in ticks [10-12].

Morphologically, *Babesia* spp. are divided into two groups that include large organisms such as *B. canis*, in which the merozoites are longer than the erythrocyte radius, and small organisms such as *B. gibsoni*, in which the merozoites are smaller than the erythrocyte radius [6]. This division into small and large *Babesia*, based on morphologic characteristics, has no clear genetic basis [3].

More *Babesia* spp. were identified based on molecular and biological data. For example, *B. canis* was previously described as a single species, but, it has now been split into at least three distinct species or subspecies, namely *B. canis canis*, *B. canis rossi*, and *B. canis vogeli*. These three subspecies have different tick hosts [1, 7, 13, 14]. More small forms of *Babesia* spp. were also identified.

Life Cycle of Babesia Species

The life cycle of *Babesia* organisms consists of asexual multiplication in vertebrate blood cells, sexual reproduction in the vector, and the development of sporozoites in the salivary glands of the vector, which is required for transmission from the vector to the vertebral host. *Babesia* organisms are transmitted from vertebrate hosts to tick vectors as gametocytes, and from the saliva of tick vectors to vertebrates as sporozoites during biting [5, 10].

Sporozoite differentiation begins only when the tick attaches to its vertebrate host [5, 10, 12]. In contrast to *Plasmodium* spp. or *Theileria* spp., sporozoites of *Babesia* spp. are endocytosed by the red blood cells (erythrocytes), in which all the parasitic stages are carried out. Once in the erythrocytes, the parasite produces two merozoites by binary fission. After erythrolysis, each merozoite invades a new erythrocyte, and successive cycles of merogony occur. Multiplication is asynchronous, and various divisional stages of the parasite can be detected in the bloodstream at the same time.

Gametocytes become gamonts after ingestion by the tick hosts. In the lumen of the tick's digestive tract, the gametes fuse to form an elongated zygote. Once the *Babesia* zygote has been internalized, it is transformed into a motile ookinete and then invades the tick's body tissues. Invasion of the ovary in female ticks results in many *Babesia*-infected eggs (transovarial transmission). Subsequent development of *Babesia* includes asexual multiplication, continuing sporogony, and the development of numerous kinetes (sporokinetes). The *Babesia* infection acquired during one life stage of the tick is passed on to the next (transtadial transmission). Some kinetes also invade the salivary glands of ticks, where the final cycle of development produces the sporozoites. Sporozoites represent the infectious stage of the parasite, as they introduce these protozoa into the mammalian host [5, 10, 12, 15].

Tick Vector for Canine Babesiosis

Babesia parasites are transmitted to a susceptible dog through the saliva of an infected tick while feeding [10, 16]. Different species of ticks have been reported to transmit different *Babesia* spp. among canine populations, leading to different epidemiological patterns. *Babesia canis canis* is transmitted by *Dermacentor reticulatus* in Europe and South Africa, *B. canis rossi* is transmitted by *Haemaphysalis leachi* (and possibly *Rhipicephalus sanguineus*, *R. evertsi evertsi,* and *Amblyomma lepidum*) in Africa and the Middle East, and *B. c. vogeli* is transmitted mainly by *R. sanguineus* in many countries worldwide [1]. *Amblyomma cajennense, Amblyomma ovale*, and *Boophilus microplus* are reported to be associated with transmission of *B. c. vogeli*, as well [17]. *Babesia gibsoni* is transmitted mainly by *R. sanguineus*, but infestation with other tick species such as *Dermacentor reticulatus*, *Dermacentor variabilis*, *Haemaphysalis longicornis*, and *Haemaphysalis bispinosa* have also been reported in association with babesiosis caused by *B. gibsoni* [18].

Non-Tick Transmission

Tick transmission is believed to be the primary route of infection, but non-tick transmission is also considered as a possible mechanism of infection. A high prevalence of canine babesiosis has been reported in certain breeds of dogs engaged in fighting, such as American Pit Bull Terrier and Tosa inu [19-23]. As these dogs are free of tick infestation in most cases, *B. gibsoni* is considered to be transmitted via transfer of infected blood during bites and fights or during blood transfusions after fights [22, 23].

Other proposed routes of non-tick transmission include biting flies, fomites and needles used to vaccinate multiple animals [20, 24]. However, controlled experiments have not been performed to establish the route of direct blood to blood transmission. Vertical transmission of *Babesia* infection via the transplacental route as well has been reported in dogs [25, 26].

Prevalence and Geographic Distribution

The geographical prevalence of canine babesiosis, isolates of *Babesia* and recognized vectors have been reported worldwide [4, 9, 17-21, 24, 27-107] (Table 1).

Geographic characteristics of canine babesiosis are closely related to vector distribution [4, 9, 17-21, 27-30, 33, 35, 37, 39-56, 58-60, 62-65, 67-98, 100-103, 105-120] (Table 2).

Increased chances of exposure to ticks, as would occur in hunting or outdoor dogs, are associated with increased canine babesiosis [38, 121, 122]. Clusters of cases have been related to a global warming-mediated increase in the aggressiveness of ticks, which leads to increasing host attacks. From a global perspective, global warming, increased globalization of humans and dogs via travel, exposure of ticks to infected dogs new to an area, and contact with wild canine populations as a reservoir of babesia parasites, is likely to spread canine babesiosis [119, 123]. Apart from exposure to tick vector, young dogs tend to be susceptible to canine babesiosis [38, 54, 63, 65, 67].

Certain breeds have been reported to be susceptible to babesiosis. These include Cocker Spaniel, Doberman, German shepherd dog, Greyhound, Griffon, Maltese terrier, Yorkshire terrier, and Pekinese. Some breeds such as Beagle, Fox terrier, Porcelain, Teckel, and mongrel dogs may be more resistant [34, 66, 85, 92]. Breeds such as American Pit Bull terrier, American Staffordshire terrier, greyhounds and Tosa, engaged in dog fighting have higher prevalence [21, 24, 34].

Pathogenesis

Parasitic and immune-mediated lysis or destruction of infected erythrocytes and systemic response to inflammation along with secondary immune reactions to infection are two major processes involved in the pathogenesis of canine babesiosis [124, 125]. The classifications of clinical manifestations, ranging from subclinical, to anemia and life-threatening hypotensive or septic shock, are based on the severity of hemolysis, number of organs involved, and intensity of inflammatory and immune responses [124-127]. Hemolysis can be induced by two mechanisms: direct damage to the erythrocytes by intra-erythrocyte multiplication of trophozoites and by the secondary immune-mediated responses.

Both intra- and extra-vascular hemolysis can develop during the course of babesiosis [124, 127-129]. Immunoglobulin G and IgM antibodies that recognize erythrocyte membrane antigens are found in dogs afflicted with babesiosis [2, 127, 130, 131]. As a result of the process activated by antibody binding, the infected erythrocytes are extravascularly destroyed, and removed by macrophages in the spleen and liver [132, 133] constituting extravascular hemolysis. Hemolysis can also be induced by oxidative damage to canine erythrocytes and the presence of a hemolytic factor in serum [128, 131, 134-136]. The hemolysis leads to anemia and hemoglobinemia.

Both regenerative and non-regenerative anemia can be observed during the course of canine babesiosis [121, 137]. The pathomechanism of non-regenerative anemia is not fully understood, but may be related to predilection of babesia parasites for reticulocytes, or decreased activity of erythrocyte 5'- nucleotidase, which leads to delayed maturation of reticulocytes in the infected dogs [138-141].

Table 1. Epidemiological surveys of canine babesiosis in the world

Geographic distribution	Method of survey	Sample size (dogs)	Incidence/prevalence	References
Africa				
Nigeria	Nested PCR 18S rRNA gene sequence analysis	400	Nested PCR: *B. canis*: 9/400 (2.3%) Sequence analysis: *B. c. rossi*: 8/400 (2%) *B. c. vogeli*: 9/400 (2.3%)	[27]
Nigeria (case report)	Blood smear PCR 18S rRNA gene sequence analysis	1	Co-infection with *B. c. canis* and *B. c. rossi*: 1/1	[28]
South Africa	Indirect fluorescent antibody tests	275	*B. canis*: Pretoria: 36/83 (43.4%) Durban: 51/78 (65.4%) Eastern Cape: 49/105 (46.7%) West Cape: 5/9 (55.6%)	[29]
South Africa	PCR Reverse line blot hybridization 18S rRNA gene sequence analysis	297	*B. c. vogeli* / *B. c. rossi* Bloemfontein 12/41 (29.3%) / 1/41 (2.4%) East Lond 0/54 / 12/54 (22.2%) Durban 0/56 / 0/56 Johannesburg 0/75 / 0/75 Keringa Kennels 0/15 / 0/15 OVAH 1/56 (1.8%) / 18/56 (32.1%)	[30]
South Africa (case report)	Blood smear PCR Reverse line blot hybridization	1	*B. gibsoni*: 1/1	[31]

Table 1. (Continued)

Geographic distribution	Method of survey	Sample size (dogs)	Incidence/prevalence	References
Tunisia	PCR Reverse line blot hybridization (RLB) 18S rRNA gene sequence analysis	180	PCR/RLB: *B. canis* Sidi Thabet 3/57(5.3%) Somâa 8/49 (16.3%) Zaga 1/29 (3.5%) Bouhajla 0/45 Sequence analysis: *B. c. vogeli* 2/3 (66.7%)	[32]
Sudan	PCR 18S rRNA gene sequence analysis	78	PCR: Babesia spp.: 7/78 (9.0%) Sequence analysis: B. c. rossi: 5/7 (71.4%) B. c. vogeli: 2/7 (28.6%)	[33]
America—North America				
USA	Indirect fluorescent antibody test	164	*B. canis* Mississippi (Aug, 1980) 11/19 (57.9%), (Oct, 1980)* 52/87 (59.8%) Mississippi (Oct, 1980) 0/21 West Virginia 1/20 (5.0%) Oklahoma 5/8(62.5%) Texas 11/20 (55.0%) Mississippi (Feb, 1981) 3/4 75.0%) Florida 2/2 *Included 2nd sample of original 17 dogs	[34]
USA	PCR	673	*B. gibsoni*: 131/673 (19.5%) *B. c. vogeli*: 10/673 (1.5%) *Babesia* DNA positive: 3/673 (0.4%) (not matched any currently described canine Babesia genotype.)	[35]
USA	Blood smear	Georgia: 39	*B. gibsoni*	[36]

Geographic distribution	Method of survey	Sample size (dogs)	Incidence/prevalence	References
	PCR 18S rRNA gene sequence analysis	(affected subjects) Alabama: 5 (affected subjects)	Blood smear Alabama: 5/5 *B. gibsoni* (Asia type) PCR Georgia: 19/39 (48.7%) Alabama: 5/5 Sequence analysis Georgia: 2/2	
USA	Nested PCR	157	*B. gibsoni* / *T. annae* Iowa 4/9 (44.4%) 0/9 Michigan 4/15 (26.7%) 0/15 Mississippi 11/47 (23.4%) 1/47 (2.1%) Ohio 12/34 (35.3%) 0/34 Pennsylvania 3/6 (50%) 0/6 Virginia 18/39 (46.1%) 0/39 Washington 1/7 (14.3%) 0/7	[37]
USA	Fluorescence resonance energy transfer PCR 18S rRNA gene sequence analysis	299	*Babesia* spp. USA 25/235(10.6%) Alabama 8/62 (12.9%) Missouri 1/62 (1.6%) Georgia 6/16 (37.5%) Virginia 0/14 Louisiana 3/12 (25.0%) Ohio 5/11(45.5%) California 0/11 Others 2/47 (4.3%)	[38]

Table 1. (Continued)

Geographic distribution	Method of survey	Sample size (dogs)	Incidence/prevalence	References
USA California	Immune fluorescence assay	971	Sep. 1989~ Dec. 1990 *B. canis* *B. gibsoni* Lancaster 0/100 0/100 Castaic 13/100 (13%) 0/100 Woodland 1/100 (1%) 0/100 El Dorado 0/100 0/100 Feb. 1990 - Dec. 1990 *B. canis* *B. gibsoni* Lancaster 5/150 (3.3%) 1/150 (0.7%) Castaic 15/150 (10%) 0/150 Downey 2/150 (1.3%) 2/150 (1.3%) Agoura 13/150 (8.7%) 4/150 (2.7%)	[39]
USA (case report) Indiana	Blood smear PCR 18S rRNA gene sequence analysis	1	*B. gibsoni:* 1/1	[19]
USA (case report) North Carolina	Blood smear Immune fluorescence assay	9	*B. gibsoni:* 9/9	[40]
USA (case report) North Carolina	Blood smear Nested PCR	1	*B. gibsoni*: 1/1	[41]
USA North Carolina	Indirect fluorescent antibody test Blood smear PCR	359 stray dogs 159 kennel dogs	IFAT seroreactive Stray dogs: 21/359 (5.8%) *B. gibsoni*: 2/21 (9.5%) *B.canis*: 15/21 (71.4%) Western piroplasm: 6/21 (28.6%) Kennel dogs: 24/159 (15.1%) *B. gibsoni*: 16/24 (66.7%) *B.canis*: 5/24 (20.8%)	[24]

Geographic distribution	Method of survey	Sample size (dogs)	Incidence/prevalence	References
			Western piroplasm: 5/24 (20.8%) PCR Stray dogs: 2/359 (5.8%) *B. gibsoni*: 1/21 (4.8%) *B.canis*: 0/21 Western piroplasm: 0/21 Kennel dogs: 24/159 (15.1%) *B. gibsoni*: 14/24 (58.3%) *B.canis*: 0/24 Western piroplasm: 0/24 Blood smear Stray dogs: 1/21 (4.8%) Kennel dogs: 13/29 (44.8%)	
USA North Carolina	PCR	158 foxes 12 coyotes (Canis latrans)	*B. microti*-like: Red fox: 50/127 (39.4%) Gray fox: 8/31 (25.8%) coyote: 0/12	[42]
USA (case report) Oklahoma	Blood smear PCR 18S rRNA gene sequence analysis	1/1	*B. gibsoni*-like: 1/1	[43]
USA Southeastern	Blood smear PCR	33 American Pit Bull Terriers 87 dogs of various other breeds	*B. gibsoni:* Blood smear: American Pit Bull Terriers 10/33 (30.3%) Other dogs 0/87 PCR: American Pit Bull Terriers 18/33 (54.5%) Other dogs 0/87	[20]

Table 1. (Continued)

Geographic distribution	Method of survey	Sample size (dogs)	Incidence/prevalence	References
North America (case report)	Blood smear	1	*B. gibsoni*: 1/1	[44]
America – South America				
Argentina	Blood smear PCR 18S rRNA gene sequence analysis	16767	Blood smear: Large piroplasms: 34/16767 (0.2%) PCR, sequence analysis: *B. c. vogeli*: 2/2	[45]
Brazil (case report)	Blood smear PCR 18S rRNA gene sequence analysis	5	*B. c. vogeli*: 5/5	[46]
Brazil (case report)	Immunohistochemistry In situ hybridization	10	*B. microti*: 10/10	[47]
Brazil	Indirect fluorescent antibody test	381	*B. c. vogeli*: 136/381 (36%)	[48]
Brazil (case report)	Blood smear PCR 18S rRNA gene sequence analysis	16	Blood smear: *B. gibsoni*: 16/16 PCR: *B. gibsoni*: 4/4 Sequence analysis: *B. gibsoni* (Asian type): 4/4	[49]
Brazil Espirito Santo	Indirect fluorescent antibody test PCR	92	Indirect fluorescent antibody test: *B. c. vogeli* Total: 31/92 (33.7%) Rural: 25/71 (46.5%) Urban: 6/14 (42.9%) School: 0/7	[50]

Geographic distribution	Method of survey	Sample size (dogs)	Incidence/prevalence	References
			PCR: *Babesia/Theileria* Total: 20/92 (21.7%) Rural: 14/ 71 (19.7%) Urban: 6/14 (42.9%) School: 0/7	
Brazil Minas Gerais	Indirect fluorescent antibody test	505	*B. c. vogeli:* 95/505 (18.8%) Vila Mato Verde 1/12 (8.3%) Vila Serranópolis 8/80 (10.0%) Saő Judas Tadeu 17/48 (35.4%) Renascenca 12/42 (28.6%)	[51]
Brazil Minas Gerais (Lavras, Belo Horizonte, Nanuque)	Indirect fluorescent antibody test PCR 18S rRNA gene sequence analysis	244	Indirect fluorescent antibody test: *B. c. vogeli* 70/244 (28.7%) Nanuque: (49%) total:102 Belo Horizonte: (34%) total: 50 Lavras: (3.3%) toal: 92 Sequence analysis: *B. c. vogeli:* 12/12	[17]
Brazil Ribeirão Preto	PCR	221	*Babesia* spp.: 18 /221 (8.1%)	[52]
Brazil São Paulo	Blood smear PCR 18S rRNA gene sequence analysis	150	Blood smear: *Babesia* spp.: 3/150 (2.0%) PCR: *Babesia* spp.: 12/150 (8.0%) Sequence analysis: *B. c. vogeli*: 4/4	[53]
Chile	Blood smear	57	*B. gibsoni*: 23/57 (40.4%)	[54]

Table 1. (Continued)

Geographic distribution	Method of survey	Sample size (dogs)	Incidence/prevalence	References
Asia				
Hong Kong	Fluorescence resonance energy transfer PCR 18S rRNA gene sequence analysis	299	*Babesia* spp.: 30/64 (46.9%)	[38]
Japan	ELISA	1206	*B. gibsoni*: 128/1206 (10.6%) Eastern part of Japan: 39/559 (7.0%) Western part of Japan: 89/647 (13.8%)	[55]
Japan Aomori	PCR	141 Tosa dogs (87 with fighting history)	*B. gibsoni*: 42/141 (29.8%) 41/42 (97.6 %) (in fighting history group)	[21]
Japan Aomori, Okinawa	Blood smear PCR	Aomori: 945 Okinawa:137	Blood smear: *B. gibsoni:* Aomori: 16/945 (1.7%) Okinawa: 12/137 (8.8%) PCR: *B. gibsoni:* Aomori: 37/945 (3.9%) All positive animals from Aomori Prefecture were male Tosa dogs (Japanese mastiff) Okinawa: 15/137 (10.9%)	[56]
Japan (case report) Miyazaki	PCR 18S rRNA gene sequence analysis	7	*B. gibsoni* Asia-1 & *B. gibsoni* Asia-2: 7/7	[57]
Japan Okayama	Indirect fluorescence assay tests	22	*B. gibsoni*: 3/22 (13.6%)	[58]
Japan	Blood smear	53	*Babesia* spp.: 10/53 (19.9 %)	[59]

Geographic distribution	Method of survey	Sample size (dogs)	Incidence/prevalence	References
Okinawa			Included: *B. canis*: 3/10 *B. gibsoni*: 6/10 Both: 1/10	
Japan Okinawa	Blood smear PCR 18S rRNA gene sequence analysis	80 (free-roaming dogs)	Blood smear: *B. canis*: 3/80 (3.7%) *B. gibsoni*: 4/80 (5.0%) PCR: *B. canis*: 5/80 (6.3%) *B. gibsoni*: 7/80 (8.8%) Sequence analysis *B. c. vogeli*: 5/5 *B. gibsoni* (Asia-1): 1/1	[60]
Japan Okinawa, Yamaguchi, Osaka China Nanjing	ELISA	Okinawa: 62 Yamaguchi: 120 Osaka: 37 Nanjing: 41	*B. gibsoni:* Japan: Okinawa: 9/62 (14.5%) Yamaguchi: 7/120 (5.8%) Osaka: 2/37 (5.4%) China: Nanjing: 2/41 (4.8%)	[61]
Japan Osaka, Hiroshima, Miyazaki, Okinawa, Akita, Fukui	PCR 18S rRNA gene sequence analysis (from ticks)	1221 (from 4122 ticks, 1136 DNA samples)	*Babesia* spp.: 13/4122 ticks *B. c. vogeli*: 6/4122 ticks Prefectures: Aomori, Nara, Hiroshima, Oita, and Okinawa *B. c. gibsoni* Asia-1: 4/4122 ticks Prefectures: Osaka, Hiroshima, Miyazaki, and Okinawa Prefectures *B. odocoilei*-like: 3/4122 ticks Prefectures: Akita and Fukui Prefectures	[18]

Table 1. (Continued)

Geographic distribution	Method of survey	Sample size (dogs)	Incidence/prevalence	References
Japan Eastern Japan	PCR ELISA	115 (affected subjects)	*B. gibsoni*: PCR or/and ELISA Positive: 35/115 (30.4%) PCR+ELISA+: 21/115 (18.3%) PCR+ELISA-: 8/115 (6.9%) PCR-ELISA+: 6/115 (5.2 %) PCR-ELISA-: 88/115 (69.9 %)	[62]
Korea Gangwon, Gyunggi	Blood smear PCR 18S rRNA gene sequence analysis	501 German shepherd dogs	Blood smear: *B. gibsoni:* 9/501 (1.8%) PCR: *B. gibsoni*: 9/501 (1.8%) Sequence analysis: *B. gibsoni*: 9/9	[63]
Korea, Japan (case report)	Indirect fluorescent antibody test ELISA	7	*B. gibsoni:* 7/7	[64]
Malaysia	Blood smear	4080	*B. gibsoni:* 726/4080 (17.7%) *B. canis:* 43/4080 (1.1%)	[65]
Taiwan North Taiwan	Blood smear	7000	*B. gibsoni*: 121/7000 (1.7%) *B. canis:* 7/7000 (0.1%)	[66]
Taiwan North Taiwan	Blood smear	101	*B. gibsoni*: 7/101 (6.3%) *B. canis*: 0/101	[67]
Taiwan Mid Taiwan	Blood smear PCR 18S rRNA gene sequence analysis	125 (affected subjects)	Blood smear: 29/125 (23.2%) PCR, sequence analysis (the one whose blood smear positive undergo PCR): *B. gibsoni* (Similar to Okinawa Island Japanese strain): 29/29	[68]

Geographic distribution	Method of survey	Sample size (dogs)	Incidence/prevalence	References
Thailand	Indirect fluorescent antibody test	49 (affected subjects)	*B. canis*: 10/49 (20.4%) *B. gibsoni*: 4/49 (8.2%) Co-infection *B. canis* and *B. gibsoni*: 3/49 (6.1%)	[69]
Australia				
Australia (case report) Victoria	Blood smear PCR 18S rRNA gene sequence analysis Indirect fluorescent antibody tests	3	PCR, sequence analysis: *B. gibsoni* (Asian type): 3/3 Indirect fluorescent antibody tests: high titre of antibody to *B. gibsoni* and low titres to *B. canis*	[70]
Australia Northern South-eastern (Victoria)	Blood smear PCR 18S rRNA gene sequence analysis	Northern:34 Victoria:3 (affected subjects)	PCR: Northern Australia: *Babesia* spp.: 14/34 (41.2%) Southern Australia (Victoria): *Babesia* spp.: 3/3 Sequence analysis: Northern Australia: *B. c. vogeli*: 4/4 Southern Australia (Victoria): *B. gibsoni* (Asian type): 3/3	[71]
Australia Aboriginal community in Central Australia	Blood smear PCR 18S rRNA gene sequence analysis	55 (free-roaming dogs)	Blood smear: *B. canis*: 22/55 (40.0%) PCR, sequence analysis (the one whose blood smear positive under PCR) *B. c. vogeli*: 22/22	[72]
Australia Aboriginal communities	PCR 18S rRNA gene sequence analysis	215	Singularly infection: *B. c. vogeli*: 22/215 (10.2%) Co-infection: *B. c. vogeli* and *A. platys*: 24/215 (11.2%)	[73]

Table 1. (Continued)

Geographic distribution	Method of survey	Sample size (dogs)	Incidence/prevalence	References
Caribbean				
British West Indies	Blood smear Immune fluorescence assay	104 (feral dogs) 83 (client-owned dogs)	No *B. canis* organisms were noted	[74]
Grenada	Nested PCR 18S rRNA gene sequence analysis	73	*B. c. vogeli*: 5/73 (6.8%)	[75]
St Kitts	Seminested PCR	12 healthy dog with thrombocytopenia (affected dog)	No *Babesia* spp. detected	[76]
Mid-East				
Iran	Blood smear	155	*B. canisi*: 1/155 (0.6%)	[4]
Israel	Immune fluorescence assay	40 (affected subjects)	*B. canis*: 36/40 (90.0%) *B. gibsoni*: 30/40 (75.0%)	[77]
Turkey (case report)	Blood smear PCR- Restriction fragment length polymorphism (RFLP) 18S rRNA gene sequence analysis	3	Blood smear: *B. canis*: 3/3 Sequence analysis: *B. c. vogeli*: 3/3	[78]
Russia				
Russia	Blood smear Hemi-nested PCR	21 (affected subjects)	*B. c. canis*: 17/21 (80.9%) Similar to B. c. canis, but with two undetermined nucleotide sequences: 4/21 (19.0%)	[79]

Geographic distribution	Method of survey	Sample size (dogs)	Incidence/prevalence	References
Europe				
Austria Albania, Kosovo	PCR Specific antibody	272 36 (stray dog)	Specific antibody: *B. canis*: (7.3%) PCR: *Babesia* spp.: (11.1%)	[80]
Belgium (case report) Import from Spain	Blood smear Indirect fluorescence immunoassays PCR	1	Indirect fluorescence immunoassays: *B. microti:* 1/1 PCR: *B. microti:* 0/1	[81]
Belgium (case report)	Blood smear	1	*B. canis*:1/1	[82]
Britain (case report)	Blood smear PCR 18S rRNA gene sequence analysis	1	*B. c. vogeli*: 1/1	[83]
Croatia	Blood smear PCR 18S rRNA gene sequence analysis	81 blood smear positive for Babesia spp. 848 asymptomatic	Blood smear Symptomatic: 81/81 (100%) PCR Asymptomatic: 29/848 (3.4 %) Sequence analysis Asymptomatic: *B. c. canis*: 20/29 (69.0%) *B. gibsoni*: 6/29 (20.7%) *B. c. vogeli:* 2/29 (6.9%) *T. anne*: 1/29 (3.4%) Symptomatic: *B. c. canis*: 78/81 (96%) *B. c. vogeli*: 1/81 (1.2%) *B. cabali:* 1/81 (1.2%) *T. equi*: 1/81 (1.2%)	[84]

Table 1. (Continued)

Geographic distribution	Method of survey	Sample size (dogs)	Incidence/prevalence	References
France	Blood smear ELISA	3090	Blood smear: *B. canis*: 511/3090 (16.5%)	[85]
France	PCR 18S rRNA gene sequence analysis	108(affected subjected)	*B. c. vogeli*: 1/108 (0.9%)	[86]
France	PCR PCR - RFLP	166(affected subjected)	*B. c. canis*: 106/166 (63.9%) *B. c. rossi:* 1 /166 (0.6%) *B. gibsoni* (California type): 1 /166 (0.6%) *T. equi*: 31/166 (18.7%) *B. caballi*: 1/166 (0.6%) *Babesia* spp. or *Theileria* spp.: 27/166 (16.3%) (did not allow differentiation)	[87]
France Southern Jura	Blood smear	437 (affected subjects)	Blood smear: *B. canis*: 73/437 (16.7%)	[88]
Germany(case report) Ravensburg	Blood smear PCR 18S rRNA gene sequence analysis	2	*B. gibsoni* (Asian type): 2/2	[89]
Hungary (case report)	Blood smear	2	*B. gibsoni*: 2/2	[90]
Hungary	Blood smear PCR 18S rRNA gene sequence analysis	44 (affected subjects)	Blood smear: *B. canis*: 27/27 PCR: *Babesia* spp.: 39/44 (88.6%) Sequence analysis: *B. c. canis*: 5/5	[91]

Geographic distribution	Method of survey	Sample size (dogs)	Incidence/prevalence	References
Hungary Budapest, Debrecen, Countryside	Indirect fluorescent antibody test	651	*B. canis*: 37/651 (5.7%)	[92]
Italy	Blood smear PCR RFLP 18S rRNA gene sequence analysis	164	PCR: 45/164 (27.4%) PCR-RFLP: *B. c. canis*: 34/ 164 (20.7%) *B. c. vogeli*: 11/164 (6.7%) 18S rRNA gene sequence analysis *B. c. canis*: 3/3 *B. c. vogeli:* 4/4	[93]
Italy Central and North-Eastern Italy	Blood smear Indirect fluorescent antibody test PCR 18S rRNA gene sequence analysis	420	Blood smear: *B. canis*: 10/416 (2.4%) Indirect fluorescent antibody test: *B. canis*: 128/376 (34.0%) PCR, sequence analysis: *B. c. canis*: 9/385 (2.3%)	[94]
Netherlands Hague, Arnhem	Blood smear Indirect fluorescent antibody test PCR—Reverse line blot hybridization 18S rRNA gene sequence analysis	Hague:10 Arnhem:13 (affected subjects)	Reverse line blot hybridization, or Indirect fluorescent antibody test or blood smear: *B. c. canis:* 23/23 18S rRNA gene sequence analysis (fatal cases): *B. c. canis*	[96]
Norway (case reports)	Blood smear PCR 18S rRNA gene sequence analysis	1	*B. c. canis*	[97]

Table 1. (Continued)

Geographic distribution	Method of survey	Sample size (dogs)	Incidence/prevalence	References
Portugal Chaves	Blood smear PCR 18S rRNA gene sequence analysis	8 (affected subjects)	*B. c. canis*: 7/8(87.5%) *B. c. vogeli*: 1/8(12.5%)	[98]
Poland Eastern Poland	Blood smear PCR RFLP 18S rRNA gene sequence analysis	76 (affected subject)	Blood smear: *B. canis*: 76/76 PCR: *B. c. canis*: 69/76 (90.8%) RFLP: *B. c. canis*	[99]
Poland Warsaw	PCR 18S rRNA gene sequence analysis	408	PCR: *B. canis*: 48/408 (11.8 %) Sequence analysis: *B. c. canis* 9/9	[100]
Poland Central Poland	PCR 18S rRNA gene sequence analysis	26 (symptomatic dog in veterinary clinics) 82 (dog kennels)	PCR: symptomatic dog in veterinary clinics: *B. canis*: 0/26 dog kennels: *B. canis*: 20/79 (25.3%) Sequence analysis: *B. c. canis* 10/10	[101]
Slovak (case report) Biel	Blood smear	3	*B. canis*: 3/3	[102]
Slovenia Ljubljana	Blood smear PCR 18S rRNA gene sequence analysis	238	Blood smear: *B. canis*: 14/238 (5.9%) PCR, sequence analysis: *B. c. canis*: 11/238 (4.6%) *B. c. vogeli*: 3/238 (1.3%)	[103]

Geographic distribution	Method of survey	Sample size (dogs)	Incidence/prevalence	References
Spain	Blood smears PCR 18S rRNA gene sequence analysis	157 (affected subjects)	Blood smear: *Babesia* spp.: 157/157 Sequence analysis: *B. microti*–like (T. annae): 1/1	[9]
Spain	PCR 18S rRNA gene sequencing analysis	10 (affected subjects)	*B. c. vogeli*: 1/10 (10.0%) *B. c. canis*: 3/10 (30.0%) *B. equi*: 1/10 (10.0%)	[104]
Spain	PCR 18S rRNA gene sequencing analysis	62 (affected subjects)	*B. microti*-like (*T. annae*): 62/62	[105]
Republic of Cape verde Praia	PCR 18S rRNA gene sequence analysis	130	PCR: *B. c. vogeli*: 5/130 (3.8%) *B. gibsoni*: 0/130 Sequence analysis: *B. c. vogeli*: 3/3	[106]
The Republic of Cape verde Praia	Blood smear PCR	57	Blood smear: *B. canis*: 0/57 *B. gibsoni*: 23/50 (46.0%)	[54]
Europe Croatia, Zagreb France, Nantes Italy, Perugia Poland, Warsaw	Blood smear PCR 18S rRNA gene sequence analysis	11 (affected subjected)	Blood smear: *B. canis*: 11/11 PCR, sequence analysis: *B. c. canis*: 10/11(90.9%) *B. c. vogeli*: 1/11(9.1%)	[107]

Table 2. Babesia species and tick vectors of dogs

Geographic distribution	Babesia species (phenotype)	Tick vectors reported	References
Africa			
Nigeria	*B. c. rossi, B. c. Vogeli, B. c. canis*	*R. sanguineus*	[27, 28, 108]
South Africa	*B. c. rossi* *B. c. vogeli* *B. c. canis* *B. gibsoni*	*D. reticulatus* *R. sanguineus* *H. leachi,* *H. leachi* (Thomas strain)	[29, 30, 31, 109, 110,111, 112]
Tunisia	*B. c. vogeli*	*R. sanguineus*	[32]
Sudan	*B. c. rossi* *B. c. vogeli*	*R. sanguineus, R. e. evertsi* *A. lepidum*	[33]
America —North America			
North America	*B. gibsoni*		[44]
USA	*B. gibsoni, B. c. Vogeli, B. microti* – like (*T. annae*)		[35, 37]
USA California	Dog: *B. canis, B. gibsoni* Coyote: *B. gibsoni, B. conradae*	*R. sanguineus* *D. variabilis*	[39, 113]
USA: Georgia, Alabama (Auburn)	*B. gibsoni* (Asian type)		[36]
USA: Indiana	*B. gibsoni* (Asian type)		[19]
USA: Oklahoma	*B. gibsoni*-like (*B. conradae*)		[43]
USA: North Carolina	*B. gibsoni, B. canis* Western piroplasm (*B. conradae*) Foxes: *B. microti*-like	*D. variabilis* *R. sanguineus* *A. aureolatum*	[24, 40, 41, 42]
USA: Southeastern	*B. gibsoni* (Asian type)		[20]
USA: Mississippi, West Virginia, Oklahoma, Texas, Florida	*B. canis*		[34]
America —South America			
Argentina	*B. c. vogeli*		[45]

Geographic distribution	Babesia species (phenotype)	Tick vectors reported	References
Brazil	*B. microti* *B. gibsoni* (Asian type) *B. c. vogeli*	*R. sanguineus* *A. aureolatum*	[46, 47, 48, 49]
Brazil Espirito Santo	*B. c. vogeli*		[50]
Brazil Minas Gerais	*B. c. vogeli*	*R. sanguineus*	[51]
Brazil Minas Gerais (Lavras, Belo Horizonte, Nanuque)	*B. c. vogeli*	*R. sanguineus* *A. cajennense* *B. microplus*	[17]
Brazil Ribeiraő Preto	*Babesia* spp.	*R. sanguineus*	[52]
Brazil São Paulo	*B. c. vogeli*	*R. sanguineus* *A. cajennense* *A. ovale*	[53]
Chile	*B. gibsoni*		[54]
Asia			
China Nanjin	*B. gibsoni*	*R. haemaphysaloides* *H. longicornis.*	[114]
Japan	*B. gibsoni*	*H. longicornis.*	[55]
Japan Aomori	*B. gibsoni*		[21] [115]
Japan Aomori Okinawa	*B. gibsoni*	*R. sanguineus*	[56]
Japan Akita Fukui	*Babesia odocoilei*-like group	*I ovatus*	[18]

Table 2. (Continued)

Geographic distribution	Babesia species (phenotype)	Tick vectors reported	References
Japan Miyazaki	*B. gibsoni*		[116]
Japan Osaka, Hiroshima Miyazaki Okinawa	*B. gibsoni* (Asian type)	*Haemaphysalis* spp. *I. ovatus* *R. sanguineus*	[18]
Japan Okayama	*B. gibsoni*	*R. sanguineus*	[58] [59]
Japan Okinawa	*B. c. vogeli* *B. gibsoni* (Asian type)	*R. sanguineus*	[60]
Japan Eastern Japan	*B. gibsoni*	*H. longicornis.* *R. sanguineus*	[62]
Korea	*B. gibsoni*		[63]
Korea, Japan	*B. gibsoni*		[64]
Malaysia	*B. gibsoni* *B. canis*		[65]
Taiwan	*B. gibsoni* (Asian type) *B. canis*	*R. sanguineus* *I. ovatus*	[67] [68]
Thailand	*B. gibsoni* *B. canis*	*R. sanguineus* *H. wellingtoni, H. Atherurus,* *H. lagrangei, H. papuana* *H. heinrichi*	[69]
Turkey	*B. c. vogeli*	*R. sanguineus*	[78]
Australia			
Australia Victoria	*B. gibsoni* (Asian type)		[70]
Australia Northern	*B. c. vogeli* *B. gibsoni* (Asian type)	*R. sanguineus* (*B. c. vogeli*)	[71]

Geographic distribution	Babesia species (phenotype)	Tick vectors reported	References
South-eastern -- Victoria			
Australia Aboriginal community in Central Australia	*B. c. vogeli*		[72]
Carribean			
British West Indies	No *B. canis* organisms were noted		[74]
Grenada	*B. c. vogeli*	*R. sanguineus*	[75]
Mid-East			
Iran	Dog: *B. canis* Fox: *B. gibsoni*	*Rhipicephalus spp.* *Haemaphysalis spp.* *Hyalomma spp.*	[4]
Israel	*B. canis* *B. gibsoni*		[77]
Turkey	*B. c. vogeli*		[78]
Russia			
Russia	*B. c. canis*	*D. reticulatus* *I. persulcatus*	[79]
Europe			
Austria Albania Kosovo	*B. canis*		[80]
Belgium	*B. canis*		[82]
Britain	*B. c. vogeli*	*R. sanguineus*	[83]
France	*B. c. vogeli* *B. c. canis* *B. c. rossi* *B gibsoni* (California type) *T. equi* *B. caballi*		[86] [87]

Table 2. (Continued)

Geographic distribution	Babesia species (phenotype)	Tick vectors reported	References
France Southern Jura	*B. canis*	*D. reticulatus* *I. ricinus*	[88, 117]
Germany	*B. canis*	*I. ricinus* *D. reticulatus* *I. hexagonus* No *B. canis* was found in the present study of a total of 135 D. reticulatus by PCR	[118]
Germany Southern Germany (Ravensburg)	*B. gibsoni* (Asian type)	*R. Sanguineus* *I. hexagonus*	[89]
Hungary	*B. gibsoni* *B. c. canis*	*D. reticulatus*	[90] [91]
Hungary Budapest Debrecen Countryside	*B. canis*	*D. reticulatus*	[92]
Italy Central and North-Eastern Italy	*B. c. canis* *B. c. vogeli* *B. microti*-like	*R. sanguineus* *I. ricinus*	[94]
Italy Southern Italy	*B. c. vogeli*		[95]
Netherlands Hague Arnhem	*B. c. canis*	*D. reticulatus*	[96]
Norway	*B. c. canis*		[97]
Poland Warsaw	*B. canis*	*D. reticulatus*	[100]

Geographic distribution	Babesia species (phenotype)	Tick vectors reported	References
Poland Central Poland	*B. c. canis*	*D. reticulatus*	[101]
Portugal Chaves	*B. c. canis* *B. c. vogeli*	*D. reticulatus* (*B. c. canis*) *R. sanguineus* (*B. c. vogeli*)	[98]
Slovak	*H. canis*	*R. sanguineus*	[119]
Slovak Biel'	*B. canis*	*D. reticulatus*	[102]
Slovenia Ljubljana	*B. c. canis* *B. c. vogeli*	*D. reticulatus* *R. sanguineus*	[103]
Spain	*B. microti*-like (*T. annae*), *B. gibsoni* (Asian type) *B. c. vogeli, B. c. canis, B. equi*	*Ixodid ticks* *I. ricinus, R. sanguineus, D. reticulatus*	[9, 81,105,120]
The Republic of Cape verde: Praia	*B. c. vogeli* *B. gibsoni*	*R. sanguineus*	[106] [54]
Europe: Croatia, Zagreb; France, Nantes; Italy, Perugia; Poland, Warsaw	*B. c. canis.*		[107]
Europe: France, Portugal, Spain	*B. c. vogeli, B. c.canis, T. annae, B. equi*		[120]

A. *Aureolatum: Amblyomma aureolatum; A. cajennense: Amblyomma cajennense; A. Lepidum: Amblyomma Lepidum; A. ovale: Amblyomma oval; B. caball: Babesia caballi; B. canis: Babesia canis; B. c. canis: Babesia canis canis; B. c. rossi: Babesia canis rossi; B. c. vogeli: Babesia canis vogeli; B. equi: Babesia equi; B. gibsoni: Babesia gibsoni; B. Microti: Babesia microti; B. microplus: Boophitus microplus; D. reticulatus: Dermacentor reticulatus; H. atherurus: Haemaphysalis atherurus; H. heinrichi: Haemaphysalis heinrichi: H. lagrangei: Haemaphysalis lagrangei; H. leachi: Haemaphysalis leachi; H. longicornis: Haemaphysalis longicornis; H. papuana: Haemaphysalis papuana; H. wellingtoni: Haemaphysalis wellingtoni; H. canis: Hepatozoon canis; I. hexagonus: Ixodes hexagonus; I. ovatus: Ixodes ovatus; I. persulcatus: Ixodes persulcatus; I. ricinus: Ixodes ricinus; R. e. evertsi: Rhipicephalus evertsi evertsi; R. haemaphysaloides: Rhipicephalus haemaphysaloides R. Sanguineus: Rhipicephalus sanguineus; T. annae: Theileria annae; T. equi: Theileria equi.*

Table 3. Morphological characteristics of Babesia species in dogs and the most frequently reported clinical signs in association with the species

Species	Size in length	Intraerythrocytic from	Clinical manifestations	References
Babesia canis *B. c. vogeli* *B. c. canis* *B. c. rossi*	Large 3.0 –5.0µm	Pyriform	Hemolytic anemia, thrombocytopenia *B. c. vogeli*: -Mild; -Inapparent clinical signs;- Severe in young or immunocompromised dogs *B. c. canis*:- Moderate; - Hyperfibrinogenemia and coagulation abnormalities; - Congestion of internal organs *B. c. rossi*:- Severe; - Fetal hemolytic syndrome, - Higher parasitemia	[14,78,93,104,137]
Babesia (large unnamed strain)	Large 2.3 – 4.0µm	Oval to pyriform or amoeboid forms	Mild; -Hemolytic anemia, thrombocytopenia; -Only reported from immunocompromised dogs	[41,152,181]
Babesia gibsoni	Small 1.5 – 2.5µm	Ring (most common), paired pyriform, petaloid, amoeboid, reticular, and dot forms	Moderate From severe hemolytic anemia and thrombocytopenia to subclinical infections	[153, 154, 155]
Babesia conradae	Small 0.5 – 3.5µm	Ring (most common), pyriform, tetrad, amoeboid, or anaplasmoid forms.	Moderate to severe Pronounced hemolytic anemia Thrombocytopenia Higher parasitemia Lymphadenopathy	[2] [156] [113]
Babesia equi (Spain isolate)	Small 1 – 2.5µm	Single small form	Mild Mostly found in asymptomatic dogs	[104]
Theileria annae (*Babesia microti* – like)	Small 1.0 – 2.0µm	Ring-form	Moderate to severe Pronounced hemolytic anemia Renal failure	[8, 9, 105,122, 151, 168]

B. canis: Babesia canis; B. c. canis: Babesia canis canis; B. c. rossi: Babesia canis rossi; B. c. vogeli: Babesia canis vogeli.

Altered hemostasis, including thrombocytopenia, coagulopathy, and change in fibrinogen concentrations, is reported commonly in canine babesiosis infections [137, 142]. The decreased platelet number is probably caused by immune-mediated sequestration and destruction of platelets in the spleen [121]. Infected erythrocytes trigger acute phase immune responses and interactions of parasitized erythrocytes with the vascular endothelial cells further activate the coagulation system [143, 144]. An increase in the activated partial thromboplastin time (aPTT) is observed from the moment babesia infection is detected [14]. Infection may also induce prolonged prothrombin time (PT) and thrombin time (TT) in dogs [145]. Acute phase responses to infection also lead to a higher fibrinogen concentration that increases steadily from day one of infection. Together with activated coagulopathy, disseminated intravascular coagulation (DIC) syndrome is induced and followed by hypotension [144, 146].

Both excessive and inadequate inflammatory responses play important role in the pathogenesis of babesiosis. The clinical manifestations and clinicopathological characteristics vary by the status of the inflammatory responses, from leukogram abnormalities, such as neutropenia and lymphopenia, to systemic inflammatory response syndrome (SIRS). Decreased packed cell volume, DIC triggered by immune reactions, and the overwhelming responses of SIRS may ultimately lead to life-threatening hypotension, multiple organ dysfunction syndromes (MODS), or septic shock [2, 125, 146, 147].

Clinical Manifestations

In general, clinical manifestations of canine babesiosis are associated with hemolytic anemia and intensity of inflammatory responses [145]. Clinically, canine babesiosis can be classified as uncomplicated or complicated, or categorized as peracute, acute, chronic and subclinical forms based on the course of development [112]. The clinical manifestations of canine babesiosis vary widely depending on the condition of the host and the specific *Babesia* spp. Young dogs and dogs with compromised immunity or splenectomized dogs may exhibit severe hemolytic anemia [26, 110, 127, 148, 149]. In general, *Babesia canis vogeli* is the least pathogenic subspecies of *B. canis*. Dogs infected with *B. c. vogeli* may present subclinical or a relatively mild disease [93, 107]. However, the clinical manifestation of infection with *B. c. vogeli* may be exacerbated in dogs with concomitant other infections or in immunocompromised dogs. Hypotensive shock has been reported in dogs infected with *B. canis* [126]. Experimentally, splenectomized dogs infected with *B. canis* and *B. gibsoni* rapidly develop parasitemia and clinical disease with high levels of parasitemia [127, 149]. Cases presenting with the most severe hemolytic anemia and complicated conditions, such as DIC or hypotensive shock, are frequently associated with *B. c. rossi* infection. Clinical signs of babesiosis due to *B. conradae* are similar to those of *B. gibsoni*; however, *B. conradae* may be more pathogenic, resulting in higher parasitemia and more pronounced anemia when compared with *B. gibsoni* [2, 150]. *Babesia annae* (now known as *T. annae*) is associated with a high risk of azotemia, consequent renal failure and mortality [8, 9, 41, 78, 105, 113, 120, 151-156] (Table 3).

In uncomplicated babesiosis, the most common signs are associated with hemolysis, such as pale mucous membrane, icterus, hemoglobinuria, splenomegaly, tachypnea, and tachycardia [124, 127, 157, 158]. Depression, anorexia, and fever are commonly reported as non-specific signs. Fever is the result of the acute phase reaction to the infection [6]. Experimentally, dogs infected with higher doses of *Babesia* may develop fever earlier [147].

Complicated babesiosis is characterized by the involvement of additional organs, and the clinical manifestations and severity of complications vary widely depending on the organs involved. The hematologic, immune, cardiovascular, metabolic, respiratory, and central nervous systems, as well as organs such as the liver, kidney and pancreas have been reported to be affected by complicated babesiosis [5, 124, 148, 151, 159].

Hematological and Immune Systems

The pale mucous membrane, bleeding tendency and petechiation of membrane are associated with anemia and secondary to altered hemostasis and dysregulated immune function, such as coagulopathy, DIC, and immune-mediated hemolytic anemia (IMHA) [6, 124, 142].

Cardiovascular System

Tachycardia reflects poor tissue perfusion as a result of anemia or hemodilution. Arrhythmia may be observed in some cases, such as sino-atrial block, ventricular premature complexes, and atrioventricular block caused by myocarditis and myocardial ischemia. The corresponding lesions of myocardia include necrosis, Haemorrhage, and fibrin microthrombi [159]. Hypotension is significantly associated with collapsed cases, and is one of the diagnostic criteria for the SIRS and consequent MODS [126].

Respiratory System

Cough, tachypnea and dyspnea may be associated with pulmonary edema, which is a result of SIRS and acute respiratory distress syndrome (ARDS) [5, 124, 148, 160].

Liver

Hepatomegaly and icterus without evidence of cholestasis are commonly observed in dogs infected with babesia [158]. Histopathologically, diffuse nonsuppurative periportal and centrilobular hepatitis has been described in affected dogs [5, 127, 148]. The actual patho-mechanism is unknown, but oxidative damage to the tissues may be one of the causes [136]. Icterus tends to be more severe in dogs with high parasitemia than those with low parasitemia, but the differences are not significant. Jaundice may also appear in some patients without hepatopathy because of hemolytic icterus [112]. Ultrasonographic examination of the liver is usually unremarkable, or merely reveals diffusely enlarged, or mild increased or decreased echogenicity of the parenchyma [124, 161].

Kidney and Urinary System

Hemoglobinuria suggests acute intravascular hemolysis. Development of acute renal failure may be related to the pathogenicity of specific *Babesia* spp. [91, 124]. The most common cause of renal failure in babesiosis is suggested to be hemoglobinuric nephrosis. However, hypoxia appears to be more important than hemoglobinuria in damaging the kidneys of experimentally- and naturally-infected dogs. The major lesions of the kidneys include vacuolar-hydropic degeneration and necrosis and detachment of the renal tubular epithelial cells of the proximal convoluted tubules, while no significant histological changes are observed in the glomeruli [5, 148]. Anemic hypoxia, hypovolemia, and consequences of SIRS and DIC may also lead to dysfunction of the kidneys [124, 125, 162]. Additionally, multifocal deposits of IgM antibodies are evident immunohistochemically within the walls of inflamed arteries and renal glomeruli, indicating immune-mediated processes [127].

Increased echogenicity of the renal cortex, enhanced corticomedullary definition, diffuse renal structure or increased cortico-medullary ratio may be revealed ultrasonographically [124, 161].

Pancreas

Pancreatitis can be a complication of canine babesiosis. Vomiting, abdominal pain, diarrhea and melena have been reported in dogs complicated with pancreatitis. Pancreatitis usually occurs concomitantly with hepatopathy or renal failure and commonly develops in older dogs [112, 124].

Central Nerve System

Intermittent loss of consciousness, seizures, stupor and coma, aggression, paddling, crying, nystagmus, anisocoria, opisthotonus, incoordination, hindquarter paresis, and muscle tremors have been reported in infected dogs with central nervous system complications [163]. Lesions are usually localized in the cerebral cortex and, occasionally, in the spinal cord. Haemorrhage is the most common type of lesion. Congestion, brain swelling or edema may also be observed. Development of lesions in the central nervous system is not related to hypoxia. Cerebellar ataxia is a rare complication in canine babesiosis [124]. Neurological signs may also result from hypoglycemia [164].

Systemic Inflammatory Response Syndrome (SIRS) and Multiple Organ Dysfunction Syndrome (MODS)

Systemic inflammatory response syndrome corresponds to the clinical expression resulting from the activation of a complex system of intrinsic mediators of the acute phase reaction. The inflammatory mediators are induced by infection and tissue hypoxia [136]. Systemic inflammatory response syndrome is defined as the occurrence of two or more of the following symptoms: tachycardia, tachypnoea (or respiratory alkalosis), hypothermia or hyperthermia, leucocytosis or leucopenia, or neutrophilic left shift. Systemic inflammatory response syndrome is a common complication of babesiosis and may lead to MODS [124]. The organs that may be involved include the liver, kidneys, muscle, lungs and central nervous system [124, 165].

DIAGNOSIS

A diagnosis of canine babesiosis is based upon history of tick exposure, along with compatible clinical signs and clinicopathological abnormalities such as pallor, intermittent fever, anemia and hemoglobinuria [157]. Further confirmation of the infection is obtained through morphological identification of *Babesia* spp. within infected erythrocytes on blood smears, positive serological results or evidence of nuclear acids [158].

Clinicopathological Findings

In general, the most relevant clinicopathological findings are related to altered hemostasis [6]. However, the abnormal clinicopathological findings are highly variable and depend on the organs or systems involved [148]. Overall, the degree of parasitemia may not correlate to the severity of clinicopathological findings.

Hematological Abnormalities

Babesia parasites may be identified in blood smears [5, 158]. The majority of infected dogs exhibit some degree of anemia, characterized by low erythrocyte counts, decreased hemoglobin concentration, or normocytic-normochromic anemia [110, 121, 148, 162]. Furthermore, both regenerative and non-regenerative anemia may be observed [137, 141].

Hemolysis and hemoglobinemia are usually secondary to inflammatory or immune reactions. The presence of hemoglobinemia and hemoglobinuria suggests acute intravascular hemolysis. In cases of developing secondary IMHA, auto-agglutination may be detected in saline-washed blood cells and/or marked spherocytosis may be identified on stained blood smears [124]. Direct Coomb's tests are positive in most dogs infected with *Babesia* spp., as well [158]. However, the severity of parasitemia is not associated with the severity of hemolysis. The degree of hemolytic anemia varies depending on age and immune status; young dogs and dogs with concomitant compromised immunity are usually more anemic [2, 26, 34]. If *Babesia* parasites are not identified in blood smears, it is very difficult to differentiate babesiosis from IMHA clinically [158, 166]. In such cases, other procedures such as *Babesia*-specific PCR, should be applied to confirm diagnosis of babesiosis [110, 112].

Paradoxical hemoconcentration refers to a normal or elevated hematocrit in association with intravascular hemolysis [124]. This phenomenon is caused by a shift of protein-rich fluid into the extravascular space in response to increased capillary permeability resulting from the inappropriate response to inflammation and rapid progression of the disease [110, 112, 163].

Thrombocytopenia is commonly detected in dogs with babesiosis and is related to altered hemostasis [93, 121, 142, 166]. Altered coagulation profiles such as prolonged PT, aPTT, and TT are associated with DIC [121, 145]. Evidence of abnormal fibrinogen degradation, including increased D-dimer concentrations and fibrin degradation products (FDPs), is found in cases of developing DIC and secondary fibrinolysis [121, 145, 146].

The leukograms of infected dogs are widely variable, and may include leukocytosis (with or without a left shift), lymphocytosis, monocytosis, neutropenia, lymphopenia or monocytopenia [121, 137]. The specific profile reflects the status of inflammation generated in response to the infections. Leukopenia is more common in dogs with uncomplicated babesiosis [148].

Biochemical Profile Abnormalities

Biochemical profile abnormalities are related to the intensity of the disease and the degree of hypoxia [121, 167]. Hypoalbuminemia and hyperglobulinemia are commonly seen in serum protein pattern [93, 168]. The detection of hyperglobulinemia suggests IgM, IgG and/or complement response to antigenic stimulation [121].

Increased activity of acute phase proteins, such as C-reactive proteins and fibrinogen, is a marker of inflammation. Even during a low-grade infection, increased C-reactive protein concentration can be detected in plasma prior to the detection of *Babesia* parasites in the blood, and indicates activation of an acute phase reaction [93, 143, 147, 169, 170].

Hypoglycemia and hyperlactatemia are often found in collapsed dogs [164, 171]. Glucose levels should be checked in collapsed dogs, in dogs with a history of neurological signs, and in young dogs [171]. Hypoglycemia is an important differential diagnosis for cerebral babesiosis [112]. The causes of hypoglycemia are speculated to include increased glucose consumption, depletion of hepatic glycogen stores, and hepatic dysfunction with impaired gluconeogenesis as a result of anemia and hypoxia [172]. Hyperlactatemia is believed to be related to hypoxia due to anemia, poor tissue perfusion or sepsis [171, 173].

Mild elevation of total bilirubin concentration and the activity of hepatic enzymes such as alanine aminotransferase, aspartate aminotransferase, alkaline phosphatase, is commonly observed in babesiosis [93, 121, 127, 136, 174].

High blood urea nitrogen (BUN) concentrations are found commonly in canine babesiosis patients. Increased degradation of hemoglobin and dehydration both may raise BUN concentrations. Severe intravascular hemolysis alone is not always associated with elevated BUN. Increased creatinine serum concentrations are indicative of renal failure and defined as complicated babesiosis [124, 125, 127, 162].

Electrolyte imbalance, mostly hypokalemia, hypochloremia and hyponatremia, may be observed as a result of renal hypotension or activation of the renin-angiotensin-aldosterone axis that leads to hypotension [146, 147]. Mixed acid-base disturbances due to hypoxia, such as metabolic acidosis and respiratory alkalosis, are common in canine babesiosis, as well [173].

Aberrant hormonal status, such as high basal and post-adrenocorticotrophic hormone (ACTH) cortisol concentrations with low serum thyroxine and free thyroxine concentrations, is associated significantly with hypoglycemia [175, 176].

Urinalysis

Hemoglobinuria, proteinuria, bilirubinuria, and increased amounts of urobilinogen may be detected in the urine [93, 122, 151].

Prognosis and Risk Factors

One of the major diagnostic aims is to identify the affected dogs at high risk for complications and death. Many clinical signs and laboratory tests have been investigated to identify the prognosis and risk of death in canine babesiosis.

Factors Associated with Higher Mortality

Old age is a risk factor for the development of multiple complications from babesiosis. Multiple organ involvement leads to poor prognosis, but the only factor associated significantly with mortality is azotemia [124, 125]. Age is the only factor associated significantly with the risk of azotemia. On average, a 4 year age increase doubles the risk of an infected dog becoming azotemic [151]. *Babesia microti–like* (*T. annae*) infection is associated with a high risk of azotemia and mortality [151, 168].

Experimentally, dogs infected with high-dose infections or severe parasitemia may not have time enough to recover and die of hypotensive shock [14, 147, 159]. High mortality rates are usually observed for fighting dogs with complicated babesiosis [110].

Collapsed and young dogs are prone to developing hypoglycemia, and development of hypoglycemia predicts a higher mortality rate [26, 112]. Cases presenting with pulmonary edema and ARDS are always given a poor prognosis [124, 165]. Central nervous system involvement increases the risk of death by 57 times [165]. Unfavorable outcome should be expected in cases of complicated babesiosis with DIC and MODS [125].

Cases characterized by paradoxical hemoconcentration, hypoglycemia hyperlactatemia, or elevated creatinine levels at admission are associated with high mortality [110, 151, 165, 177]. Higher cortisol and ACTH concentrations, with low thyroxine and free thyroxine concentrations may also predict a poor outcome [178].

Factors not Associated with Higher Mortality

Experimentally, dog given low-dose infections may survive without treatment [147]. Severe anemia, autoagglutination and degree of parasitemia are not associated with increased mortality [110, 159]. Elevation of C-reactive protein is not associated with the outcome [169]. The presence of SIRS also does not affect the outcome [165].

Controversial Factors

Severe parasitemia and elevation of C-reactive protein at admission are still controversial factors for determining prognosis [14, 110, 169]. However, C-reactive protein levels can be used as an optional clinical marker to estimate the response to treatment [170, 179].

Identification of Babesia Species

A confirmed diagnosis of babesiosis depends on the identification of *Babesia* spp. within the infected erythrocytes, positive serological results, and detection of parasite nucleic acid in blood or tissues.

Morphological Identification of Babesia Species within Infected Erythrocytes

Babesia organisms can be identified within infected erythrocytes on a stained blood smear from dogs with parasitemia [2, 8, 9, 41, 105, 113, 122, 151, 152, 156, 180, 181]. Large *Babesia* species, *Babesia canis* and subspecies *B. c. canis*, *B. c. rossi*, and *B. c. vogeli*, are characterized typically as pear-shaped with a heavy ring of cytoplasm, small vacuoles, and unusual shapes. Typical parasites are 4.5–5 μm in length and occupy more than one third of the erythrocyte area [14, 78, 93, 104, 137]. In contrast, the small *Babesia* spp., *B. gibsoni*, is characterized as a small ring with one or two chromatin dots located at the periphery of the cytoplasm [6, 153-155]. *Babesia gibsoni* is polymorphic, and six forms have been described: the ring form (2.7 ± 0.5 × 1.6 ± 0.5 μm), the paired pyriform form (2.1 ± 0.5 × 0.9 ± 0.3 μm), the petaloid form (2.4 ± 0.3 × 0.9 ± 0.2 μm), the amoeboid form (2.5–7.3 × 2.7 ± 0.8 μm), the reticular form (2.3 ± 0.6 × 1.7 ± 0.4 μm), and the dot form (0.8 ± 0.1 × 0.6 ± 0.1 μm). Among these forms, the ring form is present in the initial stage of infection and the reticulate form is present only in the terminal stage of infection [153]. *Babesia conradae*, a small form of *Babesia* spp. first isolated in California, in the United States, is also polymorphic, with the following forms described: the ring-form (2.2 ± 0.7 × 1.9 ± 0.5 μm), the pyriform (1.4 ± 0.4 × 0.7 ± 0.2 μm), the anaplasmoid form (0.5 ± 0.3 × 0.5 ± 0.3 μm), the amoeboid form (3.2 ± 0.9 × 2.4 ± 0.9 μm), and the tetrad form (2.5 ± 0.7 × 2.0 μm) (Table 3).

Microscopic identification of the parasite is suggested to be only effective for detecting acute stage infections [182]. *Babesia* parasites are difficult to detect in peripheral blood samples from dogs presenting with low-grade parasitemia or chronically infected erythrocytes. For these cases, taking blood samples from ear or toe capillaries, and separating erythrocytes by centrifugation or Percoll gradient solution may facilitate the detection of infected erythrocytes [148, 183]. Alternatively, infected erythrocytes can be seen in the liver, spleen and bone marrow at one to two weeks after infection [127, 148, 184]. Additionally, flow cytometry can be used to detect *Babesia*-infected erythrocytes [24, 183, 185-187] (Table 4).

Serodiagnosis

Several immune-based serodiagnostic methods can be used to detect *Babesia*-specific surface antigens, the corresponding proteins expressed by the parasites, and antibodies generated against the specific parasite antigens [61, 64, 188, 189, 194-201, 203-208]. These techniques include indirect hemagglutination test, indirect fluorescent antibody staining, gel-precipitation test, and corpuscular antigen [3, 189-193]. Enzyme-linked immunosorbent assays and immunochromatographic tests have been developed to detect the antibodies produced by infected dogs by using the specific antigens derived from *Babesia*-infected erythrocytes or directly to the antigenicity of *Babesia* spp. Many antigens or recombinant proteins such as p29, p32, p22, p57, RAP-1, TRAP, BgSA1, and BgSA3 have been widely applied to improve the sensitivity and specificity of the assays for identification of subclinical infections [61, 64, 194-208]. These immune-based diagnostic methods provide convenient tools to conduct large-scale sero-epidemiological surveys, to discriminate infected subjects from non-infected ones, and to identify the causative *Babesia* spp. in dogs with chronic or subclinical infection (Table 4).

Detection of Nucleic Acid

Babesia infection can be detected using sequence analysis of parasite DNA or ribosomal RNA nucleotide fragments. These methods not only provide detection and differentiation of the canine *Babesia* spp., but also describe genetic characteristics of isolated pathogens and further determine phylogenetic relationships within the established *Babesia* genus [36, 185, 209-213]. Based on the nucleotide findings, *B. microti-like* are classified as *Theileria annae* [8, 9, 105, 122, 151, 168].

Many polymerase chain reaction (PCR) techniques are developed to improve the capability to detect and differentiate *Babesia* spp., such as restriction fragment length polymorphism, seminested-PCR, loop-mediated isothermal amplification assays [13, 56, 72, 214-217]. Advanced modalities, such as high-resolution melting (HRM) analysis, real-time PCR and quantitative fluorescence resonance energy transfer (FRET)-PCR assays are even able to perform quantitative assessment of infection intensity and to monitor the response of therapy [38, 218, 219]. Multiplex quantitative real-time PCR assay is able to simultaneously detect multiple tick-borne pathogens [13, 38, 56, 72, 185, 209, 214-220] (Table 4).

Surveillance of Drug-Resistant Babesia Species

Drug-resistant *Babesia* species have gained attention of late, and the strains bearing drug-resistant genes have been identified using updated techniques for nucleotide detection. More drug-resistant genes are expected to be identified. The integration of these techniques as part of routine diagnostic procedure for canine babesiosis is strongly recommended.

Table 4. Hematological and serodiagnositc protocols and nucleotide detection of canine babesiosis

Hematological methods

Diagnostic test	Method	Target organism	Sensitivity	Specificity	Babesia detection during trial interval (days)	Refe-rence
Microscopy	Percoll gradient separation Blood smear May–Grünwald–Giemsa stain	*Babesia* spp.	16/17 (94%)	9/9 (100%)	Not reported	[183]
Microscopy	Thin blood smear Giemsa stain	*Babesia* spp.	Natural infection: PCR (+): 11/28 Blood smear (+): 9/28(81.8%) (PCR as the reference tests)	Blood smear (+): 17/17 (100%) (PCR as the reference test)	4 days postinfection	[185]
Microscopy	Thin blood smear B. canis, B. gibsoni, B. canradae PCR	*Babesia* spp.	Blood smear (+): 14/17 (82.4%) (PCR as the reference tests)	Blood smear (+): 28/28 (100%) (PCR as the reference test	Not reported	[24]
Microscopy Flow cytometry	Blood smear panoptic staining kit Flow cytometry Hydroethidine stain	quantitative assessment *B. canis*	Natural infection: HE-FC (+): 37/162 (22.8%) Blood smear (+): 3/162 (1.8%)	Not reported	HE-FC: 2-189 days Blood smear: 2-119 (under detected after 119 days)	[186]
Flow cytometry	SYTO16 stain	quantitative assessment *B. gibsoni*	Correlated well with parasitemia of in vivo samples (r=0.998)	Not reported	Not reported	[187]
Serological test/ Antigen	Target gene/ protein: molecular weight of protein	Test target	Sensitivity Test with experimental infection or natural infection dog	Specificity Test with experi-mental infection or natural infection dog	Antibody detection during trial interval (days)	Reference

Table 4. (Continued)

Diagnostic test	Method	Target organism	Sensitivity	Specificity	Babesia detection during trial interval (days)	Reference
ELSIA/ P50	p50 gene rBgP50: 50 kDa	*B. gibsoni*	Experimental infection with *B. gibsoni*: (+): 6/6 Natural infection with *B. gibsoni*: (+): 8/8	Experimental infection with *B. canis*: (+): 0/1 Healthy dog: (+): 0	14-222	[194]
ELSIA/ P50	Truncated p50 gene GST-rBgP50: 76 kDa (control GST: 26kDa)	*B. gibsoni*	Experimental infection with *B. gibsoni*: (+): 15/15 Natural infection with *B. gibsoni*: (+): 9/10 (comparative to IFA (+): 7/10 and Western blot (+): 7/10)	Experimental infection with B. canis: (+): 0/6 SPF dogs: (+): 0/24	8-442	[195]
ELSIA/ Truncated P50	Truncated p50 gene rBgP50t: 46 kDa	*B. gibsoni*	Experimental infection with *B. gibsoni*: (+): 19/19 Natural infection with *B. gibsoni*: (+): 17/17	Experimental infection with *B. canis*: (+): 0/5 Healthy dog: (+): 0/24	8-360	[196]
ELSIA/ Truncated P50	Truncated p50 gene GST-rBgP50t: 71 kDa (control GST: 26kDa)	*B. gibsoni*	Experimental infection with *B. gibsoni*: (+): 22/22	Experimental infection with *B. canis*: (+): 0/8 SPF dogs: (+): 0/30	Not reported	[61]
ELSIA/ Truncated P50	Truncated p50 gene rBgP50t	*B. gibsoni*	Acute type B. gibsoni: (+): 8/14	Not reported	Not reported	[197]

Diagnostic test	Method	Target organism	Sensitivity	Specificity	Babesia detection during trial interval (days)	Reference
			Confirmed history of *B. gibsoni* infection: (+): 14/14 higher sensitivity of ELISA in detecting a history of infection			
ELSIA/ P29	P29 gene rBgP29: 29kDa	*B. gibsoni*	Natural infection with *B. gibsoni*: (+): 15/15	Experimental infection with *B. canis*: (+): 0/7	6-207	[198]
ELSIA/ BgSA1	BgSA1 gene rBgSA1: 59 kDa	*B. gibsoni*	Experimental infection with *B. gibsoni:* (+): 14/14 BgSA1-ELISA: 48/53 (90.6%) (PCR as reference test) P50-ELISA: 33/53 (62.3%) (PCR as reference test) Sandwich ELISA: detection of circulating BgSA1 with a parasitemia of 0.2%	Experimental infection with *B. canis*: (+): 0/10 SPF dogs: (+): 0/28 BgSA1-ELISA: 838/860 (97.4%) (PCR as reference test) P50-ELISA: 870/878 (99.1%) (PCR as reference test)	8-541	[199]

Table 4. (Continued)

Diagnostic test	Method	Target organism	Sensitivity	Specificity	Babesia detection during trial interval (days)	Reference
ELSIA/ BgSA3	BgSA3 gene rBgSA3: 72.2kDa	*B. gibsoni*	Experimental infection with *B. gibsoni*: (+): 16/16 Sandwich ELISA: detection of circulating BgSA1 with a parasitemia of 0.2% in asplenic dog; ELISA titer well associated with the parasitemia	Experimental infection with *B. canis*: (+): 0/8 SPF dogs: (+): 0/27	8-541	[200]
ELSIA/ P57 specific antigen for *B. gibsoni* antibodies	BgP57 gene rBgP57: 57 kDa	*B. gibsoni*	Experimental infection with *B. gibsoni*: (+): 1/1 Natural infection: 107 dogs P57-ELISA: 68/88 (77.3%) (PCR as reference test) McNemar's chi-square test: no significant difference in relative effectiveness of rBgP57-ELISA and seminested PCR in identifying positive samples	Experimental infection with *B. canis*: (+): 0/6 SPF dogs: (+): 0/30 Natural infection: 107 dogs P57-ELISA: 9/19 (47.4%) (PCR as reference test)	10-541	[203]
ELSIA/ P32	BgP32 gene rBgP32: 32-kDa	*B. gibsoni*	Experimental infection with *B. gibsoni*: (+): 1/1 Natural infection: 47 samples P32-ELISA: 73% P50-ELISA: 71%	Experimental infection with *B. canis*: (+): 0/6 SPF dogs: (+): 0/30 Natural infection: 47 samples P32-ELISA: 100% P50-ELISA: 90%	8-541	[204]

Diagnostic test	Method	Target organism	Sensitivity	Specificity	Babesia detection during trial interval (days)	Reference
ELSIA/ Rhopy-associated protein	BgRAP-1 gene rBgRAP-1: 52-kDa	*B. gibsoni*	Experimental infection with *B. gibsoni*: (+): 12/12	Experimental infection with *B. canis*: (+): 0/8 Healthy dogs: (+): 16/16	8-207	[201]
ELSIA/ P22 No cross reaction with the closely related apicomplexan parasites	BgP22 gene rBgP22: 22-kDa	*B. gibsoni*	Experimental infection with *B. gibsoni*: (+): 10/10 Natural infection: 87 dogs Differentiate chronic subclinical carrier stage: >100 days between onset of symptoms and the date of sample collection: (+) 86.9 % <100 days between onset of symptoms and the date of sample collection: (+) 43.9%	Experimental infection with *B. canis*: (+): 0/9 Uninfected dog: 0/10	130-541	[64]
ELSIA/ P12 No cross react to related apicomplexan parasites	BgP12 gene rBgP12: 12-kDa	*B. gibsoni*	Experimental infection with *B. gibsoni*: (+): 10/10	Experimental infection with *B. canis*: (+): 0/8	10-442	[205]
ELSIA/ TRAP	BgTRAP gene rBgTRAP	*B. gibsoni*	Experimental infection with *B. gibsoni*: (+): 10/10 Natural infection: 107 dogs rBgTRAP- ELISA (+)/ PCR(+): 79.5% rBgP50- ELISA (+)/PCR(+): 59.8% rBgSA- ELISA (+)/PCR(+): 66.4% rBg32- ELISA (+)/PCR(+): 60.7%	Experimental infection with *B. canis*: (+): 0/8	8-541	[206]

Table 4. (Continued)

Diagnostic test	Method	Target organism	Sensitivity	Specificity	Babesia detection during trial interval (days)	Reference
ELSIA/ TRAP	BgTRAP gene rBgTRAP	*B. gibsoni*	Experimental infection with *B. gibsoni*: (+): 10/10 Natural infection: 107 dogs rBgTRAP- ELISA (+)/ PCR(+): 79.5% rBgP50- ELISA (+)/PCR(+): 59.8% rBgSA- ELISA (+)/PCR(+): 66.4% rBg32- ELISA (+)/PCR(+): 60.7%	Experimental infection with *B. canis*: (+): 0/8	8-541	[206]
ICT/ rP50t	Not reported	*B. gibsoni*	Experimental infection with *B. gibsoni*: (+): 10/10 Natural infection: 54dogs Sensitivity: 24/24 (100%) (ELISA as reference test)	Experimental infection with *B. canis*: (+): 0/8 Experimental infection with SPF dogs: (+): 0/25 Natural infection: 54 dogs Specificity: 30/30 (100%) (ELISA as reference test)	6~144	[207]

Diagnostic test	Method	Target organism	Sensitivity	Specificity	Babesia detection during trial interval (days)	Reference
ICT/BgSA1	BgSA1 gene rBgSA1	*B. gibsoni*	Experimental infection with *B. gibsoni*: (+): 1/1 Natural infection: (+): 24/94 Sensitivity: 87.5% (ELISA as reference test)	Experimental infection with SPF dogs: (+): 0 Natural infection: (-): 70/94 Specificity: 98.6% (ELISA as reference test)	10~2227	[208]
Immunofluorescent assay (IFA)	Fluorescent antibody stain	*B. canis*	The whole of each paired parasite is fluorescent. The margin of the parasitized erythrocyte also fluoresces.		Not reported	[189]

Target gene	Test target	Method	Feature	Sensitivity (Detection limit)	Reference	Target gene
Small unit ribosomal RNA gene	*B. c. canis* *B. c. vogeli* *B. c. rossi*	PCR- RFLP Primer: PIRO-A/PIRO-B restriction enzymes: *Hin*fI/ *Taq*I	PCR fragment: *B. canis*: 400 bp *B. divergens* (in distinguishable from *B. canis*): 400 bp *B. bovis*: 390 bp *B. bigemina*: 395 bp *B. microti*: 408 bp	PCR- RFLP fragment: *B. c. rossi* /*Hin*fI:227 and 174 bp *B. c. vogeli*/ *Taq*I:203, 171, 26 bp *B. c. canis*/ *Hin*fI, *Taq*I: non-digested	Not reported	[214]

Table 4. (Continued)

Target gene	Test target	Method	Feature		Sensitivity (Detection limit)	Reference
Small subunit ribosomal RNA gene	*Babesia* species	Nest PCR Primer: PIRO-F/ PIRO-R PIRO2-F/ PIRO2-R	Nested PCR First round	Nested PCR Second round	1.18x10^{-4} % parasitemia	[209]
			No visualized band	Natural infection: 327 bp Experimental infection: 327 bp		
18S rRNA gene	B. gibsoni (Asian genotype) B. canis	Seminested PCR Primer: 455-479F/ 793-772R 5-22F/ 1661R 455-479F/793-722R BgibAsia-F BCV-F BCC-F BCR-F GAPDH-F/ GAPDH-R	Seminested PCR First round	Seminested PCR Second round	7.3 x10^{-7} % parasitemia	[215]
			B. gibsoni (California/USA genotype): 370 bp *B. gibsoni* (Asian genotype): 340 bp *B. c. vogeli*: 340 bp *B. c. canis*: 340 bp *B. c. rossi*: 340 bp	*B. gibsoni* (Asian genotype)/ BgibAsia-F: 185 bp *B. c. vogeli*/ BCV-F: 192 bp *B. c. canis*/ BCC-F: 198 bp *B. c. rossi*/ BCR-F: 197 bp		

Target gene	Test target	Method	Feature		Sensitivity (Detection limit)	Reference
18S rRNA gene	*B. canis*	PCR Primer: *Bab-f* (novel oligonucleotide) *Babesia* common "*B. canis*"	PCR	Sequence analysis	Not reported	[72]
			B. canis: *Babesia* common/ *Bab-f*: 394 bp *B. canis*: *Babesia* common/ "*B. canis*": 208 bp	an additional 186 bp of variable region 18S rRNA for more specificity differentiation among the *B. canis* spp.		
18S rRNA gene	*B. c. canis* *B. c vogeli* *B. c rossi* *T. equi* *T. annae* *B. conradae* *B. gibsoni*	Nest PCR- RFLP with filter paper-based technologies Primer: BTF1/ BTR1 BTF2/ BTF2 Restriction digestion: *Hin*fI/ *Hin*cII/ *Bsl*I	Nest-PCR- fragment:	RFLP fragment:	DNA extracted from whole EDTA blood: 2.7×10^{-7} % DNA extracted from blood applied to both FTA™ cards and IsoCode™ PCR: 2.7×10^{-5} %	[216]
			T. equi: 830 bp	*Bsl*I 532, 297 bp		
			T. annae: 848 bp	*Bsl*I 547, 220, 79 bp		
			B. conradae:842 bp	*Bsl*I Uncut (842)		
			Babesia spp. (Coco): 781 bp	*Hin*fI 584, 102, 78, 18 bp *Hin*cII and *Bsl*I 416, 276, 98		

Table 4. (Continued)

Target gene	Test target	Method	Feature		Sensitivity (Detection limit)	Reference
			B. c. vogeli: 791 bp	*HinfI* 592, 102, 80, 18 bp *Hinc*II and *Bsl*I 514, 278		
			B. c. canis: 792 bp	*HinfI* 593, 102, 80, 18 bp *Hinc*II and *Bsl*I 327, 278, 188		
			B. c. rossi: 793 bp	*Hinf*I 303, 289, 102, 81, 18 bp		
			B. gibsoni: 791 bp	*Hinf*I 321, 270, 102, 81, 18		
			N. caninum: 870 bp	*Bsl*I 353, 222, 289, 47, 36, 3 bp		
			Negative control: no amplified			
18S rRNA gene	*B. c. canis* *B. c. vogeli* *B. c. rossi*	PCR Primer: BAB1/BAB3 BAB1/BAB4 BAB1/BAB5	PCR fragment *B. c. canis*: BAB1/BAB3: 746 bp *B. c. vogeli*: BAB1/BAB4: 590 bp *B. c. rossi*: BAB1/BAB5: 342 bp *B. gibsoni*, *B.equi*: no amplified		1 Babesia-infected blood cell/sample	[13]
P18 gene	*B. gibsoni*	Novel Oligonucleotide primer: d3/ d4	PCR fragment *B. gibsoni*: 182 bp *B. canis*: no amplified		2 x10^{-6} % parasitemia Nucleotide detection during trial interval 2-222 days	[185]

<table>
<tr><th>Target gene</th><th>Test target</th><th>Method</th><th colspan="2">Feature</th><th>Sensitivity
(Detection limit)</th><th>Reference</th></tr>
<tr><td>P18 gene</td><td>quantitative assessment
B. gibsoni
(Asian genotype)</td><td>Real-time PCR
primers targeting the P18 gene
SYBR Green</td><td colspan="2">primers targeting the P18 gene: 182 bp
intra-assay CV for Ct values: from 0.70-1.89%
inter-assay CV for Ct values: from 1.18-1.92%
Correlation coefficients— times after inoculation and P18 copy number: dogs A: 0.65, dogs B: 0.619, dogs C: 0.576
Correlation coefficients—parasitemia and P18 copy number: dogs A: 0.942, dogs B: 0.729, dogs C: 0.527</td><td>Nucleotide detection: first day after inoculation</td><td>[218]</td></tr>
<tr><td rowspan="2">18S rRNA gene</td><td rowspan="2">quantitative assessment polymorphic groups of B. c. canis</td><td rowspan="2">PCR-RFLP
Restriction digestion: HincII
Real-time HRM PCR
Primers:
GF1/ GR2
SYBR Green
Melting temperature: gradual increase from 70°C to 95°C</td><td>PCR-RFLP fragment</td><td>Real-time HRM PCR</td><td rowspan="2">Not reported</td><td rowspan="2">[219]</td></tr>
<tr><td>PCR: 559 bp
RFLP:
Group A: HincII cut
Group B: HincII uncut
Group A: guanine in position 150 and adenine in position 151
Group B: adenine in position 150 and guanine in position 151</td><td>GF1/ GR2: 559 bp
Tm:
Group A: 78°C
Group B: 81°C
assumed that substitutions in positions 2-3 in a fragment of 559 nucleotides result in a 3°C difference in melting temperatures</td></tr>
<tr><td>18S rRNA gene</td><td>quantitative assessment
B. gibsoni
B. c. canis
B. c. vogeli
B. c. rossi</td><td>FRET-PCR
Fluorescent probe:
Donor probe/
Acceptor probe
Melting curves analysis
50 to 80.8°C with a temperature transition rate of 0.2 8°C/s</td><td colspan="2">FRET-PCR
Melting curves:
B. gibsoni: 64°C
B. c. canis: 60°C
B. c. vogeli: 60°C
B. c. rossi: 54°C</td><td>1 copy/PCR</td><td>[38]</td></tr>
</table>

Target gene	Test target	Method	Feature	Sensitivity (Detection limit)	Reference	Target gene
Heat shock protein 70 gene	quantitative assessment *B. c. vogeli* *E. canis*	Multiplex real-time qPCR Primer: E. c 16S fwd E. c 16S rev B. c hsp70 fwd B. c hsp70 rev Canine actin fwd Canine actin rev Dual-labeled probes	singleplex reaction (SYBR green) ΔCq values: 0.85 efficiency: 100% detection limit: 100 copies	duplex reaction (dual-labeled probes) ΔCq values: 1.56 (increase sensitivity) efficiency: 86% (decrease) detection limit: 10 copies	1–10 copies/ul	[220]
18S ribosomal DNA	*B. gibsoni*	LAMP Primer: B18S-F3/ B18S-B3 B18S-FIP/ B18S-BIP	LAMP fragment LAMP(+)/ Microscopy(+): 37/ 16 PCR(+)/ Microscopy(+): 15/ 12 Non-specific reactions were not observed by the LAMP method.		5×10^{-4}% parasitemia	[56]
Rhoptry protein gene	*B. c. canis* *B. c. vogeli* *B. c. rossi*	LAMP Primer: BcanF3/BcanB3 BcanFIP/ BcanBIP	Specificity: Detected: *B. c. canis*, *B. c. rossi*, *B. c. vogeli* Non-detected: *B. motasi*, *B. ovis*, *B. crassa*, *B. microti* Sensitivity: 80% (RLB as the reference test) Specificity: 90% (RLB as the reference test)		25 pg gDNA (LAMP) 5pg gDNA (PCR)	[217]

B. bigemina: *Babesia bigemina*; *B. bovis*: *Babesia bovis*; *B. canis*: *Babesia canis*; *B. c. canis*: *Babesia canis canis*; *B. c. rossi*: *Babesia canis rossi*; *B. c. vogeli*: *Babesia canis vogeli*; *B. conradae*: *Babesia conradae*; *B. crassa*: *Babesia crassa*; *B. divergens: Babesia divergens*; *B. equi: Babesia equi*; *B. gibsoni: Babesia gibsoni*; *B. microti: Babesia microti; B. motasi: Babesia motasi; B. ovis: Babesia ovis*; Cq: quantification cycle; Ct: threshold cycle; CV: coefficient of variation; ELISA: Enzyme-linked immunosorbent assay; FC: flow cytometry; FRET: fluorescence resonance energy transfer; GST: glutathione S-transferase fusion protein; HE: hydroethidine stain; HRM: high-resolution melting; ICT: Immunochromatographic test; IFA: indirect immunofluorescence antibody; LAMP: loop-mediated isothermal amplification; PCR: polymerase chain reaction; PCR- RFLP: restriction fragment length polymorphism; *T. annae*: *Theileria annae*; *T. equi*: *Theileria equi*; *N. caninum*: *Neospora caninum*; rBgSA1: recombinant secreted antigen-1 of *Babesia gibsoni*; rBgSA3: recombinant secreted antigen-3 of *Babesia gibsoni*; rBgP12: recombinant P12 protein of *Babesia gibsoni*; rBgP22: recombinant P22 protein of *Babesia gibsoni*; rBgP29: recombinant P29 protein of *Babesia gibsoni*; rBgP32: recombinant P32 protein of *Babesia gibsoni*; rBgP50: recombinant P50 protein of *Babesia gibsoni*; rBgP50t: recombinant protein of *Babesia gibsoni* truncated P50 gene; rBgP57: recombinant P57 protein of *Babesia gibsoni*; rBgTRAP: recombinant thrombospondin-related adhesive protein of *Babesia gibsoni*; rBgRAP-1: recombinant rhopy-associated protein-1; RLB: reverse line blot; SPF: specific pathogen free; Tm: melting temperature; ΔCq: calculated as the delta between the last cycle and the Cq value of the examined sample.

Table 5. Antibabesial treatments of canine babesiosis

Antiprotozoal agents						
Compound (treated species)	Suggested dosage	Route	Interval	Duration	Common adverse reactions/toxicity	References
Imidocarb dipropionate (*B. canis*, *B. conradae B. gibsoni*, *T. Annae*, a large unnamed *Babesia* species)	4 mg/kg 2-7 mg/kg	IV SC/IM	- QW-Q2W	Single shot 2 shots	Pain at the injection site (not necrosis), vomiting, transient excessive salivation, anorexia, diarrhoea, weakness and ataxia Severe pain at the injection site, transient parasympathomimetic side effects Marked pulmonary congestion and oedema. Kidneys were grossly enlarged and markedly congested with extensive haemorrhage in the cortex and medulla and marked tubulonephrosis. Liver and spleen were moderately enlarged and congested Severe depression, tachycardia with premature ventricular contractions, cyanosis and hind limb tremors, and massive hepatic necrosis at 60 mg/kg	[2, 40, 41,82, 105, 124, 127, 148,167, 170, 181, 228, 229, 230, 231]
Phenamidine isethionate (*B. gibsoni*)	15-20 mg/kg	SC	QD	2 shots	Behavioral changes, deafness, disturbed vision, imbalance or ataxia, rolling movements, extensor rigidity, spasticity, opisthotonus, nystagmus, coma and terminal paralysis Moderated to severe increases in AST, ALT and BUN Prominent hemorrhagic and malacic lesions were found mainly in the cerebellum, midbrain and thalamus. Degenerative lesions may also develop in the liver, kidneys, myocardium and skeletal muscles	[227]
Diminazene aceturate (*B. canis*, *B. conradae*, *B. gibsoni*)	3.5 mg/kg 3 mg/kg on puppy 4-5 mg/kg	IM IM IM	QD or Q2W QD -	Single shot to 2 shots 2 shots Single shot	The side effects are identical to those observed with phenamidine Vocalization, convulsions, ataxia and weak Similar chemical pathological changes are encountered as for phenamidine but less severe Prominent hemorrhagic and malacic lesions mainly in the cerebellum	[26, 34, 40, 127, 227, 231, 232]

Antiprotozoal agents						
Compound (treated species)	Suggested dosage	Route	Interval	Duration	Common adverse reactions/toxicity	References
Quinoronium sulphate (*B. canis*?)	0.25-1 mg/kg/day	SC	-	5-15 days	Vomiting and diarrhoea Increase in ALT	[227]
Atovaquone (*B. gibsoni*)	30 mg/kg	PO	BID	7 days	Not reported	[115, 236]
Antibiotics						
Compound (treated species)	Dose	Route	Interval	Duration	Common side effects/toxicity	Reference
Clindamycin (*B. gibsoni*)	25 mg/kg	PO	BID	14 days	Lymphocytosis Monocytosis, but the numbers of monocytes decreased during the chronic stage	[222]
Doxycycline (*B. gibsoni*)	10 mg/kg	PO	BID	11 days	Not reported	[221]
Combination therapy						
Compound (treated species)	Dose	Route	Interval	Duration	Common adverse reactions/toxicity	Reference
Phenamidine isethionate and thiamine hydrochloride (*B. canis*?)	Thiamine 10 mg/kg/day (10 times the recommended therapeutic dosage of thiamine); Phenamidine 20 mg/kg/day	Thiamine IM, phenamidine SC	-	Thiamine single shot, followed by phenamidine until the development of nervous symptoms	Imbalance, nystagmus, coma Increases in ALT and BUN Haemorrhage and malacia in CNS, degenerative lesions may also develop in the liver, kidneys and myocardium in some dogs	[227]

Table 5. (Continued)

Antiprotozoal agents							
Compound (treated species)	Suggested dosage	Route	Interval	Duration	Common adverse reactions/toxicity		References
Phenamidine isethionate and diminazene aceturate (*B. canis*?)	Phenamidine 20 mg/kg/day; Diminazene 3.5 mg/kg		Phenamidine SC, diminazene IM	-	Phenamidine for 2 days, followed by diminazene single shot	Not reported	[227]
Atovaquone and azithromycin (*B. gibsoni*, a large unnamed *Babesia* species)	Atovaquone 13.3 mg/kg; Azithromycin 10 mg/kg		PO	Atovaquone Q8H Azithromycin Q24H	10 days	Fatigue and excessive salivation, but most cases did not show clinical side effects	[181,182, 223, 224]
Imidocarb dipropionate and clindamycin (*B. canis*)	Imidocarb 5 mg/kg; Clindamycin 10 mg/kg		Imidocarb IM, clindamycin PO	Clindamycin Q8H	Imidocarb single shot, clindamycin for 2 weeks	Urination, defecation, lacrimation, and salivation after imidocarb administration	[126]
Imidocarb dipropionate and doxycycline (*B. microti*–specific, a large unnamed *Babesia* species)	Imidocarb 3-6.6 mg/kg; Doxycycline 10-16.5 mg/kg/day		Imidocarb SC, doxycycline PO	Imidocarb Q2W, doxycycline Q12H	Imidocarb single shot or 2 shots, doxycycline for 4-8 weeks	Not reported	[81, 181]

Antiprotozoal agents							
Compound (treated species)	Suggested dosage	Route	Interval	Duration	Common adverse reactions/toxicity		References
Diminazene aceturate-Imidocarb dipropionate-Atovaquone – Azithromycin (*B. gibsoni*)	Diminazene (3.5 mg/kg)+ Imidocarb (6 mg/kg)+ Atovaquone (13.3mg/kg)+Azithromycin (10 mg/kg)		Diminazene and imidocarb SC, atovaquone and azithromycin PO	Imidocarb Q2W, atovaquone and azithromycin Q12H	Diminazene single shot, followed by 2 shots of imidocarb; Atovaquone and azithromycin for 10 days	Not reported	[31]
Diminazene aceturate-Clindamycin-Doxycycline-Metronidazole (*B. gibsoni*)	Diminazene (3mg/kg)+ Clindamycin (25mg/kg)+ Doxycycline (5 mg/kg) + Metronidazole (15 mg/kg)		Diminazene SC, others PO	Diminazene QD, others Q12H	Diminazene for 3 days, other drugs for 24-129 days (suggest 50 days)	Not reported	[179]
Doxycycline-Enrofloxacin-Metronidazole ± Diminazene diaceturate (*B. gibsoni*)	Doxycycline (7-10 mg/kg)+Enrofloxacin (2-2.5mg/kg) +Metronidazole (5-15 mg/kg) ± Diminazene (3 mg/kg)		Diminazene IM, others PO	Diminazene QW, antibiotics Q12H	Diminazene 2 shots, antibiotics for 6 weeks	Not reported	[233]
Miscellaneous							
Compound (treated species)	Dose		Route	Interval	Duration	Common side effects/toxicity	Reference
Bruceine A (*B. gibsoni*)	6.4 mg/kg		PO	QD	6 days	Not reported	[238]
C. horridus 200C (*B. gibsoni*)	4 pills		PO	QD	14 days	Not reported	[237]

B. canis: Babesia canis; B. conradae: Babesia conradae; B. gibsoni: Babesia gibsoni; B. microti–specific: Babesia microti–specific; T. annae: Theileria annae

TREATMENTS

Treatments for canine babesiosis vary depending on the condition of the dogs and complications of the infection. The treatment strategies may consist of administering antibabesial agents, managing inflammatory and immune reactions, and administering supportive and symptomatic treatments for concurrent SIRS and MODS.

Many compounds have been used for the management of canine babesiosis, and these can be categorized into two types: 1) antiprotozoal agents and 2) a combination of antibiotics and antiprotozoal agents. Additive or synergistic effects of these compounds have also been evaluated. However, no single or combination therapy has successfully eliminated *Babesia* species from infected dogs [2, 26, 31, 40, 81, 105, 115, 124, 126, 127, 148, 167, 179, 181, 182, 221-224, 227-233, 236-238] (Table 5).

Antiprotozoal agents exhibit better activities against *Babesia* spp. *in vitro* [225]. Efficacy of the specific treatment used varies based on the condition of the dog and the virulence of the parasite strain [226]. Splenectomized and immunocompromised dogs such as those with hyperadrenocorticism are prone to persistent infection and are more refractory to treatments [149, 181]. In general, *B. gibsoni* is much less responsive to treatment with conventional babesicidal agents than are the large *Babesia* spp. (*B. canis*). Responses following treatment for *B. gibsoni* infection have not always been conclusive. All in all, complete elimination should not be considered in the management of canine babesiosis [21, 40, 115, 179, 181, 182, 221-224].

Antiprotozoal Agent, Diamidines

Diamidines, a group of trypanocidal agents, were found to be effective against *Babesia* species. Within this group, several compounds, such as diminazene aceturate, diminazene diaceturate, phenamidine isethionate, pentamidine isethionate, imidocarb dipropionate and amicarbalide, are used as babesicides and trypanocides in domestic animals [2, 26, 34, 40, 41, 82, 105, 124, 127, 148, 167, 170, 227-232]. Each compound is able to decrease the severity of parasitemia; however, complete elimination of the parasites has never been achieved, and relapse occurs.

Among these compounds, diminazene and phenamidine produce similar adverse reactions. Severe neurological symptoms such as behavioral changes, deafness, disturbed vision, imbalance or ataxia, rolling movements, extensor rigidity, spasticity, opisthotonus, nystagmus, coma and terminal paralysis occur in the majority of the dogs that receive these drugs [227]. Clinicopathologically, these two compounds may also cause severe increases in hepatic enzyme activities and BUN levels [26, 227]. A single dose of diminazene (3 mg/kg) may produce weakness and ataxia in young dogs [26]. Two repeated daily doses of phenamidine (20 mg/kg) may cause brain lesions [227]. The histopathological evidence of phenamidine and diminazene poisoning include prominent hemorrhagic and malacic lesions mainly in the cerebellum, midbrain and thalamus [227]. Degenerative lesions may also develop in the liver, kidneys, myocardium and skeletal muscles in some dogs [227].

In recent years, resistance to diminazene aceturate in the control of canine babesiosis has gained attention. Concomitant use of intramuscular diminazene diaceturate may not improve the efficacy of a doxycycline-enrofloxacin-metronidazole combination in management of canine babesiosis caused by *B. gibsoni* [233]. Furthermore, the diminazene aceturate-resistant *B. gibsoni* strain appears to be resistant to other antibabesial agents *in vitro* [234].

Imidocarb, another diamidine and a derivative of carbanilide, has been shown to be an effective prophylactic agent for treating *B. canis* infections [226, 230]. However, this compound may disturb the level of protective antibodies, leaving the animals more susceptible to re-infection [230]. Adverse reactions associated with imidocarb-treatment include marked pulmonary congestion and edema, tubulonephrosis and Haemorrhage in the cortex and medulla of the kidney, enlarged and congested liver and massive hepatic necrosis, and enlarged and congested spleen [227-229].

Antiprotozoal Agent, Atovaquone

Atovaquone is an analogue of ubiquinone with broad-spectrum activities against various protozoal infections in human beings [235]. Although atovaquone has been shown to be effective against *B. gibsoni* infections in dogs, complete elimination was not achieved and sensitivity of the parasites to atovaquone decreased [115].

Single nucleotide polymorphisms in the sequence of the mitochondrial cytochrome *b* gene were found to be associated with decreased susceptibility of *Babesia* species to atovaquone [236]. The use of atovaquone alone might allow the development of atovaquone-resistant parasites [115].

Synergistic administration of atovaquone and azithromycin is extremely effective against human babesiosis and later applied as the treatment for dogs that are chronically infected with *B. gibsoni* or dogs that previously failed to respond to imidocarb diproprionate or diminazene aceturate, or both [223]. Although this combination is able to suppress the level of parasitemia below the limit of detection, relapse still occurs [182, 223, 224]. Similar analysis of the cytochrome *b* gene strongly suggests the emergence of atovaquone-resistant variants after treatment of the combination of atovaquone and azithromycin [224]. The standard 10-day course may not be effective in endemic seasons or areas, and a 3-week course is suggested to achieve a longer period of clinical remission (Table 5).

Antibiotics

The efficacies of azithromycin, clindamycin, enrofloxacin, doxycycline and metronidazole have been evaluated. In general, all these compounds, when administered individually or in combination, have been shown to improve clinical signs and to reduce levels of parasitemia, but no complete elimination of infection has been reported [115, 179, 182, 222-224, 233]. Relapse and resistance to these single therapy or combination protocols were reported [115, 179, 182, 224, 233, 236]. However, these antibiotics are suggested as alternative or synergistic treatments especially for relapse after treatment with other babesicides.

Management of Immune Reactions

The administration of immunosuppressant agents or glucocorticoids may be required to control the anemia in severe conditions of IMHA. However, long-term administration of immunosuppressant agents or glucocorticoids also increases the risk of overwhelming infection [19, 124]. Management of IMHA and eradication of babesia infection requires a balance of immunosuppressive agents to prevent the risk of persistent infection.

The spleen is important immune organ for producing antibodies in response to infections and in autoimmune reactions. Although splenectomy has been suggested for the management of IMHA that is not responding to high-dose immunosuppressant agents, once the immune reactions develop, immunity is retained, even if the spleen is subsequently removed [149, 239]. However, splenectomy increases risk of overwhelming parasitemia, as well.

Blood transfusion may be required in dogs with severe anemia as a life-saving treatment, but not in uncomplicated babesiosis. Increased tissue oxygenation by transfusion may sustain the affected dogs for the time required for other treatments to become effective. However, alloimmune reaction during blood cross-matching tests may be obscured by IMHA. Therefore, the use of blood donated from dogs with negative Dog Erythrocyte Antigen 1.1 is recommended [22]. Another oxygen-carrying substitute, bovine hemoglobin solution, for which blood typing is not required, may also improve the condition of anemia.

Supportive and Symptomatic Treatments

When dogs are hypoxic or hypotensive secondary to SIRS and MODS, rehydration is vital to improve organ perfusion. In addition, correction of acid-base and electrolyte imbalances, maintenance of the cardiovascular system and arterial blood pressure for hypovolemic or hypotensive shock [126], and antibiotic therapy for sepsis should be commenced if these complications are present.

PREVENTION

Vector (Tick) Control

Babesia persists in all tick stages, and both vertebrates and ticks serve as parasite reservoirs in the environment, which facilitates the long-term persistence of *Babesia* spp. in the ecosystem [10]. Therefore, tick control is an essential part of preventative strategies. Topical preventatives, such as spot-on and amitraz-containing collars, should be applied to both affected and non-affected dogs at regular intervals to minimize the risk of tick infestation [240]. Strict tick control measures should be applied to dogs that present with multiple or frequent relapses of parasitemia, or to dogs infected by antiprotozoal-resistant strains of *Babesia* species to prevent spreading to the local canine population. Environmental controls, such as foggers and sprays, used in tick-infested houses may also reduce the risk of tick bites.

Prophylactic Therapy

Drugs such as doxycycline have been shown to exhibit prophylactic effects [221]. Ideally, prophylactic treatment would provide protection to dogs travelling to endemic areas. However, these dogs may develop clinical signs or parasitemia when the prophylactic therapy is discontinued. Therefore, application of topical tick controls is the most preventative strategy for dogs travelling in endemic areas.

Vaccination

Vaccines against *B. canis* were developed two decades ago. Although homologous and heterologous vaccine challenges afforded protection against infection with *B. canis*, availability of the vaccines and incomplete immunity induced by specific antigens limited the clinical application of the vaccine [174, 241-245]. Several candidate antigens shared among *Babesia* species may provide a solution for universal vaccine against canine babesiosis [25, 246-253].

Public Health and Zoonosis

Babesiosis is zoonotic and potentially life threatening in human beings. Several *Babesia* species have been reported to cause human babesiosis, including *B. bovis*, *B. canis*, *B duncani*, *B. microti*, and *B. venatorum*, and a fatal case was confirmed to be caused by *B. divergens* [15, 253]. Warm weather increases the aggressiveness of *R. sanguineus*. This effect, combined with increased outdoor activities during warm seasons, leads to increased human attacks and clusters of cases caused by tick-transmitted hemoparasites [123]. Prevention of *Babesia* infection relies on avoidance of tick bites, which can be achieved by wearing appropriate clothing and applying repellents [15].

Laboratory tests for babesiosis, tick control, and prophylactic treatments should be required for dogs travelling internationally.

Conclusion

Ongoing globalization and global warming is expected to lead to greater emergence of babesiosis in the future. Tick (vector) control is the key to prevent canine babesiosis. Application of tick preventatives and environmental control in endemic areas at regular basis is strongly recommended. Future exploration of different drug-resistant genes and integrated as part of diagnostic procedure may be the cost-effective strategy for both treatments of babesiosis and surveillance drug-resistant *Babesia* species.

REFERENCES

[1] Uilenberg, G. (2006). *Babesia*–a historical overview. *Vet. Parasitol., 138*, 3-10.

[2] Conrad, P., Thomford, J., Yamane, I., Whiting, J., Bosma, L., Uno, T., Holshuh, H. J., & Shelly, S. (1991). Hemolytic anemia caused by *Babesia gibsoni* infection in dogs. *J. Am. Vet. Med. Assoc., 199*, 601-605.

[3] Ristic, M., Lykins, J. D., Smith, A. R., Huxsall, D. L., & Groves, M. G. (1971). *Babesia canis* and *Babesia gibsoni*: soluble and corpuscular antigens isolated from blood of dogs. *Exp. Parasitol., 30*, 385-392.

[4] Niak, A., Anwar, M., & Khatibi, S. (1973). Canine babesiosis in Iran. *Trop. Anim. Health. Prod., 5*, 200-201.

[5] Shortt, H. E. (1973). *Babesia canis*: the life cycle and laboratory maintenance in its arthropod and mammalian hosts. *Int. J. Parasitol., 3*, 119-148.

[6] Groves, M. G., & Dennis, G. L. (1972). *Babesia gibsoni*: field and laboratory studies of canine infections. *Exp. Parasitol., 31*, 153-159.

[7] Zahler, M., Schein, E., Rinder, H., & Gothe, R. (1998). Characteristic genotypes discriminate between *Babesia canis* isolates of differing vector specificity and pathogenicity to dogs. *Parasitol. Res., 84*, 544-548.

[8] Zahler, M., Rinder, H., Schein, E., & Gothe, R. (2000). Detection of a new pathogenic *Babesia microti*-like species in dogs. *Vet. Parasitol., 89*, 241-248.

[9] Camacho, A. T., Pallas, E., Gestal, J. J., Guitián, F. J., Olmeda, A. S., Goethert, H. K., & Telford, S. R. (2001). Infection of dogs in north-west Spain with a *Babesia microti*-like agent. *Vet. Rec., 149*, 552-555.

[10] Chauvin, A., Moreau, E., Bonnet, S., Plantard, O., & Malandrin, L. (2009). *Babesia* and its hosts: adaptation to long-lasting interactions as a way to achieve efficient transmission. *Vet. Res., 40*, 37.

[11] Martinsen, E. S., Perkins, S. L., & Schall, J. J. (2008). A three-genome phylogeny of malaria parasites (*Plasmodium* and closely related genera): evolution of life-history traits and host switches. *Mol. Phylogenet. Evol., 47*, 261-273.

[12] Mehlhorn, H., & Shein, E. (1984). The piroplasms: life cycle and sexual stages. *Adv. Parasitol., 23*, 37-103.

[13] Duarte, S. C., Linhares, G. F., Romanowsky, T. N., da Silveira Neto, O. J., & Borges, L. M. (2008). Assessment of primers designed for the subspecies-specific discrimination among *Babesia canis canis*, *Babesia canis vogeli* and *Babesia canis rossi* by PCR assay. *Vet. Parasitol., 152*, 16-20.

[14] Schetters, T. P., Moubri, K., Precigout, E., Kleuskens, J., Scholtes, N. C., & Gorenflot, A. (1997). Different *Babesia canis* isolates, different diseases. *Parasitology., 115 (Pt 5)*, 485-493.

[15] Hunfeld, K. P., Hildebrandt, A., & Gray, J. S. (2008). Babesiosis: recent insights into an ancient disease. *Int. J. Parasitol., 38*, 1219-1237.

[16] Dantas-Torres, F., Figueredo, L. A., & Faustino, M. A. G. (2004). Ectoparasitos de cães provenientes de alguns municípios da região metropolitana do Recife, Pernambuco, Brasil. *Rev. Bras. Parasitol. Vet., 13*, 151-154.

[17] Costa-Júnior, L. M., Ribeiro, M. F., Rembeck, K., Rabelo, E. M., Zahler-Rinder, M., Hirzmann, J., Pfister, K., & Passos, L. M. (2009). Canine babesiosis caused by *Babesia*

canis vogeli in rural areas of the State of Minas Gerais, Brazil and factors associated with its seroprevalence. *Res. Vet. Sci., 86*, 257-260.

[18] Inokuma, H., Yoshizaki, Y., Shimada, Y., Sakata, Y., Okuda, M., & Onishi, T. (2003). Epidemiological survey of *Babesia* species in Japan performed with specimens from ticks collected from dogs and detection of new *Babesia* DNA closely related to *Babesia odocoilei* and *Babesia divergens* DNA. *J. Clin. Microbiol., 41*, 3494-3498.

[19] Irizarry-Rovira, A. R., Stephens, J., Christian, J., Kjemtrup, A., DeNicola, D. B., Widmer, R. W., & Conrad, P. A. (2001). *Babesia gibsoni* Infection in a dog from Indiana. *Vet. Clin. Pathol., 30*, 180-188.

[20] Macinitire, D. K., Boudreaux, M. K., West, G. D., Bourne, C., wright, J. C., & Conrad, P. A. (2002). *Babesia gibsoni* infection among dogs in the southeastern United States. *J. Am. Vet. Med. Assoc., 220*, 325-329.

[21] Matsuu, A., Kawabe, A., Koshida, Y., Ikadai, H., Okano, S., & Higuchi, S. (2004). Incidence of Canine *Babesia gibsoni* Infection and Subclinical Infection among Tosa dogs in Aomori Prefecture, Japan. *J. Vet. Med. Sci., 66*, 893-897.

[22] Stegeman, J. R., Birkenheuer, A. J., Kruger, J. M., & Breitschwerdt, E. B. (2003). Transfusion-associated *Babesia gibsoni* infection in a dog. *J. Am. Vet. Med. Assoc., 222*, 959-963.

[23] Jefferies, R., Ryan, U. M., Jardine, J., Broughton, D. K., Robertson, I. D., & Irwin, P. J. (2007). Blood, Bull Terriers and Babesiosis: further evidence for direct transmission of *Babesia gibsoni* in dogs. *Aust. Vet. J., 85*, 459-463.

[24] Birkenheuer, A. J., Levy, M. G., Stebbins, M., Poore, M., & Breitschwerdt, E. (2003). Serosurvey of anti*Babesia* antibodies in stray dogs and American pit bull terriers and American staffordshire terriers from North Carolina. *J. Am. Anim. Hosp. Assoc., 38*, 551-557.

[25] Fukumoto, S., Suzuki, H., Igarashi, I., & Xuan, X. (2005). Fatal experimental transplacental *Babesia gibsoni* infections in dogs. *Int. J. Parasitol., 35*, 1031-1035.

[26] Harvey, J. W., Taboada, J., & Lewis, J. C. (1988). Babesiosis in a litter of pups. *J. Am. Vet. Med. Assoc., 192*, 1751-1752.

[27] Sasaki, M., Omobowale, O., Tozuka, M., Ohta, K., Matsuu, A., Nottidge, H. O., Hirata, H., Ikadai, H., & Oyamada, T. (2007). Molecular survey of *Babesia canis* in dogs in Nigeria. *J. Vet. Med. Sci., 69*, 1191-1193.

[28] Kamani, J., Sannusi, A., Dogo, A. G., Tanko, J. T., Egwu, K. O., Tafarki, A. E., Ogo, I. N., Kemza, S., Onovoh, E., Shamaki, D., Lombin, L. H., Catto, V., & Birkenheuer, A. J. (2010). *Babesia canis* and *Babesia rossi* co-infection in an untraveled Nigerian dog. *Vet. Parasitol., 173*, 334-335.

[29] Lewis, B. D., Penzhorn, B. L., Lopez-Rebollar, L. M., & De Waal, D. T. (1996). Isolation of a South African vector-specific strain of *Babesia canis*. *Vet. Parasitol., 63*, 9-16.

[30] Matjila, P. T., Penzhorn, B. L., Bekker, C. P., Nijhof, A. M., & Jongejan, F. (2004). Confirmation of occurrence of *Babesia canis vogeli* in domestic dogs in South Africa. *Vet. Parasitol., 122*, 119-125.

[31] Matjila, P. T., Penzhorn, B. L., Leisewitz, A. L., Bhoora, R., & Barker, R. (2007). Molecular characterisation of *Babesia gibsoni* infection from a pit-bull terrier pup recently imported into South Africa. *J. S. Afr. Vet. Assoc., 78*, 2-5.

[32] M'Ghirbi, Y., & Bouattour, A. (2008). Detection and molecular characterization of

Babesia canis vogeli from naturally infected dogs and *Rhipicephalus sanguineus* ticks in Tunisia. *Vet. Parasitol., 152*, 1-7.

[33] Oyamada, M., Davoust, B., Boni, M., Dereure, J., Bucheton, B., Hammad, A., Itamoto, K., Okuda, M., & Inokuma, H. (2005). Detection of *Babesia canis rossi, B. canis vogeli*, and *Hepatozoon canis* in dogs in a village of eastern Sudan by using a screening PCR and sequencing methodologies. *Clin. Diagn. Lab. Immunol., 12*, 1343-1346.

[34] Breitschwerdt, E. B., Malone, J. B., MacWilliams, P., Levy, M. G., Qualls Jr, C. W., & Prudich, M. J. (1983). Babesiosis in the Greyhound. *J. Am. Vet. Med. Assoc., 182*, 978-982.

[35] Birkenheuer, A. J., Correa, M. T., Levy, M. G., & Breitschwerdt, E. B. (2005). Geographic distribution of babesiosis among dogs in the United States and association with dog bites: 150 cases (2000-2003). *J. Am. Vet. Med. Assoc., 227*, 942-947.

[36] Bostrom, B., Wolf, C., Greene, C., & Peterson, D. S. (2008). Sequence conservation in the rRNA first internal transcribed spacer region of *Babesia gibsoni* genotype Asia isolates. *Vet. Parasitol., 152*, 152-157.

[37] Yeagley, T. J., Reichard, M. V., Hempstead, J. E., Allen, K. E., Parsons, L. M., White, M. A., Little, S. E., & Meinkoth, J. H. (2009). Detection of *Babesia gibsoni* and the canine small *Babesia* 'Spanish isolate' in blood samples obtained from dogs confiscated from dogfighting operations. *Am. J. Vet. Res., 235*, 535-539.

[38] Wang, C., Ahluwalia, S. K., Li, Y., Gao, D., Poudel, A., Chowdhury, E., Boudreaux, M. K., & Kaltenboeck, B. (2010). Frequency and therapy monitoring of canine *Babesia* spp. infection by high-resolution melting curve quantitative FRET-PCR. *Vet. Parasitol., 168*, 11-18.

[39] Yamane, I., Gardner, I. A., Ryan, C. P., Levy, M., Urrico, J., & Conrad, P. A. (1994). Serosurvey of *Babesia canis*, *Babesia gibsoni* and *Ehrlichia canis* in pound dogs in californai, USA. *Prev. Vet. Med., 18*, 293-304.

[40] Birkenheuer, A. J., Levy, M. G., Savary, K. C., Gager, R. B., & Breitschwerdt, E. B (1999). *Babesia gibsoni* infections in dogs from North Carolina. *J. Am. Anim. Hosp. Assoc., 35*, 125-128.

[41] Holman, P. J., Backlund, B. B., Wilcox, A. L., Stone, R., Stricklin, A. L., & Bardin, K. E. (2009). Detection of a large unnamed *Babesia* piroplasm originally identified in dogs in North Carolina in a dog with no history of travel to that state. *J. Am. Vet. Med. Assoc., 235*, 851-854.

[42] Birkenheuer, A. J., Horney, B., Bailey, M., Scott, M., Sherbert, B., Catto, V., Marr, H. S., Camacho, A. T., & Ballman, A. E. (2010). *Babesia microti*-like infections are prevalent in North American foxes. *Vet. Parasitol., 172*, 179-182.

[43] Kocan, A. A., Kjemtrup, A., Meinkoth, J., Whitworth, L. C., Murphy, G. L., Decker, L., & Lorenz, M. (2001). A genotypically unique *Babesia gibsoni*-like parasite recovered from a dog in Oklahoma. *J. Parasitol., 87*, 437-438.

[44] Anderson, J. F., Magnarelli, L. A., Donner, C. S., Spielman, A., & Piesman, J. (1979). Canine *Babesia* new to North America. *Science., 204*, 1431-1432.

[45] Eiras, D. F., Basabe, J., Mesplet, M., & Schnittger, L. (2008). First molecular characterization of *Babesia vogeli* in two naturally infected dogs of Buenos Aires, Argentina. *Vet. Parasitol., 157*, 294-298.

[46] Passos, L. M., Geiger, S. M., Ribeiro, M. F., Pfister, K., & Zahler-Rinder, M. (2005). First molecular detection of *Babesia vogeli* in dogs from Brazil. *Vet. Parasitol., 127*,

81-85.
[47] Loretti, A. P., & Barros, S. S. (2005). Hemorrhagic disease in dogs infected with an unclassified intraendothelial piroplasm in southern Brazil. *Vet. Parasitol., 134*, 193-213.
[48] Trapp, S. M., Dagnone, A. S., Vidotto, O., Freire, R. L., Amude, A. M., & de Morais, H. S. (2006). Seroepidemiology of canine babesiosis and ehrlichiosis in a hospital population. *Vet. Parasitol., 140*, 223-230.
[49] Trapp, S. M., Messick, J. B., Vidotto, O., Jojima, F. S., & de Morais, H. S. (2006). *Babesia gibsoni* genotype Asia in dogs from Brazil. *Vet. Parasitol., 141*, 177-180.
[50] Spolidorio, M. G., Labruna, M. B., Machado, R. Z., Moraes-Filho, J., Zago, A. M., Donatele, D. M., Pinheiro, S. R., Silveira, I., Caliari, K. M., & Yoshinari, N. H. (2010). Survey for tick-borne zoonoses in the state of Espirito Santo, southeastern Brazil. *Am. J. Trop. Med. Hyg., 83*, 201-206.
[51] Maia, M. G., Costa, R. T., Haddad, J. P., Passos, L. M., & Ribeiro, M. F. (2007). Epidemiological aspects of canine babesiosis in the semiarid area of the state of Minas Gerais, Brazil. *Prev. Vet. Med., 79*, 155-162.
[52] Santos, F., Coppede, J. S., Pereira, A. L., Oliveira, L. P., Roberto, P. G., Benedetti, R. B., Zucoloto, L. B., Lucas, F., Sobreira, L., & Marins, M. (2009). Molecular evaluation of the incidence of *Ehrlichia canis*, *Anaplasma platys* and *Babesia* spp. in dogs from Ribeirão Preto, Brazil. *Vet. J., 179*, 145-148.
[53] O'Dwyer, L. H., Lopes, V. V., Rubini, A. S., Paduan Kdos, S., & Ribolla, P. E. (2009). *Babesia* spp. infection in dogs from rural areas of São Paulo State, Brazil. *Rev. Bras. Parasitol. Vet., 18*, 23-26.
[54] Kirchner, M., Brunner, A., Edelhofer, R., & Joachim, A. (2008). Vector-borne parasites of dogs on the Islands of Cabo Verde. *Wien. Klin. Wochenschr., 120*, 49-53.
[55] Konishi, K., Sakata, Y., Miyazaki, N., Jia, H., Goo, Y. K., Xuan, X., & Inokuma, H. (2008). Epidemiological survey of *Babesia gibsoni* infection in dogs in Japan by enzyme-linked immunosorbent assay using *B. gibsoni* thrombospondin-related adhesive protein antigen. *Vet. Parasitol., 155*, 204-208.
[56] Ikadai, H., Tanaka, H., Shibahara, N., Matsuu, A., Uechi, M., Itoh, N., Oshiro, S., Kudo, N., Igarashi, I., & Oyamada, T. (2004). Molecular evidence of infections with *Babesia gibsoni* parasites in Japan and evaluation of the diagnostic potential of a loop-mediated isothermal amplification method. *J. Clin. Microbiol., 42*, 2465-2469.
[57] Ano, H., Makimura, S., & Harasawa, R. (2001). Comparison of partial ribosomal DNA sequences of *Babesia gibsoni* occurring in Miyazaki prefecture, Japan. *J. Vet. Med. Sci., 63*, 561-562.
[58] Inokuma, H., Yamamoto, S., & Morita, C. (1998). Survey of tick-borne diseases in dogs infested with *Rhipicephalus sanguineus* at a kennel in Okayama Prefecture, Japan. *J. Vet. Med. Sci., 60*, 761-763.
[59] Yonamine, H., Ichiki, H., Hamakawa, M., Shimabukuro, T., Sugiyama, M., & Isoda, M. (1984). Studies on canine babesiosis in Okinawa Island. *Nihon. Juigaku. Zasshi., 46*, 511-518.
[60] Inokuma, H., Yoshizaki, Y., Matsumoto, K., Okuda, M., Onishi, T., Nakagome, K., Kosugi, R., & Hirakawa, M. (2004). Molecular survey of *Babesia* infection in dogs in Okinawa, Japan. *Vet. Parasitol., 121*, 341-346.
[61] Verdida, R. A., Hara, O. A., Xuan, X., Fukumoto, S., Igarashi, I., Zhang, S., Dong, J.,

Inokuma, H., Kabeya, H., Sato, Y., Moritomo, T., Maruyama, S., Claveria, F., & Nagasawa, H. (2004). Serodiagnosis of *Babesia gibsoni* infection in dogs by an improved enzyme-linked immunosorbent assay with recombinant truncated P50. *J. Vet. Med. Sci., 66*, 1517-1521.

[62] Miyama, T., Sakata, Y., Ogino, S., Watanabe, M., Itamoto, K., Okuda, M., Verdida, R. A., Xuan, X., Nagasawa, H., & Inokuma, H. (2005). Epidemiological survey of *Babesia gibsoni* Infection in dogs in eastern Japan. *J. Vet. Med. Sci., 67*, 467-471.

[63] Song, K. H., Kim, D. H., & Hayasaki, M. (2004). The PCR-based detection of *Babesia gibsoni* infection in dogs (German shepherds) reared in South Korea. *Ann. Trop. Med. Parasitol., 98*, 149-153.

[64] Goo, Y. K., Jia, H., Terkawi, M. A., Aboge, G. O., Yamagishi, J., Nishikawa, Y., Kim, S., Jang, H. K., Fujisaki, K., & Xuan, X. (2009). *Babesia gibsoni*: identification, expression, localization, and serological characterization of a *Babesia gibsoni* 22-kDa protein. *Exp. Parasitol., 123*, 273-276.

[65] Rajamanickam, C., Wiesenhutter, E., Zin, F. M., & Hamid, J. (1985). The incidence of canine haematozoa in Peninsular Malaysia. *Vet. Parasitol., 17*, 151-157.

[66] Chen, Y. P., & Huang, H. P. (2003). Prevalence and laboratory findings of caninc tick-borne diseases in the National Taiwan University Veterinary Hospital. *Taiwan. Vet. J., 29*, 97-103.

[67] Wu, T. J., Sun, H. J., Wu, Y. C., & Huang, H. P. (2009). Prevalence and risk factors of canine ticks and tick-borne diseases in Taipei, Taiwan. *J. Vet. Clin. Sci., 2*, 75-78.

[68] Lee, C., Hsieh, Y., Huang, C., Tsang, C., & Chung, Y. (2010). Sequence and phylogenetic analysis of the thrombospondin-related adhesive protein (TRAP) gene of *Babesia gibsoni* Isolates from dogs in Taiwan. *J. Vet. Med. Sci., 72*, 1329-1335.

[69] Suksawat, J., Xuejie, Y., Hancock, S. I., Hegarty, B. C., Nilkumhang, P., & Breitschwerdt, E. B. (2001). Serologic and molecular evidence of coinfection with multiple vector-borne pathogens in dogs from Thailand. *J. Vet. Intern. Med., 15*, 453-462.

[70] Muhlnickel, C. J., Hefferies, R., Morgan-ryan, U. M., & Irwin, P. J. (2002). *Babesia gibsoni* infection in three dogs in Victoria. *Aust. Vet. J., 80*, 606-610.

[71] Jefferies, R., Ryan, U. M., Muhlnickel, C. J., & Irwin, P. J. (2003). Two species of canine *Babesia* in Australia: detection and characterization by PCR. *J. Parasitol., 89*, 409-412.

[72] Martin, A. R., Dunstan, R. H., Roberts, T. K., & Brown, G. K. (2006). *Babesia canis vogeli*: a novel PCR for its detection in dogs in Australia. *Exp. Parasitol., 112*, 63-65.

[73] Brown, G. K., Canfield, P. J., Dunstan, R. H., Roberts, T. K., Martin, A. R., Brown, C. S., & Irving, R. (2006). Detection of *Anaplasma platys* and *Babesia canis vogeli* and their impact on platelet numbers in free-roaming dogs associated with remote Aboriginal communities in Australia. *Aust. Vet. J., 84*, 321-325.

[74] Hoff, B., McEwen, B., & Peregrine, A. S. (2008). A survey for infection with *Dirofilaria immitis*, *Ehrlichia canis*, *Borrelia burgdorferi*, and *Babesia canis* in feral and client-owned dogs in the Turks and Caicos Islands, British West Indies. *Can. Vet. J., 49*, 593-594.

[75] Yabsley, M. J., McKibben, J., Macpherson, C. N., Cattan, P. F., Cherry, N. A., Hegarty, B. C., Breitschwerdt, E. B., O'Connor, T., Chandrashekar, R., Paterson, T., Perea, M. L., Ball, G., Friesen, S., Goedde, J., Henderson, B., & Sylvester, W. (2008). Prevalence

of *Ehrlichia canis, Anaplasma platys, Babesia canis vogeli, Hepatozoon canis, Bartonella vinsonii berkhoffii*, and *Rickettsia* spp. in dogs from Grenada. *Vet. Parasitol., 151*, 279-285.

[76] Kelly, P. J., & Lucas, H. (2009). Failure to demonstrate *Babesia, Anaplasma* or *Ehrlichia* in thrombocytopenic dogs from St Kitts. *J. Infect. Dev. Ctries., 3*, 561-563.

[77] Baneth, G., Breitschwerdt, E. B., Hegarty, B. C., Pappalardo, B., & Ryan, J. (1998). A survey of tick-borne bacteria and protozoa in naturally exposed dogs from Israel. *Vet. Parasitol., 74*, 133-142.

[78] Gülanber, A., Gorenflot, A., Schetters, T. P., & Carcy, B. (2006). First molecular diagnosis of *Babesia vogeli* in domestic dogs from Turkey. *Vet. Parasitol., 139*, 224-230.

[79] Rar, V. A., Maksimova, T. G., Zakharenko, L. P., Bolykhina, S. A., Dobrotvorsky, A. K., & Morozova, O. V. (2005). *Babesia* DNA detection in canine blood and *Dermacentor reticulatus* ticks in southwestern Siberia, Russia. *Vector. Borne. Zoonotic. Dis., 5*, 285-287.

[80] Lazri, T., Duscher, G., Edelhofer, R., Bytyci, B., Gjino, P., & Joachim, A. (2008). Infektionen mit arthropodenübertragenen Parasiten bei Hunden im Kosovo und in Albanien unter besonderer Berücksichtigung der Leishmanieninfektionen. *Wien. Klin. Wochenschr., 120*, 54-58.

[81] Heyman, P., Duh, D., Kuylen, V. D. K., Cochez, C., Esbroeck, M. V., Vandenvelde, C., & Avsič-Županc, T. (2007). Molecular and serological evidence for *Anaplasma platys* and *Babesia* sp. Infection in a dog, imported in Belgium, from southern Spain. *J. Vet. Med. A. Physiol. Pathol. Clin. Med., 54*, 276-279.

[82] Van de Maele, I., Savary-Bataille, K., Gielen, I., & Daminet, S. (2008). An unusual form of canine babesiosis. *Can. Vet. J., 49*, 283-286.

[83] Holm, L. P., kerr, M. G., Trees, A. J., McGarry, J. W., Munro, E. R., & Shaw, S. E. (2006). Fatal babesiosis in an untravelled British dog. *Vet. Rec., 159*, 179-180.

[84] Beck, R., Vojta, L., Mrljak, V., Marinculić, A., Beck, A., Živičnjak, T., & Cacciò, S. M. (2009). Diversity of *Babesia* and *Theileria* species in symptomatic and asymptomatic dogs in Croatia. *Int. J. Parasitol., 39*, 843-848.

[85] Martinod, S., Laurent, N., & Moreau, Y. (1986). Resistance and immunity of dogs against *Babesia canis* in an endemic area. *Vet. Parasitol., 19*, 245-254.

[86] Criado-Fornelio, A., Buling, A., Pingret, J. L., Etievant, M., Boucraut-Baralon, C., Alongi, A., Agnone, A., & Torina, A. (2009). Hemoprotozoa of domestic animals in France: prevalence and molecular characterization. *Vet. Parasitol., 159*, 73-76.

[87] Fritz, D. (2010). A PCR study of piroplasms in 166 dogs and 111 horses in France (March 2006 to March 2008). *Parasitol. Res., 106*, 1339-1342.

[88] Martinod, S., & Gilot, B. (1991). Epidemiology of canine babesiosis in relation to the activity of *Dermacentor reticulatus* in southern Jura (France). *Exp. Appl. Acarol., 11*, 215-222.

[89] Hartelt, K., Rieker, T., Oehme, R. M., Brockmann, S. O., Müller, W., & Dorn, N. (2007). First evidence of *Babesia gibsoni* (Asian genotype) in dogs in Western Europe. *Vector. Borne. Zoonotic. Dis., 7*, 163-166.

[90] Farkas, R., Földvári, G., Fenyves, B., Kótai, I., A., S., & Hegedüs, G. Y. (2004). First detection of small babesiae in two dogs in Hungary. *Vet. Rec., 154*, 176-178.

[91] Földvári, G., Hell, É., & Farkas, R. (2005). *Babesia canis canis* in dogs from Hungary:

detection by PCR and sequencing. *Vet. Parasitol., 127*, 221-226.

[92] Hornok, S., Edelhofer, R., & Farkas, R. (2006). Seroprevalence of canine babesiosis in Hungary suggesting breed predisposition. *Parasitol. Res., 99*, 638-642.

[93] Solano-Gallego, L., Trotta, M., Carli, E., Carcy, B., Caldin, M., & Furlanello, T. (2008). *Babesia canis canis* and *Babesia canis vogeli* clinicopathological findings and DNA detection by means of PCR-RFLP in blood from Italian dogs suspected of tick-borne disease. *Vet. Parasitol., 157*, 211-221.

[94] Cassini, R., Zanutto, S., Frangipane di Regalbono, A., Gabrielli, S., Calderini, P., Moretti, A., Tampieri, M. P., & Pietrobelli, M. (2009). Canine piroplasmosis in Italy: epidemiological aspects in vertebrate and invertebrate hosts. *Vet. Parasitol., 165*, 30-35.

[95] Otranto, D., Testini, G., Dantas-Torres, F., Latrofa, M. S., Diniz, P. P., de Caprariis, D., Lia, R. P., Mencke, N., Stanneck, D., Capelli, G., & Breitschwerdt, E. B. (2010). Diagnosis of canine vector-borne diseases in young dogs: a longitudinal study. *J. Clin. Microbiol., 48*, 3316-3324.

[96] Matjila, T. P., Nijhof, A. M., Taoufik, A., Houwers, D., Teske, E., Penzhorn, B. L., de Lange, T., & Jongejan, F. (2005). Autochthonous canine babesiosis in The Netherlands. *Vet. Parasitol., 131*, 23-29.

[97] Øines, Ø., Storli, K., & Brun-Hansen, H. (2010). First case of babesiosis caused by *Babesia canis canis* in a dog from Norway. *Vet. Parasitol., 171*, 350-353.

[98] Cardoso, L., Costa, Á., Tuna, J., Vieira, L., Eyal, O., Yisaschar-Mekuzas, Y., & Baneth, G. (2008). *Babesia canis canis* and *Babesia canis vogeli* infections in dogs from northern Portugal. *Vet. Parasitol., 156*, 199-204.

[99] Adaszek, L., & Winiarczyk, S. (2008). Molecular characterization of *Babesia canis canis* isolates from naturally infected dogs in Poland. *Vet. Parasitol., 152*, 235-241.

[100] Zygner, W., Górski, P., & Wędrychowicz, H. (2009). Detection of the DNA of B*orrelia afzelii*, *Anaplasma phagocytophilum* and *Babesia canis* in blood samples from dogs in Warsaw. *Vet. Rec., 164*, 465-467.

[101] Welc-Faleciak, R., Rodo, A., Sinski, E., & Bajer, A. (2009). *Babesia canis* and other tick-borne infections in dogs in Central Poland. *Vet. Parasitol., 166*, 191-198.

[102] Chandoga, P., Goldová, M., Baranová, D., & Kozák, M. (2002). First cases of canine babesiosis in the Slovak Republic. *Vet. Rec., 150*, 82-84.

[103] Duh, D., Tozon, N., Petrovec, M., Strašek, K., & Avšič-Županc, T. (2004). Canine babesiosis in Slovenia: molecular evidence of *Babesia canis canis* and *Babesia canis vogeli*. *Vet. Res., 35*, 363-368.

[104] Criado-Fornelio, A., Martinez-Marcos, A., Buling-Saraña, A., & Barba-Carretero, J. C. (2003). Molecular studies on *Babesia*, *Theileria* and *Hepatozoon* in southern Europe Part I. Epizootiological aspects. *Vet. Parasitol., 113*, 189-201.

[105] García, A. T. (2006). Piroplasma infection in dogs in northern Spain. *Vet. Parasitol., 138*, 97-102.

[106] Götsch, S., Leschnik, M., Duscher, G., Burgstaller, J. P., Wille-Piazzai, W., & Joachim, A. (2009). Ticks and haemoparasites of dogs from Praia, Cape Verde. *Vet. Parasitol., 166*, 171-174.

[107] Cacció, S. M., Antunovic, B., Moretti, A., Mangili, V., Marinculic, A., Baric, R. R., Slemenda, S. B., & Pieniazek, N. J. (2002). Molecular characterisation of *Babesia canis canis* and *Babesia canis vogeli* from naturally infected European dogs. *Vet. Parasitol., 106*, 285-292.

[108] Leeflang, P. (1977). Tick-borne diseases of domestic animals in northern Nigeria. I. Historical review. *Trop. Anim. Health. Prod., 9*, 147-152.

[109] Uilenberg, G., Franssen, F. F., Perié, N. M., & Spanjer, A. A. (1989). Three groups of *Babesia canis* distinguished and a proposal for nomenclature. *Vet. Q., 11*, 33-40.

[110] Reyers, F., Leisewitz, A. L., Lobetti, R. G., Milner, R. J., Jacobson, L. S., & van Zyl, M. (1998). Canine babesiosis in South Africa: more than one disease. Does this serve as a model for falciparum malaria? *Ann. Trop. Med. Parasitol., 92*, 503-511.

[111] Depoix, D., Carcy, B., Jumas-Bilak, E., Pages, M., Precigout, E., Schetters, T. P. M., Ravel, C., & Gorenflot, A. (2002). Chromosome number, genome size and polymorphism of European and south African isolates of large *Babesia* parasites that infect dogs. *Parasitology., 125*, 313-321.

[112] Jacobson, L. S. (2006). The South African form of severe and complicated canine babesiosis: clinical advances 1994-2004. *Vet. Parasitol., 138*, 126-139.

[113] Kjemtrup, A. M., Wainwright, K., Miller, M., Penzhorn, B. L., & Carreno, R. A. (2006). *Babesia conradae*, sp. Nov., a small canine *Babesia* identified in California. *Vet. Parasitol., 138*, 103-111.

[114] Shen, Y., Gao, J., Xu, K., Xue, L., Zhang, Y., Shi, B., Li, D., Wei, X., & Higuchi, S. (1997). Babesiasis in Nanjing area, China. *Trop. Anim. Health. Prod., 29*, 19S-22S.

[115] Matsuu, A., Koshida, Y., Kawahara, M., Inoue, K., Ikadai, H., Hikasa, Y., Okano, S., & Higuchi, S. (2004). Efficacy of atovaquone against *Babesia gibsoni* in vivo and *in vitro*. *Vet. Parasitol., 124*, 9-18.

[116] Ano, H., Makimura, S., & Harasawa, R. (2001). Comparison of partial ribosomal DNA sequences of *Babesia gibsoni* occurring in Miyazaki *J. Vet. Med. Sci., 63*, 561-562.

[117] Martinod, S., Brossard, M., & Moreau, Y. (1985). Immunity of dogs against *Babesia canis*, its vector tick *Dermacentor reticulatus*, and *Ixodes ricinus* in endemic area. *J. Parasitol., 71*, 269-273.

[118] Dautel, H., Dippel, C., Oehme, R., Hartelt, K., & Schettler, E. (2006). Evidence for an increased geographical distribution of *Dermacentor reticulatus* in Germany and detection of *Rickettsia* sp. RpA4. *Int. J. Med. Microbiol., 296 Suppl 40*, 149-156.

[119] Majláthová, V., Hurníkova, Z., Majláth, I., & Peťko, B. (2007). *Hepatozoon canis* infection in Slovakia: imported or autochthonous? *Vector. Borne. Zoonotic. Dis., 7*, 199-202.

[120] Criado-Fornelio, A., Gónzalez-del-Río, M. A., Buling-Saraña, A., & Barba-Carretero, J. C. (2003). Molecular characterization of a *Babesia gibsoni* isolate from a Spanish dog. *Vet. Parasitol., 117*, 123-129.

[121] Furlanello, T., Fiorio, F., Caldin, M., Lubas, G., & Solano-Gallego, L. (2005). Clinicopathological findings in naturally occurring cases of babesiosis caused by large form *Babesia* from dogs of northeastern Italy. *Vet. Parasitol., 134*, 77-85.

[122] Guitian, F. J., Camacho, A. T., & Telford, S. R. (2003). Case-control study of canine infection by a newly recognised *Babesia microti*-like piroplasm. *Prev. Vet. Med., 61*, 137-145.

[123] Parola, P., Socolovschi, C., Jeanjean, L., Bitam, I., Fournier, P. E., Sotto, A., Labauge, P., & Raoult, D. (2008). Warmer weather linked to tick attack and emergence of severe rickettsioses. *PLoS. Negl. Trop. Dis., 2*, e338.

[124] Máthé, Á., Vörös, K., Papp, L., & Reiczigel, J. (2006). Clinical manifestations of canine babesiosis in Hungary (63 cases). *Acta. Vet. Hung., 54*, 367-385.

[125] Matijatko, V., Kiš, I., Torti, M., Brkljačić, M., Kučer, N., Rafaj, R. B., Grden, D., Živičnjak, T., & Mrljak, V. (2009). Septic shock in canine babesiosis. *Vet. Parasitol., 162*, 263-270.

[126] Freeman, M. J., Kirby, B. M., Panciera, D. L., Henik, R. A., Rosin, E., & Sullivan, L. J. (1994). Hypotensive shock syndrome associated with acute *Babesia canis* infection in a dog. *J. Am. Vet. Med. Assoc., 204*, 94-96.

[127] Wozniak, E. J., Barr, B. C., Thomford, J. W., Yamane, I., McDonough, S. P., Moore, P. F., Naydan, D., Robinson, T. W., & Conrad, P. A. (1997). Clinical, anatomic, and immunopathologic characterization of *Babesia gibsoni* infection in the domestic dog (*Canis familiaris*). *J. Parasitol., 83*, 692-699.

[128] Onishi, T., Ueda, K., Horie, M., Kajikawa, T., & Ohishi, I. (1990). Serum hemolytic activity in dogs infected with *Babesia gibsoni. J. Parasitol.*, 76, 564-567.

[129] Gneno, R., Azzar, G., Got, R., & Roux, B. (1986). Permeability of membrane of *Babesia canis* infected erythrocytes-influence of an external electric field. *Int. J. Biochem., 18*, 1151-1154.

[130] Adachi, K., Yoshimoto, A., Hasegawa, T., Shimizu, T., Goto, Y., & Makimura, S. (1992). Anti-erythrocyte membrane antibodies detected in sera of dogs naturally infected with *Babesia gibsoni. J. Vet. Med. Sci., 54*, 1081-1084.

[131] Morita, T., Saeki, H., Imai, S., & Ishii, T. (1995). Reactivity of anti-erythrocyte antibody induced by *Babesia gibsoni* infection against aged erythrocytes. *Vet. Parasitol., 58*, 291-299.

[132] Murase, T., & Maede, Y. (1990). Increased erythrophagocytic activity of macrophages in dogs with *Babesia gibsoni* infection. *Jap. J. Vet. Sci., 52*, 321-327.

[133] Otsuka, Y., Yamasaki, M., Yamato, O., & Maede, Y. (2002). The effect of macrophages on the erythrocyte oxidative damage and the pathogenesis of anemia in *Babesia gibsoni*-infected dogs with low parasitemia. *J. Vet. Med. Sci., 64*, 221-226.

[134] Chaudhuri, S., Varshney, J. P., & Patra, R. C. (2008). Erythrocytic antioxidant defense, lipid peroxides level and blood iron, zinc and copper concentrations in dogs naturally infected with *Babesia gibsoni. Res. Vet. Sci., 85*, 120 124.

[135] Otsuka, Y., Yamasaki, M., Yamato, O., & Maede, Y. (2001). Increased generation of superoxide in erythrocytes infected with *Babesia gibsoni. J. Vet. Med. Sci., 63*, 1077-1081.

[136] Crnogaj, M., Petlevski, R., Mrljak, V., Kis, I., Torti, M., Kucer, N., Matijatko, V., Sacer, I., & Stokovic, I. (2010). Malondialdehyde levels in serum of dogs infected with *Babesia canis. Vet. Med-Czech., 55*, 163-171.

[137] Zygner, W., Gójska, O., Rapacka, G., Jaros, D., & Wędrychowicz, H. (2007). Hematological changes during the course of canine babesiosis caused by large *Babesia* in domestic dogs in Warsaw (Poland). *Vet. Parasitol., 145*, 146-151.

[138] Hossain, M. A., Yamato, O., Yamasaki, M., Jeong, J. R., Chang, H. S., & Maede, Y. (2003). Serum from dogs infected with *Babesia gibsoni* inhibits maturation of reticulocytes and erythrocyte 5'nucleotidase activity *in vitro. J. Vet. Med. Sci., 65*, 1281-1286.

[139] Yamasaki, M., Yamato, O., Hossain, M. A., Jeong, J. R., Chang, H. S., Satoh, H., & Maede, Y. (2002). *Babesia gibsoni*: preferential multiplication in reticulocytes is related to the presence of mitochondria and a high concentration of adenosine 5′-triphosphate in the cells. *Exp. Parasitol., 102*, 164-169.

[140] Yamasaki, M., Hossain, M. A., Jeong, J. R., Chang, H. S., Satoh, H., Yamato, O., & Maede, Y. (2003). *Babesia gibsoni*-specific isoenzymes related to energy metabolism of the parasite in infected erythrocytes. *J. Parasitol., 89*, 1142-1146.

[141] Hossain, M. A., Yamato, O., Kim, G., Yamasaki, M., & Maede, Y. (2007). Suppressive effect of culture supernatant of erythrocytes and serum from dogs infected with *Babesia gibsoni* on the morphological maturation of canine reticulocytes *in vitro*. *J. Vet. Sci., 8*, 169-174.

[142] Kettner, F., Reyers, F., & Miller, D. (2003). Thrombocytopenia in canine babesiosis and its clinical usefulness. *J. S. Afr. Vet. Assoc., 74*, 63-68.

[143] Ulutas, B., Bayramli, G., Ulutas, P. A., & Karagenc, T. (2005). Serum concentration of some acute phase proteins in naturally occurring canine babesiosis: a preliminary study. *Vet. Clin. Pathol., 34*, 144-147.

[144] Schetters, T. (2005). Vaccination against canine babesiosis. *Trends. Parasitol., 21*, 179-184.

[145] Ruiz de Gopegui, R., Peñalba, B., Goicoa, A., Espada, Y., Fidalgo, L. E., & Espino, L. (2007). Clinico-pathological findings and coagulation disorders in 45 cases of canine babesiosis in Spain. *Vet. J., 174*, 129-132.

[146] Schetters, T. P., Kleuskens, J., Scholtes, N., & Gorenflot, A. (1998). Parasite localization and dissemination in the *Babesia*-infected host. *Ann. Trop. Med. Parasitol., 92*, 513-539.

[147] Schetters, T. P., Kleuskens, J. A., Van De Crommert, J., De Leeuw, P. W., Finizio, A. L., & Gorenflot, A. (2009). Systemic inflammatory responses in dogs experimentally infected with *Babesia canis*; a haematological study. *Vet. Parasitol., 162*, 7-15.

[148] Irwin, P. J., & Hutchinson, G. W. (1991). Clinical and pathological findings of *Babesia* infection in dogs. *Aust. Vet. J., 68*, 204-209.

[149] Kawamura, M., Maede, Y., & Namioka, S. (1987). Mitogenic responsibilities of lymphocytes in canine babesiosis and the effects of splenectomy on it. *Jpn. J. Vet. Res., 35*, 1-10.

[150] Kjemtrup, A. M., & Conrad, P. A. (2006). A review of the small canine piroplasms from California: *Babesia conradae* in the literature. *Vet. Parasitol., 138*, 112-117.

[151] Camacho, A. T., Guitian, F. J., Pallas, E., Gestal, J. J., Olmeda, A. S., Goethert, H. K., Telford, S. R., III, & Spielman, A. (2004). Azotemia and mortality among *Babesia microti*-like infected dogs. *Vet. Intern. Med., 18*, 141-146.

[152] Birkenheuer, A. J., Neel, J., Ruslander, D., Levy, M. G., & Breitschwerdt, E. B. (2004). Detection and molecular characterization of a novel large *Babesia* species in a dog. *Vet. Parasitol., 124*, 151-160.

[153] Namikawa, K., Sunaga, F., & Kanno, Y. (1988). Morphology of *Babesia gibsoni* in canine erythrocytes. *Nihon. Juigaku. Zasshi., 50*, 936-938.

[154] Walter, S., Mehlhorn, H., Zweygarth, E., & Schein, E. (2002). Electron microscopic investigations on stages of dog piroplasms cultured *in vitro*: Asian isolates of *Babesia gibsoni* and strains of *B. canis* from France and Hungary. *Parasitol. Res., 88*, 32-37.

[155] Trotta, M., Carli, E., Novari, G., Furlanello, T., & Solano-Gallego, L. (2009). Clinicopathological findings, molecular detection and characterization of *Babesia gibsoni* infection in a sick dog from Italy. *Vet. Parasitol., 165*, 318-322.

[156] Kjemtrup, A. M., Kocan, A. A., Whitworth, L., Meinkoth, J., Birkenheuer, A. J., Cummings, J., Boudreaux, M. K., Stockham, S. L., Irizarry-Rovira, A., & Conrad, P. A.

(2000). There are at least three genetically distinct small piroplasms from dogs. *Int. J. Parasitol., 30*, 1501-1505.

[157] Casapulla, R., Baldi, L., Avallone, V., Sannino, R., Pazzanese, L., & Mizzoni, V. (1998). Canine piroplasmosis due to *Babesia gibsoni*: clinical and morphological aspects. *Vet. Rec., 142*, 168-169.

[158] Inokuma, H., Okuda, M., Yoshizaki, Y., Hiraoka, H., MIyama, T., Itamoto, K., Une, S., Nakaichi, M., & Taura, Y. (2005). Clinical observations of *Babesia gibsoni* infection with low parasitaemia confirmed by PCR in dogs. *Vet. Rec., 156*, 116-118.

[159] Lobetti, R., Dvir, E., & Pearson, J. (2002). Cardiac troponins in canine babesiosis. *J. Vet. Intern. Med., 16*, 63-68.

[160] Sweers, L., Kirberger, R. M., Leisewitz, A. L., Dormehl, I. C., Killian, E., & Naude, F. (2008). The scintigraphic evaluation of the pulmonary perfusion pattern of dogs hospitalised with babesiosis. *J. S. Afr. Vet. Assoc., 79*, 76-83.

[161] Fraga, E., Barreiro, J. D., Goicoa, A., Espino, L., Fraga, G., & Barreiro, A. (2011). Abdominal ultrasonographic findings in dogs naturally infected with babesiosis. *Vet. Radiol. Ultrasound., 52*, 323-329.

[162] Meinkoth, J. H., Kocan, A. A., Loud, S. D., & Lorenz, M. D. (2002). Clinical and hematologic effects of experimental infection of dogs with recently identified *Babesia gibsoni*-like isolates from Oklahoma. *J. Am. Vet. Med. Assoc., 220*, 185-189.

[163] Pardini, A. D. (2000). The pathology and pathogenesis of canine cerebral babesiosis (dissertation). Faculty of Veterinary Science, University of Pretoria. *Onderstepoort.*, 1-64.

[164] Keller, N., Jacobson, L. S., Nel, M., de Clerq, M., Thompson, P. N., & Schoeman, J. P. (2004). Prevalence and risk factors of hypoglycemia in virulent canine babesiosis. *J. Vet. Intern. Med., 18*, 265-270.

[165] Welzl, C., Leisewitz, A. L., Jacobson, L. S., Vaughan-Scott, T., & Myburgh, E. (2001). Systemic inflammatory response syndrome and multiple-organ damage/dysfunction in complicated canine babesiosis. *J. S. Afr. Vet. Assoc., 72*, 158-162.

[166] Tuttle, A. D., Birkenheuer, A. J., Juopperi, T., Levy, M. G., & Breitschwerdt, E. B. (2003). Concurrent bartonellosis and babesiosis in a dog with persistent thrombocytopenia. *J. Am. Vet. Med. Assoc., 223*, 1306-1310.

[167] Bourdoiseau, G. (2006). Canine babesiosis in France. *Vet. Parasitol., 138*, 118-125.

[168] Camacho, A. T., Guitian, F. J., Pallas, E., Gestal, J. J., Olmeda, S., Goethert, H., Telford, S., III, & Spielman, A. (2005). Serum protein response and renal failure in canine *Babesia annae* infection. *Vet. Res., 36*, 713-722.

[169] Köster, L. S., Van Schoor, M., Goddard, A., Thompson, P. N., Matjila, P. T., & Kjelgaard-Hansen, M. (2009). C-reactive protein in canine babesiosis caused by *Babesia rossi* and its association with outcome. *J. S. Afr. Vet. Assoc., 80*, 87-91.

[170] Matijatko, V., Mrljak, V., Kiš, I., Kučer, N., Foršek, J., Živičnjak, T., Romić, Ž., Šimec, Z., & Ceron, J. J. (2007). Evidence of an acute phase response in dogs naturally infected with *Babesia canis*. *Vet. Parasitol., 144*, 242-250.

[171] Jacobson, L. S., & Lobetti, R. G. (2005). Glucose, lactate, and pyruvate concentrations in dogs with babesiosis. *Am. J. Vet. Res., 66*, 244-250.

[172] Rees, P., & Schoeman, J. P. (2008). Plasma insulin concentrations in hypoglycaemic dogs with *Babesia canis rossi* infection. *Vet. Parasitol., 152*, 60-66.

[173] Leisewitz, A. L., Jacobson, L. S., de Morais, H. S., & Reyers, F. (2001). The mixed

acid-base disturbances of severe canine babesiosis. *J. Vet. Intern. Med., 15*, 445-452.

[174] Vercammen, F., De Deken, R., & Maes, L. (1997). Duration of protective immunity in experimental canine babesiosis after homologous and heterologous challenge. *Vet. Parasitol., 68*, 51-55.

[175] Schoeman, J. P., & Herrtage, M. E. (2007). The response of the pituitary-adrenal and pituitary-thyroidal axes to the plasma glucose perturbations in *Babesia canis rossi* babesiosis. *J. S. Afr. Vet. Assoc., 78*, 215-220.

[176] Schoeman, J. P., & Herrtage, M. E. (2008). Adrenal response to the low dose ACTH stimulation test and the cortisol-to-adrenocorticotrophic hormone ratio in canine babesiosis. *Vet. Parasitol., 154*, 205-213.

[177] Nel, M., Lobetti, R. G., Keller, N., & Thompson, P. N. (2004). Prognostic value of blood lactate, blood glucose, and hematocrit in canine babesiosis. *J. Vet. Intern. Med., 18*, 471-476.

[178] Schoeman, J. P., Rees, P., & Herrtage, M. E. (2007). Endocrine predictors of mortality in canine babesiosis caused by *Babesia canis rossi*. *Vet. Parasitol., 148*, 75-82.

[179] Suzuki, K., Wakabayashi, H., Takahashi, M., Fukushima, K., Yabuki, A., & Endo, Y. (2007). A Possible treatment strategy and clinical factors to estimate the treatment response in *Bebesia gibsoni* infection. *J. Vet. Med. Sci., 69*, 563-568.

[180] Camacho, A. T. (2005). Do eosinophils have a role in the severity of *Babesia annae* infection? *Vet. Parasitol., 134*, 281-282.

[181] Sikorski, L. E., Birkenheuer, A. J., Holowaychuk, M. K., McCleary-Wheeler, A. L., Davis, J. M., & Littman, M. P. (2010). Babesiosis caused by a large *Babesia* species in 7 immunocompromised dogs. *J. Vet. Intern. Med., 24*, 127-131.

[182] Jefferies, R., Ryan, U. M., Jardine, J., Robertson, I. D., & Irwin, P. J. (2007). *Babesia gibsoni*: detection during experimental infections and after combined atovaquone and azithromycin therapy. *Exp. Parasitol., 117*, 115-123.

[183] Comazzi, S., Paltrinieri, S., Manfredi, M. T., & Agnes, F. (1999). Diagnosis of canine babesiosis by Percoll gradient separation of parasitized erythrocytes. *J. Vet. Diagn. Invest., 11*, 102-104.

[184] Mimori, T., Kono, I., Sakamoto, T., & Yasuda, N. (1982). Morphological studies of multiplication of *Babesia gibsoni* in canine erythrocytes. *Nihon. Juigaku. Zasshi., 44*, 699-701.

[185] Fukumoto, S., Xuan, X., Shigeno, S., Kimbita, E., Igarashi, I., Nagasawa, H., Fujisaki, K., & Mikami, T. (2001). Development of a polymerase chain reaction method for diagnosing *Babesia gibsoni* infection in dogs. *J. Vet. Med. Sci., 63*, 977-981.

[186] Bicalho, K. A., Ribeiro, M. F., & Martins-Filho, O. A. (2004). Molecular fluorescent approach to assessing intraerythrocytic hemoprotozoan *Babesia canis* infection in dogs. *Vet. Parasitol., 125*, 221-235.

[187] Yamasaki, M., Hwang, S. J., Ohta, H., Yamato, O., Maede, Y., & Takiguchi, M. (2008). Flow cytometry to evaluate the level of *Babesia gibsoni* parasitemia in vivo and *in vitro* by using the fluorescent nucleic acid stain SYTO16. *Jpn. J. Vet. Res., 55*, 129-136.

[188] Hauschild, S., Shayan, P., & Schein, E. (1995). Characterization and comparison of merozoite antigens of different *Babesia canis* isolates by serological and immunological investigations. *Parasitol. Res., 81*, 638-642.

[189] Ludford, C. G. (1969). Fluorescent antibody staining of four *Babesia* species. *Exp.*

Parasitol., 24, 327-335.

[190] Wanduragala, L., Kakoma, I., Clabaugh, G. W., Abeygunawardena, I., Levy, M. G., & Ristic, M. (1987). Development of Dot-enzyme immunoassay for diagnosis of canine babesiosis. *Am. J. Trop. Med. Hyg., 36*, 20-21.

[191] Dell'Porto, A., Oliveira, M. R., & Miguel, O. (1993). *Babesia canis* in stray dogs from the city of São Paulo. Comparative studies between the clinical and hematological aspects and the indirect fluresecent antibody test. *Rev. Bras. Parasitol. Vet., 2*, 37-40.

[192] Furuta, P. I., Oliveira, T. M., Theixeira, M. C., Rocha, A. G., Machado, R. Z., & Tinucci-Costa, M. G. (2009). Comparison between a soluble antigen-based ELISA and IFAT in detecting antibodies against *Babesia canis* in dogs. *Rev. Bras. Parasitol. Vet., 18*, 41-45.

[193] Hirose, Y., Matsuhira, S., Batanova, T. A., Goo, Y. K., Terkawi, M. A., Nishikawa, Y., Kitoh, K., Xuan, X., & Takashima, Y. (2009). Detection of *Babesia gibsoni* by reaction of phage display single chain antibodies with P50 proteins. *J. Vet. Med. Sci., 71*, 1133-1136.

[194] Fukumoto, S., Xuan, X., Nishikawa, Y., Inoue, N., Igarashi, I., Nagasawa, H., Fujisaki, K., & Mikami, T. (2001). Identification and expression of a 50-kilodalton surface antigen of *Babesia gibsoni* and evaluation of its diagnostic potential in an enzyme-linked immunosorbent assay. *J. Clin. Microbiol., 39*, 2603-2609.

[195] Fukumoto, S., Sekine, Y., Xuan, X., Igarashi, I., Sugimoto, C., Nagasawa, H., Fujisaki, K., Mikami, T., & Suzuki, H. (2004). Serodiagnosis of canine *Babesia gibsoni* infection by enzyme-linked immunosorbent assay with recombinant P50 expressed in *Escherichia coli*. *J. Parasitol., 90*, 387-391.

[196] Fukumoto, S., Xuan, X., Kadota, K., Igarashi, I., Sugimoto, C., Fujisaki, K., Nagasawa, H., Mikami, T., & Suzuki, H. (2003). High-level expression of truncated surface antigen P50 of *Babesia gibsoni* in insect cells by baculovirus and evaluation of its immunogenicity and antigenicity. *Clin. Diagn. Lab. Immunol., 10*, 596-601.

[197] Miyama, T., Inokuma, H., Itamoto, K., Okuda, M., Verdida, R. A., & Xuan, X. (2006). Clinical Usefulness of Antibodies against *Babesia gibsoni* Detected by ELISA with Recombinant P50. *J. Vet. Med. Sci., 68*, 1371-1371.

[198] Fukumoto, S., Xuan, X., Inoue, N., Igarashi, I., Sugimoto, C., Fujisaki, K., Nagasawa, H., Mikami, T., & Suzuki, H. (2003). Molecular characterization of a gene encoding a 29-kDa cytoplasmic protein of *Babesia gibsoni* and evaluation of its diagnostic potentiality. *Mol. Biochem. Parasitol., 131*, 129-136.

[199] Jia, H., Zhou, J., Ikadai, H., Matsuu, A., Suzuki, H., Igarashi, I., Fujisaki, K., & Xuan, X. (2006). Identification of a novel gene encoding a secreted antigen 1 of *Babesia gibsoni* and evaluation of its use in serodiagnosis. *Am. J. Trop. Med. Hyg., 75*, 843-850.

[200] Jia, H., Terkawi, M. A., Aboge, G. O., Goo, Y. K., Ma, L., Zhou, J., Nishikawa, Y., Igarashi, I., Fujisaki, K., & Xuan, X. (2009). Identification of secreted antigen 3 from *Babesia gibsoni*. *Clin. Vaccine. Immunol., 16*, 944-948.

[201] Zhou, J., Jia, H., Nishikawa, Y., Fujisaki, K., & Xuan, X. (2007). *Babesia gibsoni* rhoptry-associated protein 1 and its potential use as a diagnostic antigen. *Vet. Parasitol., 145*, 16-20.

[202] Terkawi, M. A., Amornthep, A., Ooka, H., Aboge, G., Jia, H., Goo, Y. K., Nelson, B., Yamagishi, J., Nishikawa, Y., Igarashi, I., Kawazu, S. I., Fujisaki, K., & Xuan, X. (2009). Molecular characterizations of three distinct *Babesia gibsoni* rhoptry-associated

protein-1s (RAP-1s). *Parasitology., 136*, 1147-1160.

[203] Aboge, G. O., Jia, H., Terkawi, M. A., Goo, Y., Kuriki, K., Nishikawa, Y., Igarashi, I., Suzuki, H., & Xuan, X. (2007). A novel 57-kDa merozoite protein of *Babesia gibsoni* is a prospective antigen for diagnosis and serosurvey of canine babesiosis by enzyme-linked immunosorbent assay. *Vet. Parasitol., 149*, 85-94.

[204] Aboge, G. O., Jia, H., Kuriki, K., Zhou, J., Nishikawa, Y., Igarashi, I., Fujisaki, K., Suzuki, H., & Xuan, X. (2007). Molecular characterization of a novel 32-kDa merozoite antigen of *Babesia gibsoni* with a better diagnostic performance by enzyme-linked immunosorbent assay. *Parasitology., 134*, 1185-1194.

[205] Goo, Y. K., Jia, H., Aboge, G. O., Terkawi, M. A., Lee, E. G., Yamagishi, J., Nishikawa, Y., Jang, H. K., Sunaga, F., Namikawa, K., Fujisaki, K., & Xuan, X. (2009). Characterization of the *Babesia gibsoni* 12-kDa protein as a potential antigen for the serodiagnosis. *Parasitol. Int., 58*, 55-60.

[206] Goo, Y. K., Jia, H., Aboge, G. O., Terkawi, M. A., Kuriki, K., Nakamura, C., Kumagai, A., Zhou, J., Lee, E. G., Nishikawa, Y., Igarashi, I., Fujisaki, K., & Xuan, X. (2008). *Babesia gibsoni*: Serodiagnosis of infection in dogs by an enzyme-linked immunosorbent assay with recombinant BgTRAP. *Exp. Parasitol., 118*, 555-560.

[207] Verdida, R. A., Xuan, X., Fukumoto, S., Huang, X., Zhou, J., Igarashi, I., Claveria, F. G., & Nagasawa, H. (2005). Development of a practical immunochromatographic test with recombinant P50 for the diagnosis of *Babesia gibsoni* infection in dogs. *Parasitology., 131*, 769-774.

[208] Jia, H., Liao, M., Lee, E., Nishikawa, Y., Inokuma, H., Ikadai, H., Matsuu, A., Igarashi, I., & Xuan, X. (2007). Development of an immunochromatographic test with recombinant BgSA1 for the diagnosis of *Babesia gibsoni* infection in dogs. *Parasitol. Res., 100*, 1381-1384.

[209] Ano, H., Makimura, S., & Harasawa, R. (2001). Detection of *Babesia* species from infected dog blood by polymerase chain reaction. *J. Vet. Med. Sci., 63*, 111-113.

[210] Tani, H., Tada, Y., Sasai, K., & Baba, E. (2008). Improvement of DNA extraction method for dried blood spots and comparison of four PCR methods for detection of *Babesia gibsoni* (Asian genotype) infection in canine blood samples. *J. Vet. Med. Sci., 70*, 461-467.

[211] Yamasaki, M., Inokuma, H., Sugimoto, C., Shaw, S. E., Aktas, M., Yabsley, M. J., Yamato, O., & Maede, Y. (2007). Comparison and phylogenetic analysis of the heat shock protein 70 gene of *Babesia* parasites from dogs. *Vet. Parasitol., 145*, 217-227.

[212] Yamasaki, M., Tajima, M., Lee, K. W., Jeong, J. R., Yamato, O., & Maede, Y. (2002). Molecular cloning and phylogenetic analysis of *Babesia gibsoni* heat shock protein 70. *Vet Parasitol., 110*, 123-129.

[213] Jia, H., Terkawi, M. A., Aboge, G. O., Goo, Y. K., Zhou, J., Lee, E. G., Nishikawa, Y., Igarashi, I., Fujisaki, K., & Xuan, X. (2008). *Babesia gibsoni*: identification of an immunodominant, interspersed repeat antigen. *Exp. Parasitol., 118*, 146-149.

[214] Carret, C., Walas, F., Carcy, B., Grande, N., Précigout, É., Moubri, K., Schetters, T. P., & Gorenflot, A. (1999). *Babesia canis canis*, *Babesia canis vogeli*, *Babesia canis rossi*: differentiation of the three subspecies by a restriction fragment length polymorphism analysis on amplified small subunit ribosomal RNA genes. *J. Eukaryot. Microbiol., 46*, 298-303.

[215] Birkenheuer, A. J., Levy, M. G., & Breitschwerdt, E. B. (2003). Development and

evaluation of a seminested PCR for detection and differentiation of *Babesia gibsoni* (Asian genotype) and *B. canis* DNA in canine blood samples. *J. Clin. Microbiol., 41*, 4172-4177.

[216] Jefferies, R., Ryan, U. M., & Irwin, P. J. (2007). PCR-RFLP for the detection and differentiation of the canine piroplasm species and its use with filter paper-based technologies. *Vet. Parasitol., 144*, 20-27.

[217] Müller, H., Aysul, N., Liu, Z., Salih, D. A., Karagenc, T., Beyer, D., Kullmann, B., Ahmed, J. S., & Seitzer, U. (2010). Development of a loop-mediated isothermal amplification (LAMP) assay for rapid diagnosis of *Babesia canis* infections. *Transbound. Emerg. Dis., 57*, 63-65.

[218] Matsuu, A., Ono, S., Ikadai, H., Uchide, T., Imamura, S., Onuma, M., Okano, S., & Higuchi, S. (2005). Development of a SYBR green real-time polymerase chain reaction assay for quantitative detection of *Babesia gibsoni* (Asian genotype) DNA. *J. Vet. Diagn. Invest., 17*, 569-573.

[219] Adaszek, L., & Winiarczyk, S. (2010). Application of the SYBR Green real-time HRM PCR technique in the differentiation of the *Babesia canis canis* protozoa isolated in the areas of eastern Poland. *Parasitol. Res., 106*, 1253-1256.

[220] Peleg, O., Baneth, G., Eyal, O., Inbar, J., & Harrus, S. (2010). Multiplex real-time qPCR for the detection of *Ehrlichia canis* and *Babesia canis vogeli*. *Vet. Parasitol., 173*, 292-299.

[221] Vercammen, F., De Deken, R., & Maes, L. (1996). Prophylactic treatment of experimental canine babesiosis (*Babesia canis*) with doxycycline. *Vet. Parasitol., 66*, 251-255.

[222] Wulansari, R., Wijaya, A., Ano, H., Horii, Y., Nasu, T., Yamane, S., & Makimura, S. (2003). Clindamycin in the treatment of *Babesia gibsoni* infections in dogs. *J. Am. Anim. Hosp. Assoc., 39*, 558-562.

[223] Birkenheuer, A. J., Levy, M. G., & Breitschwerdt, E. B. (2004). Efficacy of combined atovaquone and azithromycin for therapy of chronic *Babesia gibsoni* (Asian genotype) infections in dogs. *J. Vet. Intern. Med., 18*, 494-498.

[224] Sakuma, M., Setoguchi, A., & Endo, Y. (2009). Possible emergence of drug-resistant variants of *Babesia gibsoni* in clinical cases treated with atovaquone and azithromycin. *J. Vet. Intern. Med., 23*, 493-498.

[225] Matsuu, A., Yamasaki, M., Xuan, X., Ikadai, H., & Hikasa, Y. (2008). *In vitro* evaluation of the growth inhibitory activities of 15 drugs against *Babesia gibsoni* (Aomori strain). *Vet. Parasitol., 157*, 1-8.

[226] Vercammen, F., De Deken, R., & Maes, L. (1996). Prophylactic activity of imidocarb against experimental infection with *Babesia canis*. *Vet. Parasitol., 63*, 195-198.

[227] Naudé, T. W., Basson, P. A., & Pienaar, J. G. (1970). Experimental diamidine poisoning due to commonly used babecides. *Onderstepoort. J. Vet. Res., 37*, 173-184.

[228] Abdullah, A. S., Sheikh-Omar, A. R., Baggot, J. D., & Zamri, M. (1984). Adverse effects of imidocarb dipropionate (Imizol®) in a dog. *Vet. Res. Commun., 8*, 55-59.

[229] Kock, N., & Kelly, P. (1991). Massive hepatic necrosis associated with accidental imidocarb dipropionate toxicosis in a dog. *J. Comp. Pathol., 104*, 113-116.

[230] Brandão, L. P., Hagiwara, M. K., & Myiashiro, S. I. (2003). Humoral immunity and reinfection resistance in dogs experimentally inoculated with *Babesia canis* and either treated or untreated with imidocarb dipropionate. *Vet. Parasitol., 114*, 253-265.

[231] Brosey, B. P. (2003). *Babesia gibsoni*: a clinical perspective from Southeast Asia. *Proc. 21st ACVIM*, 720-721.

[232] Miller, D. M., Swan, G. E., Lobetti, R. G., & Jacobson, L. S. (2005). The pharmacokinetics of diminazene aceturate after intramuscular administration in healthy dogs. *J. S. Afr. Vet. Assoc., 76*, 146-150.

[233] Lin, M. Y., & Huang, H. P. (2010). Use of a doxycycline-enrofloxacin-metronidazole combination with/without diminazene diaceturate to treat naturally occurring canine babesiosis caused by *Babesia gibsoni*. *Acta. Vet. Scand. Suppl., 52*, 27.

[234] Hwang, S. J., Yamasaki, M., Nakamura, K., Sasaki, N., Murakami, M., Wickramasekara Rajapakshage, B. K., Ohta, H., Maede, Y., & Takiguchi, M. (2010). Development and characterization of a strain of *Babesia gibsoni* resistant to diminazene aceturate *in vitro*. *J. Vet. Med. Sci., 72*, 765-771.

[235] Hudson, A. T. (1993). Atovaquone - a novel broad-spectrum anti-infective drug. *Parasitol. Today., 9*, 66-68.

[236] Matsuu, A., Miyamoto, K., Ikadai, H., Okano, S., & Higuchi, S. (2006). Short report: cloning of the *Babesia gibsoni* cytochrome B gene and isolation of three single nucleotide polymorphisms from parasites present after atovaquone treatment. *Am. J. Trop. Med. Hyg., 74*, 593-597.

[237] Chaudhuri, S., & Varshney, J. P. (2007). Clinical management of babesiosis in dogs with homeopathic *Crotalus horridus* 200C. *Homeopathy., 96*, 90-94.

[238] Nakao, R., Mizukami, C., Kawamura, Y., Bawm, S., Yamasaki, M., Maede, Y., Matsuura, H., Nabeta, K., Nonaka, N., & Oku, Y. (2009). Evaluation of efficacy of bruceine A, a natural quassinoid compound extracted from a medicinal plant, *Brucea javanica*, for canine babesiosis. *J. Vet. Med. Sci., 71*, 33-41.

[239] Eling, W. M. (1982). Chronic, patent Plasmodium berghei malaria in splenectomized mice. *Infect. Immun., 35*, 880-886.

[240] Last, R. D., Hill, J. M., Matjila, P. T., & Réme, C. A. (2007). A field trial evaluation of the prophylactic efficacy of amitraz-impregnated collars against canine babesiosis (*Babesia canis rossi*) in South Africa. *J. S. Afr. Vet. Assoc., 78*, 63-65.

[241] Moreau, Y., Vidor, E., Bissuel, G., & Dubreuil, N. (1989). Vaccination against canine babesiosis an overview of field observation. *Trans. R. Soc. Trop. Med. Hyg., 83*, 95-96.

[242] Schetters, T. (1995). Vaccine development from a commercial point of view. *Vet. Parasitol., 57*, 267-275.

[243] Schetters, T. P., Kleuskens, J. A., Scholtes, N. C., Pasman, J. W., & Goovaerts, D. (1997). Vaccination of dogs against *Babesia canis* infection. *Vet. Parasitol., 73*, 35-41.

[244] Schetters, T. P., Kleuskens, J. A., Scholtes, N. C., Gorenflot, A., Moubri, K., & Vermeulen, A. N. (2001). Vaccination of dogs against heterologous *Babesia canis* infection using antigens from culture supernatants. *Vet. Parasitol., 100*, 75-86.

[245] Schetters, T. P., Strydom, T., Crafford, D., Kleuskens, J. A., van de Crommert, J., & Vermeulen, A. N. (2007). Immunity against *Babesia rossi* infection in dogs vaccinated with antigens from culture supernatants. *Vet. Parasitol., 144*, 10-19.

[246] Fukumoto, S., Xuan, X., Takabatake, N., Igarashi, I., Sugimoto, C., Fujisaki, K., Nagasawa, H., Mikami, T., & Suzuki, H. (2004). Inhibitory effect of antiserum to surface antigen P50 of *Babesia gibsoni* on growth of parasites in severe combined immunodeficiency mice given canine red blood cells. *Infect. Immun., 72*, 1795-1798.

[247] Fukumoto, S., Tamaki, Y., Shirafuji, H., Harakawa, S., Suzuki, H., & Xuan, X. (2005).

Immunization with recombinant surface antigen P50 of *Babesia gibsoni* expressed in insect cells induced parasite growth inhibition in dogs. *Clin. Diagn. Lab. Immunol., 12*, 557-559.

[248] Zhou, J., Yang, J., Zhang, G., Nishikawa, Y., Fujisaki, K., & Xuan, X. (2006). *Babesia gibsoni*: An apical membrane antigen-1 homologue and its antibody response in the infected dogs. *Exp. Parasitol., 114*, 329-333.

[249] Zhou, J., Zhang, G., Nishikawa, Y., Fujisaki, K., & Xuan, X. (2006). A 38-kDa protein from *Babesia gibsoni* and its antibody response in an experimentally infected dog. *Vet. Parasitol., 141*, 345-348.

[250] Zhou, J., Fukumoto, S., Jia, H., Yokoyama, N., Zhang, G., Fujisaki, K., Lin, J., & Xuan, X. (2006). Characterization of the *Babesia gibsoni* P18 as a homologue of thrombospondin related adhesive protein. *Mol. Biochem. Parasitol., 148*, 190-198.

[251] Fukumoto, S., Tamaki, Y., Okamura, M., Bannai, H., Yokoyama, N., Suzuki, T., Igarashi, I., Suzuki, H., & Xuan, X. (2007). Prime-boost immunization with DNA followed by a recombinant vaccinia virus expressing P50 induced protective immunity against *Babesia gibsoni* infection in dogs. *Vaccine., 25*, 1334-1341.

[252] Terkawi, M. A., Jia, H., Gabriel, A., Goo, Y. K., Nishikawa, Y., Yokoyama, N., Igarashi, I., Fujisaki, K., & Xuan, X. (2007). A shared antigen among *Babesia* species: ribosomal phosphoprotein P0 as a universal babesial vaccine candidate. *Parasitol. Res., 102*, 35-40.

[253] Fukumoto, S., Tamaki, Y., Igarashi, I., Suzuki, H., & Xuan, X. (2009). Immunogenicity and growth inhibitory efficacy of the prime-boost immunization regime with DNA followed by recombinant vaccinia virus carrying the P29 gene of *Babesia gibsoni* in dogs. *Exp. Parasitol., 123*, 296-301.

[254] Homer, M. J., Bruinsma, E. S., Lodes, M. J., Moro, M. H., Telford, S. r., Krause, P. J., Reynolds, L. D., Mohamath, R., Benson, D. R., Houghton, R. L., Reed, S. G., & Persing, D. H. (2000). A polymorphic multigene family encoding an immunodominant protein from *Babesia microti*. *J. Clin. Microbiol., 38*, 362-368.

SUGGEST READINGS

[1] DeClue, A. (2009). Sepsis and the systemic inflammatory response syndrome. In *Ettinger S. J., & Feldman, E. C. Textbook of Veterinary internal medicine, 7th edition. St Lois Missouri, Saunders Elsevier*. Pp 523-527.

[2] Boller, E. M., & Otto, C. M. (2009). Shock. In *Ettinger S. J., & Feldman, E. C. Textbook of Veterinary internal medicine, 7th edition. St Lois Missouri, Saunders Elsevier*. Pp 528-531.

[3] Hohenhaus, A. E. (2009). Blood transfusions, component therapy, and oxy-carrying solutions. In *Ettinger S. J., & Feldman, E. C. Textbook of Veterinary internal medicine, 7th edition. St Lois Missouri, Saunders Elsevier*. Pp 537-544.

[4] Mitchell, K. & Kruth, S. (2009). Immune-mediated hemolytic anemia and other regenerative anemia. In *Ettinger S. J., & Feldman, E. C. Textbook of Veterinary internal medicine, 7th edition. St Lois Missouri, Saunders Elsevier*. Pp 761-772.

In: Ticks: Disease, Management and Control
Editor: Moges Woldemeskel
ISBN: 978-1-62081-136-8

Chapter 6

THE ROLE OF TICKS AND TICK CONTROL IN THE MANAGEMENT OF CUTANEOUS DERMATOPHILOSIS IN DOMESTIC ANIMALS

Moges Woldemeskel*
The University of Georgia, USA, Tifton Veterinary Diagnostic and Investigational laboratory, 43 Brighton road, Tifton, GA, US

ABSTRACT

Cutaneous dermatophilosis is a severe skin disease of domestic animals caused by the bacterium *Dermatophilus congolensis* (*D. congolensis*). Although the disease is caused by *D. congolensis*, several risk factors are associated with the pathology and pathogenesis of the disease. Tick infestation is among the major risk factors, which play a very important role in the establishment, transmission and progression of the disease. Dermatophilosis and tick infestation entail heavy economic losses in animal production due to treatment costs, management of the disease and condemnation of skin and hides in various parts of the world, particularly in the tropical countries. In this chapter, the role ticks play in the pathology and pathogenesis of dermatophilosis and the effect of tick control in the management of the disease will be covered.

INTRODUCTION

Ticks are external parasites of domestic and wild animals and humans. They cause various health hazards through direct skin damage, feeding on blood, and transmission of diseases. They transmit large numbers of bacterial, parasitic and viral diseases in animals and humans. Ticks and several tick-borne diseases are among the major stumbling blocks to the development of livestock industry in the tropics and the sub tropics. Ticks are among the major risk factors associated with the infection and establishment of clinical cutaneous dermatophilosis in cattle and horses. Cutaneous dermatophilosis is a bacterial disease, in

* E-mail: mwoldem@uga.edu.

which transmission and pathogenesis is associated with tick infestation and other immunosuppressive conditions. The disease is a severe skin disease of domestic animals caused by *Dermatophilus congolensis* (*D. congolensis*). Several risk factors are associated with the infection, establishment, pathology and pathogenesis of the disease. These include mechanical skin damage, prolonged wetting of the skin, and arthropod bites, which disrupt the protective or mechanical skin barrier and immunosuppression due to malnutrition, immunosuppressive therapy, and debilitating diseases due to other concomitant infections. Among the risk factors, tick infestation plays a very important role in the establishment, progression and transmission of the disease. Dermatophilosis and tick infestation entail heavy economic losses in animal production due to treatment costs, management of the disease and downgrading the quality of skin and hides in various parts of the world, particularly in the tropical countries. A number of tick resistant breeds of cattle are reported to exist in tropical countries. However, introduction of exotic breeds to these areas and cross breeding with the tick-resistant, but low-producing local breeds of cattle is widely practiced to increase milk production. Tick infestation and dermatophilosis pose severe problem to the introduction of the high milk producing exotic breeds of cattle and their crosses into the tropical environment, thereby hindering the development and productivity of the livestock sector. Therefore, understanding the impact of tick infestation in favoring the establishment of the infection and aggravating clinical disease in cutaneous dermatophilosis is very important to envisage, devise and implement a sound control strategy against the disease. In this chapter, the role ticks play in the pathology and pathogenesis of dermatophilosis and the effect of tick control in the management of the disease will be covered.

Tick Infestation and Cutaneous Dermatophilosis

Ticks are obligatory blood-sucking external parasites of mammals, birds and reptiles and are among the most important ectoparasites affecting the health of man and domestic and wild animals. They serve as reservoirs and vectors of many organisms pathogenic to humans and other animals and cause direct harm by their bites, including irritation, anemia, toxemia, allergic sensitization and paralysis [1]. Ticks comprise veterinary problem because they transmit diseases, produce paralysis or toxicosis, and cause physical damage to livestock [2]. One characteristic of ticks which greatly enhances their importance as vectors and reservoirs of pathogens is their ability to transmit microorganisms from one generation of ticks to the next one transovarially and from one developmental instar to the next [1].Tick-borne diseases of livestock constitute a complex of several diseases whose etiological agents may be protozoal, rickettsial, bacterial or viral; their single common feature is that they can all be transmitted by ticks [3].

Ticks belonging to various genera and species and the consequent pathology they induce in domestic animals have negatively impacted global animal production. Tick-borne diseases are present throughout the world, but are most numerous and exert their greatest impact in the tropical and subtropical regions [3]. Tick-borne diseases are the cause of banning of exports of livestock and livestock products from some tropical countries. Ticks suppress the host immune system and predispose infested animals to various infectious diseases.

In recent years, interest in crossbreeding and demand for exotic cattle in developing countries have increased. In catering for this demand, some international and national bodies have underestimated the risks associated with the presence of pathogenic ticks in countries utilizing exotic cattle [4]. Tick-borne diseases of cattle make havoc of the most ambitious plans for importing exotic breeds into tick-infested countries, or for upgrading a local breed by distributing improved animals from well-managed ranches that adopt good tick control [3]. Exotic breeds of cattle introduced into tick-infested tropical regions are highly susceptible to ticks and tick-borne diseases. For example, *Bos taurus* × Sahiwal calves have died from theileriosis in Asia under circumstances in which a purebred Sahiwal might have survived [4]. Tick infestation among others is also the major risk factor associated with the development of clinical cutaneous dermatophilosis in domestic ruminants, particularly cattle, in the tropics.

Cutaneous dermatophilosis is an acute, subacute, or chronic superficial exudative dermatitis and affects several species of animals. It is known under different names including cutaneous streptothricosis, mycotic dermatitis, cutaneous actinomycosis, lumpy wool, strawberry foot rot, rain scald, rain rot, Kirchi, Gasin-Gishu, Senkobo disease, Drodo-Boka, Savi, and Ambarr-Madow [5]. The disease is caused by the filamentous actinomycete bacterium *Dermatophilus congolensis* [5, 6]. Dermatophilus is included in the order Actinomycetales, consitituting separate genus in the family of Polyseptaceae. It is a gram positive non-acidfast organism, which divides transversally and longitudinally and thus forms coccoid elements, which give rise to motile spores (zoospores) [7]. The disease is of worldwide in distribution and has a wide host range but it is most commonly seen in the hot humid tropics and subtropics and in areas with heavy prolonged rains. Cattle are the primary animals affected in hot monsoon climates and some parts of Africa. In more temperate climates, sheep and goats are involved primarily [5].

Cutaneous dermatophilosis is seen occasionally in horses and rarely in pigs, and humans. Cases of dermatophilosis have also been reported in various species of animals including camels, mules, donkeys, zebras, giraffes, Thompson's gazelles, woodchucks, striped skunks, raccoons, hedgehogs, gerbils, foxes, ground squirrels, seals, owl monkeys, captive polar bears, chamois, tortoises, and Australian bearded lizards [5]. *Dermatophilus congolensis* was also isolated from a cat [8] and dogs [9] with dermatitis, although it is seen rarely in these species [5].

Ambrose et al. [6] reported that the disease in cattle results in severe production losses and an average mortality rate of 15%. Cutaneous dermatophilosis is one of the major causes of rejection of hides and skin from tropical countries due to poor quality [10].

The disease commonly affects malnourished cattle, particularly exotic breeds, kept under poor management [6]. There is a complex interplay of several factors including management, and the breed affected in the pathogenesis of dermatophilosis [11]. The multicomponent system underlying the establishment of dermatophilosis involves the bacterium *Dermatophilus congolensis*, transmission by various routes including flies, host genetic factors and immunosuppression by *Amblyomma variegatum* ticks [6]. In the tropics, where dermatophilosis is of major economic importance, mage mites, ticks, and biting flies, which damage the skin barrier and facilitate penetration by infective zoospores, immunosuppression due to other infections and other stressors such as poor nutrition, which lower the protective immunity are associated with clinical dermatophilosis. However, ticks are among the most important risk factors for the establishment and progress of the disease in ruminant livestock. In the tropical environment 36% and 45.2% to 76% of clinical cutaneous dermatophilosis in goats and cattle, respectively, were associated with infestation of the animals with various tick species [10, 11, 12]. *Amblyomma variegatum, Boophilus decoloratus* and unidentifiable aggregates of nymphal stages were observed in infected goats showing clinical dermatophilosis [12]. In one report, out of the 76% tick-infested cattle with dermatophilosis, the respective infestation with *Amblyomma variegatum*, *Boophilus decoloratus* and *Rhipicephalus evertsi,* was 51.6%, 28.1% and 20.3% [10]. An outbreak of dermatophilosis in horse was also reported in association with tick infestation [13]. *Amblyomma variegatum, A. cohaerens, A. gemma, Rhipicephalus (Rh.) evertsi, Rh. Parvus, Boophilus decoloratus,* and *Hyaloma excavatum* were collected and identified from dermatophilosis infected cattle, particularly from the scrotal areas in male and udder in females. Although the main lesions were not confined to these sites, the lesion of dermatophilosis in cattle around the scrotum in males and in the udder areas in the females with the collection of ticks from these sites would suggest establishment of the disease in association with tick bites [14].The occurrence of the disease in cattle in the tropics is closely associated with infestation by the tick *Amblyomma variegatum* [15]. The clinical disease in cattle is aggravated by the presence of the tick *Amblyomma variegatum* [6, 16-18]. Animals that are more susceptible to the disease also tend to carry higher tick burdens [6].

THE TICK: AMBLYOMMA VARIEGATUM

The tick *Amblyomma variegatum* is a three-host hard tick that feeds on a number of domestic animals including cattle, sheep, goats, horses and dogs, as well as humans [19]. Although it infests several species of animals, cattle are regarded to be the preferred hosts [20]. *A. variegatum* is a host for a number of microbial pathogens including *Ehrlichia ruminantium* (formerly *Cowdria ruminantium*), the agent of heartwater, and *Rickettsia africae*, the agent of African tick-bite fever, which is an emerging zoonosis in rural sub-Saharan Africa and the Caribbean [19].

This tick requires moistures and warmth for its survival and is not found in open grassland [20]. Activity of adult tick commences when spring rains begin (19, 20), while the immature ticks feed primarily during the dry season [19]. The adult ticks are usually found on the relatively hairless parts of the body; most are located on the ventral body surface and the genitalia, or under the tail [19]. In cattle, most *Amblyoma variegatum* ticks attach on the lower extremities, the dewlap (16.5%), axillary areas (25%), mammary gland-scrotal areas, (13.6%) and in the perineum (including the vulva and anal areas, 24%). On goats, adult ticks are frequently found attached between the digits of the feet [21].

A definite association between infestation with the tick *Amblyomma variegatum* and occurrence of dermatophilosis in cattle has been found [5]. Although several risk factors play role in the establishment of bovine cutaneous dermatophilosis, the disease in most cases, particularly in the tropical environment, is associated with the tick *A. variegatum.* Kassaye et al. [17] reported that *Amblyomma variegatum*-infested cattle were at seven times higher risk (Odds Ratio (OR) = 6.8) of acquiring the disease than non-infested ones. It was reported that recent outbreaks and spread of dermatophilosis in cattle in Zimbabwe were preceded by the appearance of *Amblyomma variegatum* ticks within the area. The report further indicated that, in all the infected areas, the spread of dermatophilosis was consequent to reduced tick control on cattle [18].

The tick Amblyomma variegatum has been shown by field and controlled laboratory experiments to have a direct causal relationship with severe clinical dermatophilosis. The tick does not transmit the pathogen in the usual sense, either biologically or mechanically. There are other factors such as malnutrition involved with the progression of dermatophilosis to clinical disease [22]. Two factors that appear to be the most important ones are trauma to the skin and prolonged wetting. Zoospores are unable to overcome the protective barriers of the hair, surface lipid film, and stratum corneum, and their entry is facilitated by breaks in the skin surface. Trauma from ectoparasites, shearing, dipping, barb wire injuries, sharp stones, and scratches from sharp vegetation can act as portals of entry for the organisms. Besides producing skin trauma, external parasites such as flies, mites, lice, ticks, and mosquitoes also act as mechanical vectors [5]. The long mouthparts of *A. variegatum* leave large wounds, and make this tick difficult to remove manually. Its bite is severe and painful, and can result in significant damage to the skin [19]. Therefore, the skin trauma induced by the long mouthparts of *A. variegatum* might partly contribute to the role the tick plays in the establishment of the infection with *Dermatophilus congolensis*. In addition to the skin trauma, prolonged moisture is needed for the activation, proliferation, and spread of the zoospores. Wetting may also act to breach the skin barriers by dissolving the surface lipid film and softening the stratum corneum [5]. An outbreak of dermatophilosis in sheep and goats was recorded two weeks after dipping the animals in acaricide dipping fluid against ticks, and the animals were held in a draining pen for about 10-15 minutes after dipping to avoid pasture contamination with the acaricide used. The asymptomatic carriers among the dipped animals during and after dipping are suggested to be the source of the contamination and spread to the previously non-infected sheep. Similarly, clinical dermatophilosis was reported in association with frequent washing (wetting) with rough brushing in cart-horses [15].

Figure 1. Cutaneous dermatophilosis in the left dorsolateral thorax in a Holestein-Fresian x Zebu cross adult cow.

Pathogen, host and environmental factors must be considered in order to understand the pathogenesis of dermatophilosis. A frequently cited sequence of events involves physical damage to the skin, bacterial multiplication in the epidermis, repeated cycles of invasion by the hyphae, infiltration by neutrophils and exudate, regeneration of the epidermis and reinvasion [23]. Since the lesions of dermatophilosis do not correspond to the sites of tick attachment, a systemic effect on the animal's immune system has been suspected [5]. Dermatophilosis in most cases affects animals with compromised immunity [24]. Ticks suppress the immune system of the host, for unknown reasons, and this appears to enable the existing dermatophilosis to become a severe disease [22]. Ambrose [23] indicated that primary infections of calves under controlled conditions at clipped, cleaned, defatted sites result in characteristic dermatophilosis crusts, illustrating that *D. congolensis* itself is pathogenic. Calves infected in this way develop protective immune responses to subsequent infections. Immunity to dermatophilosis might involve non-classic responses mediated by CD1 antigen presentation and gammadelta T- cells [6]. First and second infections of calves simultaneously infested with *D. congolensis* and *Amblyomma variegatum* result in more severe lesions that take significantly longer to heal than lesions on tick free calves. Immunity to *D. congolensis* is isolate specific. However, the antigens that elicit immunity have not been defined; they might include the pathogenic or virulence factors. Hyphae are the life-cycle stage most closely associated with the living epidermis. During the in-vitro growth, the hyphae secrete antigens and proteolytic enzymes into the culture medium. The proteolytic activity has been linked to the virulence of *D. congolensis*. The characteristics of proteases released into the culture medium vary between isolates. This raises the possibility that proteases, with as yet undefined functions, act as pathogenic and virulence factors, that they are the targets for protective immune responses and that *A. variegatum* infestation interferes with host immune responses which normally inhibit their activity [23]. Rebeiro et al. [25] quoting several authors indicated that, among the adaptations found in ticks for successful blood feeding, their salivary glands have compounds that counteract host hemostasis and inflammation, including anticlotting, antiplatelet, vasodilatory, antihistaminic, antileukotriene, anticomplement, antichemokine, and immune-modulatory compounds. A proteome study by these authors using salivary gland homogenates of *A. variegatum* identified 19 putative secreted proteins within a total of 211 matches. Most of the proteins described have no known function but, if secreted into their hosts, they are suggested to have antihemostatic, antiinflammatory, anti-angiogenic, or immunomodulatory function. Oliveira et al. [26] suggest that tick saliva induces regulatory dendritic cells (DCs), which secrete IL-10 and low levels of IL-12 and TNF-α when stimulated by Toll-like receptor (TLR) ligands. Such regulatory DCs are associated with expression of TLR-2 and inhibition of ERK and p38, which promotes the production of IL-10 and thus down-modulates the host's immune response, possibly favoring susceptibility to tick infestations.

Logistic regression analysis of the overall prevalence of dermatophilosis in one study showed a significant difference ($p < 0.05$) between the prevalence in the indigenous and the cross breeds of cattle; the cross breeds being at about three times higher risk (OR = 2.68) than the indigenous breeds [17]. Breed-based differences in resistance to dermatophilosis are probably related to immunity to ticks or resistance to the immunosuppressive effects of the ticks [6].

TICK CONTROL IN THE MANAGEMENT OF CUTANEOUS DERMATOPHILOSIS

Control of tick infestations and the transmission of tick-borne diseases remain a challenge for the cattle industry in tropical and subtropical areas of the world [2]. Although complex mechanisms underlie the establishment of dermatophilosis, an exudative and proliferative skin disease of ruminants [6], ticks are known to be the most important risk factors associated with the disease. Recently, dermatophilosis associated with tick infestation is reported to be the major threat to the prime cattle production zone in Zimbabwe [18]. Infected animals may develop a protective immune response to the infection. As previously indicated, immunity to dermatophilosis might involve non-classic responses mediated by CD1 antigen presentation and gammadelta T cells. However, progress towards vaccination is complicated by the strain-specific acquired immunity to *D. congolensis* [6]. Therefore, the control of dermatophilosis should be based on the control and management of the predisposing factors. Among the predisposing factors, the role of ticks is so distinct that it is likely that appropriate tick control using acaricide, when combined with other measures to improve the husbandry of the cattle, will reduce the incidence and severity of dermatophilosis [22]. In herds with effective tick prevention, there is a much lower incidence or diminished severity of the disease [5].

Morrow et al. [27] reported that fortnightly treatment of cattle with Amitraz reduced the level of infestation with *A. variegatum* and the prevalence of dermatophilosis dropped to a low level. Furthermore, the authors indicated that the use of deltamethrin-based pour-on acaricide similarly lowered the prevalence of dermatophilosis.

Susceptibility to *A. variegatum* associated dermatophilosis varies between breeds of cattle; therefore, the level of tick control required is breed-dependent. Rigorous tick control is necessary for the prevention of dermatophilosis in highly susceptible exotic breeds of cattle. The limited use of acaricides applied either at the predilection feeding sites of *A. variegatum* and/or at selected times when the level of challenge increases is sufficient to control the occurrence of the disease in indigenous cattle [12]. Immature ticks of *A. variegatum* feed on small mammals, ground-feeding birds and reptiles, as well as cattle, sheep and goats. Adult ticks prefer cattle, but can also be found on sheep, goats, horses, camels, dogs and some large wildlife including antelope [19]. Therefore, due attention should be given to all potential hosts (including wild and feral animals) of the important tick species involved in a given area under consideration. Three-host ticks such as *A. variegatum* spend at least 90% of their life cycle in the environment rather than on the host animal; ticks in the environment must also be controlled to prevent their spread. If ticks are already widespread in a region, eradication can be difficult due to re-infestation of the animals from the infested environment. In countries where *A. variegatum* is already present, acaricides can eliminate the ticks from the animal, but do not prevent re-infestation and must be repeated periodically [19]. *A. variegatum* has affinity for the protected areas on the body of the host (axillary areas, under tail, dewlap, mammary-scrotal areas). This requires special care in the application of acaricides. In an eradication program where hosts are sprayed with a non-systemic acaricide, considerable effort by the applicator must be given to treat those areas of the host body where ticks are found [21].

Generally, the method of tick control in a given area depends upon the epidemiology of tick and tick-borne diseases in that particular area, the species of animals (indigenous or exotic) and whether acaricide resistant ticks have developed due to inappropriate dosage and use. To counteract the adverse effects of ticks and tick-borne diseases, a variety of tick-control programs have been integrated into modern livestock management practices. Foremost among these is chemical tick-control, which not only decreases the detrimental effects of the ticks themselves, but also interrupts the transmission cycle of pathogenic agents and, thus, reduces the incidence of tick-borne diseases. Depending on the tick species involved, the livestock-management practices followed, the tick-borne diseases present, and the environmental conditions of an area, cattle are treated at periodical intervals with any one of over 30 available acaricides [3].

Disease caused by ticks *per se,* often underestimated in developing countries, is the least likely to be controllable by chemical or immunological means in the future. The effectiveness of acaricides can be expected to decline as ticks become tolerant to chemicals, and in some countries the increasingly high cost of dips and sprays will militate against their continued use. Biological control using tick-resistant cattle, especially *Bos indicus* breeds, should eventually succeed (Wharton, 1976, cited by [4]), but the prospect is long-term, particularly for dairy cattle, and will depend on continuing endeavor by workers in animal production, breeding and parasitology in both developed and developing countries. The possibility of eradicating ticks or holding populations at very low levels with minimal use of acaricides should be considered [4]. The eradication of tick-borne livestock diseases and their vectors is a viable alternative to the control of ticks and tick-borne disease. However, the assessment of this alternative deserves the consideration of a variety of interrelated parameters that will eventually influence the success or failure of a campaign [3]. The issue of enzootic instability should be taken into account in case of a trial to completely eradicate ticks from a given specific region.

Conclusion

Effective control of ticks using acaricides and other biological methods coupled with management of other risk factors would help to alleviate the incidence and impact of ticks and tick-borne diseases including dermatophilosis on livestock industry. The public officials, livestock owners, as well as the general public should be adequately informed of the impact of tick infestation causing severe health hazards to the livestock and the public, and also of the consequent damage to the livestock industry. Appropriately programmed surveillance, and effective tick control regimens should be devised and implemented based on the epidemiology, host range and habitat of a given area. If necessary, quarantines and movement restrictions should be put in effect during the implementation of the tick control measures. Introduction and maintenance of tick-resistant breeds into tick infested areas would help in sustaining livestock productivity.

References

[1] Furman, D.P. & Loomis, E.C. (1984). The Ticks of California (Acari: Ixodida). *University of California Publications, Bulletin of the California Insect Survey, Vol. 25.* University of California Press, California, p.1-239.

[2] Rajput, Z.I., Hu S.H., Chen, W.J., Arijo, A.G. & Xiao C.W. (2006). Importance of ticks and their chemical and immunological control in livestock. A Review. *Journal of Zhejiang University* SCIENCE B. ISSN 1673-1581 (Print); ISSN 1862-1783 (Online)

[3] Bram, R.A., & Gray, J.H. (1983). Eradication - an alternative to tick and tick-borne disease control. In *FAO, Ticks and tick-borne diseases selected articles from the World Animal Review.* http://www.fao.org/DOCREP/004/X6538E/X6538E00.HTM

[4] Callow, L.L. (1983). Ticks and tick-borne diseases as a barrier to the introduction of exotic cattle to the tropics. *In FAO, Ticks and tick-borne diseases selected articles from the World Animal Review.* http://www.fao.org/DOCREP/004/X6538E/X6538E00.HTM

[5] Ginn, P. E., Mansell, J.E. K. L. & Rakich, P.M. (2007). Skin and appendages. In: *Maxie M.G. (editor), Jubb, Kennedy and Palmer's pathology of domestic animals.* New York, NY, Elsevier Saunders, 5th ed., vol. 1, pp. 680-684.

[6] Ambrose, N., Lloyd, D. & Maillard J-C. (1999). Immune responses to Dermatophilus congolensis infections. *Parasitol Today* 15:295–300.

[7] El-Nageh, M.M. (1971). Comparison of strains of Dermatophilus congolensis Van Saceghem 1915 isolated from different species of animals. *Ann Soc Belge Méd Trop.* 51, (2), 239-246.

[8] Kaya, O., Kirkan, S. & Unal, B. (2000). Isolation of Dermatophilus congolensis from a cat. *J Vet Med B Infect Dis Vet Public Health.* Mar; 47 (2), 155-157.

[9] Chastain, C.B., Carithers, R.W., Hogle, R.M., Abou-Gabal, M., Graham, C.L. & Branstetter, D. (1976). Dermatophilosis in two dogs. *J Am Vet Med Assoc.* Nov 15; 169 (10):1079-80.

[10] Woldemeskel, M. & Taye, G. (2002). Prevalence of bovine dermatophilosis in a tropical highland region of Ethiopia. *Trop Anim Health Prod.* 2002 May; 34(3), 189-94.

[11] Berhanu, D. & Woldemeskel, M. (1999). Bovine dermatophilosis and its influencing factors in central Ethiopia. *Zentralbl Veterinarmed A.* 1999 Dec; 46 (10), 593-7.

[12] Woldemeskel, M. & Mersha, G. (2010). Study on caprine and ovine dermatophilosis in Wollo, Northeast Ethiopia. *Trop Anim Health Prod.* 42(1), 41-4.

[13] Shannon, D. (1997). Dermatophilosis and abscessation of lymph nodes in a group of tick-infested horses in Uganda. *Trop Anim Health Prod.* 29, 102-104.

[14] Woldemeskel, M. (2000). Dermatophilosis: A threat to livestock production in Ethiopia. Dtsch. *Tieraerztl.Wschr.* 107, 144-146.

[15] Morrow, A.N., Koney, E.B. & Heron, I.D. (1996). Control of Amblyomma variegatum and dermatophilosis on local and exotic breeds of cattle in Ghana. *Trop Anim Health Prod.* 28 (2 Suppl):44S-49S; discussion 74S-86S.

[16] Morrow, A.N., Compton, E.A.E. (1991).The occurrence of streptothricosis and its association with A. variegatum ticks in St Lucia. *J Vet Med Ser B.* 38, 635–638.

[17] Kassaye, E., Moser, I. & Woldemeskel, M. (2003). Epidemiological study on clinical bovine dermatophilosis in northern Ethiopia. *Dtsch Tierarztl Wochenschr.* 110 (10), 422-5.

[18] Chatikobo, P., Choga, T., Ncube, C. & Muzenda-Mutambara, J. (2009). Bovine dermatophilosis, a re-emerging pandemic disease in Zimbabwe. *Trop Anim Health Prod.* 41(7), 1289-97.

[19] *Anonymus. (2009). Amblyoma variegatum, Tropical bont tick. Online:* www.cfsph.iastate.edu/Factsheets/pdfs/amblyomma_variegatum.pdf. *P1-2.*

[20] Pegram, R.G., Hoogstraal, H., & Wassef, H.Y. (1981). Ticks (Acari: Ixodoidea) of Ethiopia. I. Distribution, ecology and host relationship of species infesting livestock. *Bull Ent Res,* 71, 339-359.

[21] Barre, N., &. Garris, G.I. (1990). Biology and Ecology of Amblyomma variegatum (Acari: Ixodidae) in the caribean: Implication for a regional eradication program. *J Agric Entomol.* 7(1), 1-9.

[22] Walker, A.R. (1996). Amblyomma tick feeding in relation to host health. *Trop Anim Health Prod.* 28 (2 Suppl): 26s-28s.

[23] Ambrose, N.C. (1996). The pathogenesis of dermatophilosis. *Trop Anim Health Prod.* 28 (2 Suppl): 29S-37S.

[24] Woldemeskel, M. and Ashenafi, H. (2003). Study on skin diseases in sheep from northern Ethiopia. *Deut Tierarztl Woch* 110 (1), 20-22.

[25] Ribeiro, J.M.C. Anderson, J. M., Manoukis1, N. C., Meng, Z. & Francischetti, I.M.B. (2011). A further insight into the sialome of the tropical bont tick, *Amblyomma variegatum. BMC Genomics.* 12, 136.

[26] Oliveira, C.J. F., Carvalho, W.A., Garcia, G.R., Gutierrez, F.R.S., De Miranda Santos, I.K.F., Silva, J.S. & Ferreira, B.R. (2010). Tick saliva induces regulatory dendritic cells: MAP-kinases and Toll-like receptor-2 expression as potential targets. *Veterinary Parasitology,* 67 (2-4), 288-297.

[27] Morrow, A.N., Arnott, J.L., Heron, I.D., Koney, E.B., &Walker, A.R. (1993). The effect of tick control on the prevalence of dermatophilosis on indigenous cattle in Ghana. *Rev Elev Med Vet Pays Trop.* 46 (1-2), 317-22.

In: Ticks: Disease, Management and Control
Editor: Moges Woldemeskel ISBN: 978-1-62081-136-8

Chapter 7

ACARICIDE RESISTANCE OF SINGLE AND MULTI-HOST TICKS AND COMPARISON OF *IN VITRO* LARVAL AND ADULT TICK BIOASSAY METHODS

***Sileshi Mekonnen*[*,1], *N. R. Bryson*[2] *and Moges Woldemeskel*[3]**
[1]Food and Agriculture Organization (FAO) Office, P.O. Box 5536, Addis Ababa, Ethiopia
[2]Department of Veterinary Tropical Diseases, Faculty of Veterinary Science, University of Pretoria, South Africa
[3]The University of Georgia, College of Vet. Med., Dept. of Pathology, Tifton Vct. Diagnostic & Investigational Lab., P. O. Box 1389, Tifton, GA, US

1. ABSTRACT

Tick resistance to acaricides is an increasing problem in tropical Africa. It poses a real economic threat to the livestock and veterinary pharmaceutical industries. New acaricides are extremely expensive to develop. Therefore, the present acaricides should be seen as an ever-diminishing resource, which should be protected by all means possible. To alleviate tick resistance problems through resistance management regimens, it is very important to determining the pattern of resistance to ticks in a given area. A field survey was carried out at selected communal and commercial farms in the Eastern Cape and Northwest Provinces of South Africa to monitor the levels of field ticks (B. decoloratus, R. evertsi evertsi, R. appendiculatus and A. hebraeum) resistance to Amitraz, Cypermethrin and Chlorfenvinphos.

On the communal farms high levels of tick resistance were detected to Cypermethrin as well as partial resistance to Chlorfenvinphos whilst no resistance was detected against Amitraz. On the commercial farms, however, ticks were equally resistant to Amitraz, Cypermethrin and Chlorfenvinphos. The populations of B. decoloratus on these farms had developed higher levels of resistance to the test acaricides than the populations of R. evertsi evertsi, R. appendiculatus and A. hebraeum. Higher level of tick resistance to Amitraz was observed on commercial farms than on communal farms. However, there

[*] Corresponding author. Sileshimekonnen99@yahoo.com; Sileshi.Mekonnen@fao.org.

was no significant difference in tick resistance to Chlorfenvinphos and Cypermethrin at both the commercial and communal farms. It was summarized that inappropriate use of acaricides on the commercial as well as the communal farms might have resulted in higher tick resistance to the currently available acaricides.

This chapter describes the levels of tick resistance to acaricides at selected commercial and communal farms in South Africa, compares the *in vitro* adult and larval test methods and discusses the acaricide management strategies, which may increase the lifespan of the presently used acaricides.

2. Acaricide Resistance of Single and Multi-Host Ticks

2.1. Introduction

Ixodid ticks are one of the most economically important external parasites of livestock in tropical and sub-tropical parts of the world [1]. Heavy infestation can cause loss of blood, reduce the rate of live-weight gain, lower milk yield, and downgrade the quality of the hides [2]. It is estimated that 80% of the world's cattle population suffer to some extent from the deleterious effects caused by ticks [3]. Because of the direct and indirect effects on their hosts, ticks are considered to be not only a serious threat to successful livestock production but also seriously interfere with the national economy, particularly in the African continent.

Ticks undergo a one-, two-, or three-host lifecycle [4]. One-host ticks spend all three stages of their development on one host. In the case of two host ticks, the larvae develop into nymph on the first host and then drops to the ground where they moult into adults, which then seek a second host. In three host-ticks, each of the larval, nymph and adult stages feed and engorges on a separate host before moulting to the next stage [5].

Throughout most of the 20th century, tick infestation on cattle were mainly controlled by chemicals, administered by plunge dipping, spraying in spray races or by hand-spraying [6], and more recently, in the form of pour-ons, spot-ons, parenteral injections and intra-ruminal boluses. Aresenic was first used for tick control in 1893 in South Africa [7] and this organic compound was followed by a range of orgainic acarcicides, which include organochlorines, organophosphates, carbamates, amidines, and the pyrethroids primarily by economic demands but also tempered by the development of acaricide resistance to these chemicals.

Tick resistance to acaricides used for tick control is an increasing problem in the world including South Africa. The first published report on the development of tick resistance was to arsenic [8]. The resistance was detected in *B. decoloratus* after controlled field trials [8] and laboratory tests [9, 10] confirmed the resistance. Later, resistance to arsenic was also detected in *R. evertsi evertsi* [11, 12], *R. appendiculatus* [13], *A. hebraeum* [12] and *B. microplus* [14]. Distinction

Acaricide resistance is more widespread and diverse in one-host tick [15] and one of the main reasons for the more rapid selection for resistance in one-host ticks is the short generation time [16]. Generally, it is hypothesized that the rate of development of resistance would be quicker in the single-host ticks compared to the multi-host ticks. The blue tick, *B. decoloratus* has developed resistance to a variety of chemicals used at many different localities in southern Africa [17]. In South Africa, published reports indicate a wide spread resistance of *B. decoloratus* all over the country (Spickett, Pers. Commin., 2001). The detection of resistance at an early stage, combined with an understanding of the factors which enhance its development, could result in the formulation of better resistance management strategies. This chapter deals with acaricide resistance profiles in one- and multi-host ticks on cattle in eastern Cape and Northwest provinces of South Africa, and compares the *in vitro* adult and larval test methods used to detect resistance in populations of *B. decoloratus*.

2.2. Materials and Methods

The study was conducted at selected communal (n=6) and commercial (n=6) farms (each with 100-500 head of cattle) located in the Northwest and Eastern Cape Provinces of South Africa. Ticks collected from these farms were used for acaricide resistance tests. Breeds of cattle in the commercial farms were mostly exotic-crosses with Simmental breeds mixed with locally developed Bonsmara (Afrikaner x Shorthorn).

The animals were examined and ticks were collected in a crash and the collected ticks were identified according to the descriptions given by various authors [18-22]. Ticks were collected from each farm during the peak occurrence of the adult ticks in summer (January-April, 2001). One-host ticks were collected from any part of the body of the animal while multi-host ticks were collected from their specific predilection sites. Only fully engorged adult female *A. hebraeum, B. decoloratus, R. appendiculatus* and *R. evertsi evertsi* were removed manually before cattle were treated with an acaricide.

The identified ticks were washed on a sieve using clean tap water. All damaged and undersized ticks were discarded. The collected ticks were then air-dried in absorbent paper, placed in a glass flask and incubated in a glass jar incubator at 27°C and 80-90% relative humidity (R.H) in a temperature-controlled room equipped with a humidifier and a fan heater. The ticks were kept and maintained under these conditions until egg-laying and larval hatching were completed. The egg laying period by *Amblyoma* species was approximately seven weeks, and that of *Rhipicephalus* and *Boophilus* species took four and three weeks, respectively. The hatching period of the *Amblyoma, Rhipicephalus* and *Boophilus* species was about three weeks (Strydom, Pers. Comm, 2001).

The acricide used for the study were Amitraz, 12.5% m/v (Triatix®- Intervet South Africa Pty. Ltd.), Chlorfenvinphos 30% m/v (Supona 30®- Fort Dodge, Bayer Animal Health Pty Ltd.) and Cypermethrin 15% m/v (Curatik Dip®- Fort Dodge, Bayer Animal Health Pty Ltd.).

Acarcicide Resistance Testing Procedures

The larvae for the test were originally obtained from engorged female *A. hebraeum, B. decoloratus, R. appendiculatus* and *R. evertsi evertsi* and tested against different concentrations of Amitraz, Chlorfenvinphos and Cypermethrin using the Shaw Larval Immersion Test (SLIT) originally described by [23] and later modified to include a longer holding period for larval ticks after treatment [24]. The acaricide concentration, tick species, code and testing dates were noted and labeled. Serial dilutions of the emulsified commercial acaricides (Amitraz, 12.5% or Chlorfenviphos, 30% or Cypermethrin, 15%) were prepared for each sample using seven different concentrations for each chemical. Batches of between

14- to 21- day-old larvae were dipped in the different acaricide concentrations according to the published protocol of acaricide resistance testing [23] and the tests were read.

Responses of field ticks exposed to acaricides were compared with baseline data obtained from susceptible strains on the basis of the LC_{50} value (an estimate of the acaricide concentration which kills 50% of the population). A Factor of Resistance (FOR) was calculated by dividing the LC_{50} value obtained with a field strain for that obtained with a susceptible reference strain. The degree of resistance is the number of times the LC_{50} value of a field strain exceeded that of the susceptible tick strain [25].

Susceptible reference strain values were included as controls. The larvae obtained from engorged female ticks collected from the field were considered to be resistant when the FOR values were more than 100 for Amitraz and Cypermethrin or more than five for Chlorfenvinphos. They were considered emerging resistant when FOR values were between 50 and 100 for Amitraz and Cypermethrin and between 2.5 and five for Chlorfenvinphos. The cut-off points for the tests were determined based on previous field trials (Taylor, pers. Comm., 2001).

B. decoloratus susceptible reference strains were obtained from the University of the Free State (Fourie, Pers. Comm. 2001). A. hebraeum, R. appendicuklatus and R. evertsi evertsi susceptible strains were obtained from Dr. R.J. Taylor. Corrections for mortality were also made using the formula given below [26]:

Corrected mortality (%) = (% test mortality- % control mortality) x 100 100 - % control mortality.

1.3. Results

The LC_{50} values of Amitraz, Chlorfenvinphos and Cypermethrin tested against the four different tick species, as well as factors of resistance (FOR) for each tick species, and each acaricide tested are illustrated in Tables 1-4. A. hebraeum larvae were susceptible to both Amitraz and Cypermethrin, however, there was partial resistance to Chlorfenvinphos (Table 1).

Two Amitraz, four Cypermethrin, and five Chlorfenvinphos resistant strains of *B. decoloratus* were detected (Table 2). One, two and four emerging resistant strains of *B. decoloratus* were detected to Amitraz, Cypermethrin and Chlorfenvinphos, respexctively (Table 3).

All *R. appendiculatus* larvae tested were susceptible to the three acricides used (Table 3) and only Chlorfenvinphos resistant strain of *R. evertsi evertsi* was found (Table 4).

The percentage corrected mortality (% CM) of the different concentrations of the acaricide tested against the different tick species calculated from the dead/alive tick count, showed a good kill in relation to increases in concentration except where there is a resistant strain.

Table 1. *In vitro* larval bioassay: Results of the susceptibility of A. hebraeum larvae to Amitraz, Cypermethrin and Chlorfenvinphos

Farm Name	A. heb.strain	Active compound								
		Amitraz			Cypermethrin			Chlorfenvinphos		
		LC50	FOR	Comment	LC50	FOR	Comment	LC50	FOR	Comment
	Ref. Strain	0.0001	NC		0.000015	NC		0.00052	NC	
Mabeleni	S-3	0.000026	0.260	S	0,000015	1.000	S	0,00024	0.462	S
Mozana	S-6	0,000007	0,070	S	0.000031	2.067	S	0.00064	1.231	S
ColleyWable	S-7	0.0000063	0.063	S	0.000018	1.200	S	0.00026	0.500	S
Cizele	S-1.1	0.00013	1.300	S	0,0001 1	7.333	S	0.0013	2.500	ER
Ntubeni	S-12	0.000043	0.430	S	0.000036	2.4	S	0.00022	0.423	S

	Amitraz	Cypermethrin	Chlorfenvinphos
	FOR	FOR	FOR
R	>100	>100	>5
ER	50-100	50-100	2.5-5
S	<50	<50	<2.5

R= Resistant; ER= Emerging resistance; S= Susceptible; FOR= Factor of resistance; NC=Not Calculated; A. heb. = A. hebraeum.

Table 2. ***In vitro*** **larval bioassay: Results of the susceptibility of B. decoloratus larvae to Amitraz, Cypermethrin and Chlorfenvinphos**

Farm Name	B. dec strain	Active compound								
		Amitraz			Cypermethrin			Chlorfenvinphos		
		LC_{50}	FOR	Comment	LC_{50}	FOR	Comment	LC_{50}	FOR	Comment
	Ref. Strain	0.000042	NC		0.000015	NC		0.00052	NC	
Mabeleni	S-1	0.0000002	0.005	S	0.016	280.9187	R	0,0015	3.511	ER
Mozana	S-4	0.000016	0.376	S	0,028	501.7667	R	0.0012	2.976	ER
ColleyWable	S-8-1	0.0000072	0.172	S	0.0032	56.53710	ER	0.00036	0.872	S
Cizele	S-IO	0.0000079	0.189	S	0.0051	90.10600	ER	0,0025	6.053	R
Ntubeni	S-15	0.00000068	0.016	S	0,00032	5.653710	S	0.0013	3,148	ER.
Basfontein	S-19	0.097	2326	R	0.00013	2.29689	S	0.00044	1.065	S
Wood stock	S-20	0.000041	0.981	S	0.00024	4.240282	S	0.00032	0.775	S
Mddelfontein	S-23	0.013	311.01	R	0.012	212.0141	R	0,0009	2.179	S
Brycedale	S-100	0.0033	77.75	ER	0.0024	42.049	S	0,0026	6.199	R
Sunny Grove	S-101	0,0013	31.34	S	2.0E+01	>200	R	0.0020	4.860	ER
Welgevind	S-102	0.000041	0.971	S	0.00068	12.032	S	0,0024	5.920	R

	Amitraz FOR	Cypermethrin FOR	Chlorfenvinphos FOR
R	>100	>100	>5
ER	50-100	50-100	2.5-5
S	<50	<50	<2.5

R= Resistant; ER= Emerging resistance; S= Susceptible; FOR= Factor of resistance; NC=Not Calculated; B. dec = B. decoloratus.

Table 3. ***In vitro*** **larval bioassay: Results of the susceptibility of R. appendiculatus larvae to Amitraz, Cypermethrin and Chlorfenvinphos**

Farm Name	R.app strain	Active compound								
		Amitraz			Cypermethrin			Chlorfenvinphos		
		LC_{50}	FOR	Comment	LC_{50}	FOR	Comment	LC_{50}	FOR	Comment
	Ref. Strain	0.00005	NC		0.0002	NC		0.0006	NC	
Ciko	S-18	0,0000056	0.11	S	0.00006	0.3	S	0.00031	0.517	S
Cizele	S-9	0.00000064	0.013	S	0.000045	0.225	S	0.00038	0.633	S
Ntubeni	S-16	0.0000069	0.14	S	0.000038	0.19	S	0,00017	0.283	S
Middelfontein	S-26	IL	NC		0.00019	0.95	S	0.0007	1.167	S

	Amitraz	Cypermethrin	Chlorfenvinphos
	FOR	FOR	FOR
R	>100	>100	>5
ER	50-100	50-100	2.5-5
S	<50	<50	<2.5

R= Resistant; ER= Emerging resistance; S= Susceptible; FOR= Factor of resistance; NC=Not Calculated; R. app = R. appendiculatus.

Table 4. ***In vitro*** **larval bioassay: Results of the susceptibility of R. evertsi evertsi larvae to Amitraz, Cypermethrin and Chlorfenvinphos7**

Farm Name	R. ever strain	Active compound								
		Amitraz			Cypermethrin			Chlorfenvinphos		
		LC_{50}	FOR	Comm.	LC_{50}	FOR	Comm.	LC_{50}	FOR	Comm.
	Ref. Strain	0.00005	NC		0,00002	NC		0.0005	NC	
Mozana	S-5	1.5×10^{-9}	0.00003	S	0.000017	0.850	S	0.000066	0.132	S
Ntubeni	S-17	1.3×10^{-9}	0.000026	S	0.0000065	0.325	S	0.00024	0.480	S
Woodstock	S-21	1.2×10^{-7}	0.0024	S	0.00001	0.500	S	0,00015	0.300	S
Middelfontein	S-24	4.0×10^{-7}	0.008	S	0.000017	0,850	S	0.0029	5.800	R

	Amitraz FOR	Cypermethrin FOR	Chlorfenvinphos FOR
R	>100	>100	>5
ER	50-100	50-100	2.5-5
S	<50	<50	<2.5

R= Resistant; ER= Emerging resistance; S= Susceptible; FOR= Factor of resistance; NC=Not Calculated; Comm.= comments; R. ever = R. evertsi evertsi.

Table 5. Summary of the tick resistance data collected from the communal farms/dip tanks (n = 6) in the Eastern Cape Province

District	Dip tank	Tick species	Amitraz		Cypermethrin		Chlorfenviophos	
			FOR	Comment	FOR	Comment	FOR	Comment
East London	Mabeleni	A. hebraeum	0.260	S	1.000	S	0.462	S
		B. decoloratus	0.005	S	280.9	R	3.5	ER
	Mozana	A. hebraeum	0.070	S	2.067	S	1.231	S
		B. decoloratus	0.376	S	501.8	R	3.0	ER
		R. evertsi evertsi	0.00003	S	0.850	S	0.132	S
Idutywa	Colley-wable	A. hebraeum	0,063	S	1.200	S	0.500	S
		B. decoloratus	0.172	S	56.5	ER	0.872	S
	Ctzele	A. hebraeum	1.300	S	7,333	S	2.50	ER
		B. decoloratus	0.189	S	90.1	ER	6.053	R
		R. appendiculatus	0.013	S	0.225	S	0.633	S
Willow-vale	NtubenI	A. hebraeum	0.43	S	2,4	S	0.423	S
		B. decoloratus	0.016	S	5.65	S	3,14	ER
		R. appendiculatus	0.0014	S	0.190	S	0.283	S
		R. evertsi evertsi	0,00003	S	0.325	S	0.480	S
	CIko	R. appendiculatus	0.011	S	0.030	S	0.517	S

R= Resistant; ER= Emerging resistance; S= Susceptible; FOR= Factor of resistance.

Table 6. Summary of the tick resistance data collected from the commercial farms (n = 6) in the Eastern Cape and North-West Province

District	Dip tank	Tick species	Amitraz		Cypermethrin		Chlorfenviophos	
			FOR	Comment	FOR	Comment	FOR	Comment
East London (Eastern Cape Province)	Brycedale Sunny Grove	B. decoloratus	77.75	ER	42.05	S	6.20	R
Koster (North West Province)	Sunny Groove	B. decoloratus	31.34	S	3508	R	4.86	ER
	Welgevind	B. decoloratus	0.97	S	12.03	S	5.92	R
Koster (NWP)	Middelfontein	B. decoloratus	311.005	R	212.01	R	2,179	S
		R. appendiculatus	IL	NC	0.950	S	1.167	S
		R. evertsi evertsi	0.008	S	0.850	S	5.800	R
	Basfoatein	B. decoloratus	>200	R	2.297	S	1.065	S
Swartruggens (NWP) (North West Province)	Woodstock	B. decoloratus	0.981	S	4.240	S	0.775	S
		R. evertsi evertsi	0.0024	S	0.500	S	0300	S

R= Resistant; ER= Emerging resistance; S= Susceptible; FOR= Factor of resistance; IL= Insufficient larvae; NC=Not calculated; NWP =Northwest province.

Aacricde Resistance of Ticks on the Communal Farm Dip-Tanks at Mabeleni, Mozana and Ntubeni

Laboratry results obtained from the larval progeny of *B. decoloratus* collected at the communal farms/dip tanks (Table 5) demonstrated high level of resistance to both Cypermethrin (two resistant and two emerging resistant farms), and Chlorfenvinphos (one resistant and four emerging resistant forms) and no resistance to Amitraz (all susceptible). *B. decoloratus* from dip thanks at two sites (Mabelini and Mozana) in East London district showed considerable resistance to Cypermethrin.

2.4. Discussion

Resistance in ticks to pyrethroids was reported after only 18 months of use in the field [27] and in Australia cross-resistance between DDT and the pyrethroids has been reported [28]. Coetzee *et al.* [27] observed the association in *B. decoloratus* between Fenvalerate resistance, which is a pyrethroid, and DDT resistance in South Africa. DDT was extensively used from 1949 to 1955 [29] in East London. Therefore, the possibility that the current outbreak of pyrethroid resistance may be associated with the previous DDT resistance has to be considered. There are, however, no grounds to imply that the cypermethrin resistance evolved from DDT resistant strains of *B. decoloratus* at the communal dip tanks.

One of the most important factors, which affects the efficacy of an acaricide, is the use of an incorrect acaricide concentration and this is one of the prime causes of tick control failure at communal dipping tanks [30]. Use of incorrect acaricide concentration was recorded at the Mozana dip tank near East London, where there was no assigned dip tank manager and farmers were responsible for the replenishment of acaricides in the dip (Magadla, pers, comm., 2001). The farmers believed that they needed to increase the concentration of the acaricides in the dip tank during the peak tick season to control the excessive tick burdens on their cattle. Such increased dip acaricide concentration might have undoubtedly lead to a higher selection pressure for tick resistance [31] since the high acaricide concentration would effectively kill all susceptible ticks, leaving only a residue of highly resistant individuals in the population [31]. Each successive dipping would be a selective process, which would concentrate the genes responsible for the resistance. Eventually, the majority of the ticks in the population would be resistant to the acaricide applied [11].

At both the Mabeleni and Mozana communal dip tanks, high levels of tick resistance to Cypermethrin were observed. These dip tanks were close to the coast, where there was high humidity and high temperatures for most of the year. These were ideal conditions for tick population to rapidly expand. Under such favorable climatic conditions the ticks were able to thrive all the year round necessitating a tick control regimen at a two-week-interval dipping all year round. This intensive dipping program probably also contributed to an increased selection pressure for resistance. Zeropar®, which is a mixture of alphamethrin and Chlorfenvinphos, had been used at these dip tanks for nearly 6 years (Magadla, pers. comm., 2001). The ticks in this area had, therefore, been exposed to high levels of both pyrethroid and organophosphate dips in the field.

R. evertsi evertsi which were also collected from the Mozana and Ntubeni communal dip tanks were found to be susceptible to all acaricides tested (Table 5). A. hebraeum and R. appendiculatus, which are both three-host ticks, were all susceptible to the acaricides tested.

Only at dip tank in one site (Cizele) did A. hebraeum showed emerging resistance to Chlorfenvinphos (Table 5). Baker et al. [32] previously reported that A. hebraeum was fully susceptible to Chlorfenvinphos in dip tanks in the East London and Willowvale districts.

Breakdown in dipping is a common phenomenon in the Eastern Cape and it is often due, not to acaricide failure to control ticks, but rather to incorrect acaricide application [23]. During the current study period, acaricide failure at the Ntubeni communal dip tank in the Willovale district (Table 5) resulted in loss of tick control. The communal farmers complained that the ticks were not killed after dipping due to the use of inefficient acaricide. This coincided with heavy tick burdens of *A. hebraeum, B. decoloratus, R. appendicidatus* and *R. evertsi evertsi* on cattle. *In vitro* laboratory tests, however, indicated that all the ticks at this dip tank were susceptible to the test acaricides, although emerging resistance to Chlorfenvinvos was recorded in *B. decoloratus* (Table 5). This may be because of the improper replenishing of the tank at the recommended concentration, practiced at the site (Ntubeni) resulting in poor tick control.

Aacricde Resistance of Ticks on the Commercial Farms

The *B. decoloratus* population on the commercial farms was quite resistant to Chlorfenvinphos and moderately resistant to Amitraz and Cypermethrin (Table 6). Preliminary results from the National Tick Resistance Survey (NTRS), however, indicated that B. decoloratus was very resistant to Cypermethrin, partially resistant to Chlorfenvinphos and more susceptible to Amitraz (Fourie, pers. comm., 2001). The resistance status of ticks on a particular farm will, however, depend on the historical pattern of acaricide use.

B. decoloratus demonstrated multiple resistance to both Amitraz and Cypemiethrin (Table 6) in North-West Province of South Africa (Middelfontein), where Amitraz and Cypermethrin mixed with citric or olive oil as a pour-on were in use by the farmers, a practice not recommended by the manufacturers.

B. decoloratus was resistant to Amitraz, at a farm (Basfontein) where, only four concentrations were used per test as opposed to the standard seven. Many commercial dairy farmers in the East London district use Ektoban®, which is a mixture of Cymiazol and Cypermethrin. *B. decoloratus* resistance to Amitraz on these farms might have been due to cross-resistance with Cymiazol, an active ingredient belonging to the same chemical group as Amitraz. Nolan [33] reported high levels of B. microplus resistance to Amitraz with cross-resistance to a closely related Amidine, Cymiazol, in Australia. Amitraz-resistant strains have been reported in B. *decoloratus* [34] and more recently in *A. hebraeum* and *B. microplus* (Taylor, pers. comm., 2001), in South Africa,. In the current study, only *B. decoloratus* was collected for resistance testing, as *B. microplus* was not found during the study period. Chlorfenvinphos resistant *B. decoloratus* was found on two (Brycedale, Welgevind) of the three dairies (Brycedale, Welgevind and Sunny Grove) in East London district.

Differences in Selection Pressure in One- and Multi-Host Ticks

It was clear from the results that *B. decoloratus* (one-host tick) had developed significant resistance to all the currently used acaricides. There was much less resistance in *A. hebraeum, R. appendiculatus* and *R. evertsi evertsi*, in which much less contact with acaricides decreased the selection pressure for resistance.

The reasons for the more rapid selection for resistance in one-host ticks are numerous and include a shorter generation time [16] which allows more frequent exposure to the acaricides

[35]. The shorter life cycle also leads to a quick succession of generations of ticks exposed to the chemicals, resulting in the selective elimination of the majority of susceptible individuals in the tick population [12]. B. decoloratus appears to have as many as four generations per year [36], while the three-host ticks have a single generation which may extend over one, two or even three years [35].

Other factor which can delay the development of resistance is the availability of alternative hosts [37]. Two- and three-host ticks spent a greater length of their life cycle off the host and have a broader host range, such as wild animal hosts [15]. This would help to reduce the selection pressure from acaricides on the multi-host ticks compared to the one-host ticks. The extended period of the use of acaricides coupled with the high frequency of acaricide application might also have increased the resistance problems on these farms. Solomon [38] reported that high dipping frequencies removes susceptible ticks, leaving only resistant males to mate with resistant females, leading to the proliferation of resistant ticks.

Boophilus decoloratus is a one-host tick and all stages occur mainly on cattle at the same time, taking three weeks from the time the larvae attach to when the adults detach [39]. Due to short lifecycle, this tick species is able to pass through two to four generations per year [36] remaining for 42 63 days on the host each year and can potentially be exposed nine times during the year if weekly dipping is practiced (Horak, pers. comm., 2001). The one-host tick lifecycle implies that all the three parasitic stages of development are exposed to acaricides at each dipping, hence increasing the likelihood of selection for resistance.

Rhipicephalus evertsi evertsi is a two-host tick [40]. It takes 14-18 days from the time the larvae attaches to the time the engorged nymphs drop. The adult tick remains attached for seven days. This tick has a continuous lifecycle in warmer months and can probably complete two or more life cycles in a year. Consequently it is on the host for 42 days or more during the year [19]. Larvae and nymphs may share the same host as the adults [41], which feed on cattle, sheep, goats, horses, zebras and eland [42] and consequently may not be exposed to acaricide at each dipping. However, if only cattle are present at a locality, all developmental stages could be exposed to an acaricide applied at weekly intervals on six occasions during a year and hence the pressure to select for resistance would be high.

Amblyomma hebraeum is a three-host tick [43] with one generation per year [44] in which the larvae, nymphs and adults feed on separate hosts [40]. Cattle are the preferred host of the adult ticks [45], which also feed on a wide range of other species including sheep, goats, horses and donkeys [46]. The immature stages feed on a wide range of hosts including birds and small and large mammals and consequently might not be exposed to acaricides at all [47]. The larvae and nymphs of this tick species spend one week each on an animal while adult females and males spend 1-2 and 1-10 weeks, respectively on an animal. The total time spent per annum on hosts is approximately 21 days or more.

Rhipicephalus appendiculatus is a three-host tick [19] with cattle as the preferred domestic host of all stages of development [48]. However, sheep, goats, horses, donkeys and mules are also parasitized to a lesser extent. The larvae and nymphs can utilize a number of host species [16]. Total time on hosts is 15-21 clays in a year and consequently there are only five days a year within a particular tick's life cycle when the adults can potentially be exposed to acaricides on the host, hence the low level of resistance in this tick species. In conclusion the current study supports the hypothesis that single-host ticks develop resistance faster than multi-host ticks.

3. *In Vitro* Larval and Adult Bioassay Methods to Determine Acaricide Resistance at Commercial Dairy Farms

3.1. Introduction

A number of laboratory and field test methods have been developed for the detection of resistance of ticks to acaricides. These techniques usually involve the use of either larval or adult ticks [38]. The acaricide resistance tests used in this study were the Shaw Larval Immersion Test (SLIT) and the Adult Immersion Test (AIT) which include both the Reproductive Estimate Test (RET) and the Egg Laying Test (ELT).

The SLIT is based on using the larval stage of ticks to detect resistance. This method was first described by Shaw [23] and subsequently modified to include a longer holding period for larval ticks after treatment [24].

The RET uses field collected engorged female ticks, which are then tested directly in the laboratory for resistance [49]. The RET was first described by [49]., [49], who used it for evaluating the efficacy of new acaricides, acaricide resistance testing and toxicity studies [50]. Engorged adult female ticks are sometimes preferred to the larvae for testing for acaricide resistance because of their greater tolerance to chemicals [51]. With multi-host ticks, when it is possible to obtain sufficient adult ticks, at the same stage of engorgement, then the bioassay on adult ticks has certain advantages. The test can be used directly on the adult ticks and quicker response is obtained when screening for tick resistance [38]. One disadvantage of this test is that it is difficult to obtain enough ticks and difficult to assess the mortality in the relatively immobile engorged female ticks (Spickett, pers. comm., 2001). To overcome this, Drummond, et al. [49] suggested that a ratio of reproductive efficiency be used as an assay criterion. This method is being assessed for resistance testing at several laboratories in the world but has not been standardized (Nari, pers. comm., 2001).

The ELT is an assessment of oviposition and is the same as the initial stage of the RET. The ELT is based on a comparison of the number of eggs laid by engorged treated field ticks compared with engorged control female ticks. The method is similar to the Adult Immersion Test-Discriminating Dose (AIT-DD) which is in the process of development by the Food and Agriculture Organization (FAO) of the United Nations (Nari, pers. comm., 2001).

3.2. Materials and Methods

The study was made in dairy farms at Brycedale, Sunny Grove and Welgevind in the East London District, Eastern Cape Province, South Africa. Ektoban®, (a mixture of Cymiazol and Cypermethrin) was being used by all farmers during tick collection for the current study. However, tick control failure has been reported on most of these dairies.

Adult and larval resistance tests were then carried out to investigate the breakdown in tick control. Engorged female B. decoloratus were collected from each dairy farm and comparative laboratory tests on the larval progeny and the adults of B. decoloratus were made. A comparison of the susceptibility of B. decoloratus to different acaricides was done by using the SLIT, RET and ELT.

Acaricide resistance testing procedures: The commercial acaricides used to test the susceptibility of B. decoloratus in the three tests were the same as described earlier (Amitraz, Cypermethrin, and Chlorfenvinphos). The SLIT was conducted at the Acaricide Resistance Laboratory (ARL) of the Department of Zoology and Entomology, the University of the Free State, South Africa following the method previously described by Shaw [23]. The data was analyzed as previously described [25].

For the RET, engorged female B. decoloratus of uniform size and free from visible abnormalities were collected from the dairy farms and tested by measuring the RET. The laboratory protocol which was first described by Drummond et al [49] had been slightly modified by the South African Bureau of Standards (SABS) in East London, South Africa (Strydom, pers. comm., 2001). Briefly, field collected engorged female ticks were washed in water and air dried [52] and divided into groups according to size. Groups of ten ticks were weighed and randomly allocated into two and four replicates for each treatment and control groups. Concentrations of the acaricide used were the recommended field concentrations, i.e. 0.025%, 0.05% and 0.015% for Amitraz, Chlorfenvinphos and Cypermethrin, respectively.

The treatment groups were immersed in field concentrations of the three test acaricides and the control group was immersed in water. These were incubated at 27°C and 80 - 90% R.H. At the end of the hatching period, which normally takes 42 days for B. decoloratus, the Reproductive Estimate (RE) was calculated by estimating the number of larvae which had hatched using a numerical scale of 0 to 4 as follows:

Percentage (%) hatch

0	1	2	3	4
0	<25	25-50	50-75	75-100

The RE for each treatment group was calculated as follows:

$$RE = \frac{m_1 \times n \times h}{m_2 \times s \times 4}$$

m_1 = mass of eggs per treatment group (mg); m_2 = mass of engorged female ticks per treatment; group (mg); n = number of ticks per treatment group; h = hatchability of the eggs (scale of 0 to 4); s = number of female ticks alive after seven days of incubation.

The percentage Reproductive Estimate (% RE) was calculated by dividing the RE of female ticks treated with test acaricide by the RE of untreated (control) female ticks multiplied by one hundred:

$$\% RE = \frac{\text{RE of female ticks treated with test acaricide X 100}}{\text{RE of untreated (control) female ticks}}$$

The tick population was considered resistant if the % RE was greater than 80 % and susceptible if less than 80 %.

In the ELT, fully engorged female B. decoloratus of uniform size and free from any visible abnormalities were collected from the commercial dairies in the East London district, Eastern Cape Province, South Africa. The ticks were initially washed in distilled water and air dried at room temperature. They were then divided into groups (ten ticks per group) according to size with ten ticks randomly allocated to two or four replicates for each treatment and control groups. Concentrations of acaricide were 0.025%, 0.05% and 0.015% for Amitraz, Chlorfenvinphos and Cypermethrin, respectively. The AIT-DD being tested by FAO uses a Discriminating Dose (DD) of acaricide and this has only been worked out for B. microplus (Nari, pers. comrn., 2001). The initial procedures for this test were the same as that already described for the AIT.

After seven days of incubation at 27°C and 80 - 90% R.H., the number of engorged female B. decoloratus, which had laid eggs, was assessed. With the ELT test, engorged female B. decoloratus which have been immersed in water (control ticks) should lay a normal quantity of eggs within seven days in an incubator. Ticks treated with an acaricide, which still lay as many eggs as the control ticks, are considered to be resistant. Those treated with acaricide and do not lay eggs, are considered susceptible. In this study, the ticks were considered resistant if the percentage resistance (% R) was greater than 80% on the seventh day after incubation. They were considered susceptible if they lay less than 80% when compared with the eggs laid by the control. The results of the ELT are summarized in Table 9.

The percentage resistance (% R) obtained with the different test acaricides was calculated by dividing the number of treated engorged female ticks laying eggs (egg laying response) with the number of untreated (water control) ticks laying eggs multiplied by a hundred as follows:

$$\% R = \frac{\text{No. of treated engorged female ticks laying eggs}}{\text{No. of untreated control engorged female ticks laying eggs}} \times 100$$

3.3. Results

Shaw Larval Immersion Test (Slit)

The results of the SLIT used to determine the susceptibility of the B. decoloratus larvae, from the three commercial dairy farms, to Amitraz, Chlorfenvmphos and Cypermethrin are summarized in Table 7.

All B. decoloratus populations from Brycedale, Sunny Grove and Welgevind commercial dairies showed resistance to the test acaricides. There was no resistance at two of the sites (Sunny Grove and Welgevind) for Amitraz, and an emerging resistance was observed at one site (Brycedale). There was emerging resistance for Chlorfenvinphos at Sunny Grove and full resistance at Brycedale and Welgevind, which resulted in the known control problems. For Cypermethrin, there was no resistance at Brycedale and Welgevind but a high FOR was recorded at Sunny Grove.

Reproductive Estimate Test (Ret)

The mean engorgement weights of the engorged female ticks (m_2), the total mass of the eggs (m_1), the hatchability (h) of the eggs to larvae, the RE, and the % RE of *B. decoloratus* immersed at recommended field concentrations of acaricides are illustrated in Table 8.

B. decoloratus is resistant if %RE is greater than 80% and susceptible if %RE is less than 80%.

Table 7. *In vitro* larval bioassay: Results of the susceptibility of B. decoloratus larvae from three commercial dairy farms in East Londion District, Eastern Cape Province to Amitraz, Cypermethrin and Chlorfenvinphos

Farm Name	Active compound								
	Amitraz			Cypermethrin			Chlorfenvinphos		
	LC_{50}	FOR	Comm.	LC_{50}	FOR	Comm.	LC_{50}	FOR	Comm.
Ref. strain	4.2×10^{-5}			4.1×10^{-4}			5.7×10^{-5}		
Brycedale	3.3×10^{-3}	77.751	ER	2.6×10^{-3}	6.199	R	2.4×10^{-3}	42.049	S
Sunny Grove	1.3×10^{-3}	31.340	S	2.0×10^{-3}	4.860	ER	2.0×10^{1}	>200	R
Weigevind	4.1×10^{-5}	0.971	S	2.4×10^{-3}	5.920	R	6.8×10^{-4}	12.032	S

	Amitraz FOR	Cypermethrin FOR	Chlorfenvinphos FOR
R	>100	>100	>5
ER	50-100	50-100	2.5-5
S	<50	<50	<2.5

R= Resistant; ER= Emerging resistance; S= Susceptible; FOR= Factor of resistance; Comm.= comments; LC_{50} is percent concentration.

B. decoloratus was resistant to Amitraz (% RE = 84.83), and Chlorfenvinphos (% RE = 80.34) at one farm (Brycedale). Poor percentage control at this farm against all the three acaricides tested supported the field observations, which recorded difficulty to control heavy burdens of B. decoloratus. No resistance was detected at either Sunny Grove or Welgevind daily farms, which further supports the field observations.

The mean weight of the egg masses produced by the water treated control females (0.109g) was almost 2.2 times greater than the egg masses produced by the acaricide treated females (0.049g). Similarly, the hatching capacity (h) of the eggs in the water treated control females was higher than the hatching capacity of the eggs obtained from the treated females. It was additionally observed that, the RE of all water treated females was greater than those of the acaricide treated females, despite no significant difference between the mean weight of the female ticks allocated per treatment group (0.215g) and mean weight of female ticks allocated per control group (0.218g).

Egg Laying Test

At one of the three dairy farms (Brycedale), acaricide treated female B. decoloratus were resistant to Amitraz (84.2%) and Chlorfenvinphos (94.7%) but susceptible to Cypermethrin (73.7%) (Table 9). The population of B. decoloratus at one other farm (Sunny Grove) was susceptible to Amitraz, Chlorfenvinphos and Cypermethrin. At the remaining farm (Welgevind), resistance was detected only to Cypermethrin (82.1%), whilst Amitraz and Chlorfenvinphos still successfully controlled B. decoloratus.

$$\% R = \frac{\text{No. of treated ticks laying eggs (C)}}{\text{No. of untreated ticks (control) laying eggs (D)}} \times 100$$

B. decoloratus is resistant if % R is greater than 80%.

Table 8. Percentage resistance, percentage reproductive estimate, reproductive estimate and percent control of female B. decoloratus from three commercial dairies after treatment with different acarieides (Amitraz, Chlorfenvinphos and Cypermethrin) compared with a water control

Daily farms	Acaricide	NO. of ticks immersed & tested	Survivors after 7 days	*Mass of female (mg)	+Mass of eggs (mg)	h	RE	%RE	°R
Brycedale	Amitraz	20	20	3.790	1.503	4.00	0.397	84.83	R
	Chlorfenvinphos	20	20	3.771	1.620	3.50	0.376	80.34	R
	Cypennethrin	20	18	3.822	1.406	3.00	0.307	65.60	S
	D* = Control	20	19	3.791	1.687	4.00	0.468		
Sunny Grove	Amitraz	20	20	4.573	1.154	2.00	0.126	25.45	S
	Chlorfenvinphos	20	20	4.446	0.003	0.00	0.000	0.00	S
	Cypennethrin	20	20	4.397	1.629	3.00	0.278	56.16	S
	D* = Control	20	20	4.543	2.250	4.00	0.495		
Welge-vind	Amitraz	40	40	8.623	0.015	0.00	0.000	0.00	S
	Chlorfenvinphos	40	37	8.546	0.721	1.75	0.040	7.46	S
	Cypennethrin	40	40	8.496	3.545	3.00	0.313	58.40	S
	D* = Control	40	39	8.670	4.527	4.00	0.536		

* = Engorgement weight of treated female ticks; + = Total mass of eggs per treatment group; h= hatchability estimate (0-4); RE= Reproductive Estimate; °R= Degree of resistance; % RE = Percentage Reproductive Estimate.

Table 9. Egg-laying response of engorged adult female B. decoloratus after immersion in three different acaricide groups and then incubated for seven days

Daily farms	Acaricide	A	B	C	%R	°R
Brycedale	Amitraz (250 ppm)	20	20	16	84.2	R
	Chlorfenvinphos (500 ppm)	20	20	18	94.7	R
	Cypennethrin (150 ppm)	20	19	14	73.7	S
	D* = Control	20	19	19		
Sunny Grove	Amitraz (250 ppm)	20	20	7	38.8	S
	Chlorfenvinphos (500 ppm)	20	20	1	5.6	S
	Cypennethrin (150 ppm)	20	20	13	72.2	S
	D* = Control	20	20	18		
Welgevind	Amitraz (250 ppm)	40	40	0	0.0	S
	Chlorfenvinphos (500 ppm)	40	37	5	12.8	S
	Cypennethrin (150 ppm)	40	40	32	82.1	R
	D* = Control	40	39	39		

ppm = parts per million; A= Number of engorged female B. decoloratus treated with acaricide (treatment group); B = Number of female ticks alive after seven days of incubation; C = Number of females laying eggs after seven days of incubation; % R = Percentage resistance.

3.4. Discussion

The resistance of B. decoloratus on the three farms, to Chlorfenvinphos was greater than of either Amitraz or Cypermethrin. This is surprising, since with the exception of Brycedale, organophosphate (OP) acaricides had not been used for the past ten years, on any of the three farms. As resistance to Chlorfenvinphos was still detected, one can deduce that once OP resistance has become established in a tick population, reversing back to susceptibility is either very slow or does not occur [53].

Other OP acaricides, which had been used earlier, might have produced cross-resistance to the OP compounds, since resistance to one member of a group of chemically similar acaricides can result in a degree of resistance to the other members of the same group [10, 17, 54]. Cross-resistance to B. decoloratus within groups of chemically related acaricides has been previously documented in the area [32].

Comparison of the Shaw Larval Immersion Test (SLIT), the Reproductive Estimate Test (RET) and the Egg Laying Test (ELT)

The susceptibility of B. decoloratus to various acaricides was determined using the SLIT [23], the RET and the ELT [49]. The period required to detect resistance in B. decoloratus was seven, 42 and 60 days of incubation using the ELT, RET and SLIT, respectively. The ELT was far quicker than either the RET and SLIT.

B. decoloratus from Brycedale dairy farm showed resistance to Amitraz, and Chlorfeuviuphos when tested with the ELT and RET. However, it showed resistance to Chlorfenvinphos, and emerging resistance to Amitraz.

Table 10. Comparison of the results of the three tests used on the different farms

Farms	Tick species	Test methods		
		SLIT	RET	ELT
		A Ch Cy	A Ch Cy	A Ch Cy
Bryceclale	B. decoloratus	ER R S	R R S	R R S
Sunny Grove	B. decoloratus	S ER R	S S S	S S S
Welgevind	B. decoloratus	S R S	S S S	S S R

SLIT = Shaw Larval Immersion Test; RET = Reproductive Estimate Test; ELT = Egg Laying Test; A = Amitraz ; Ch =Chlorfenvinphos; Cy = Cypermethrin; S =Susceptible; ER = Emerging resistance; R = Resistance.

The population *of B. decoloratus* from Sunny Grove dairy, was susceptible to Amitraz and Chlorfenvinphos with the RET and susceptible to Cypermethrin (% RE=72%) with ELT, while it showed resistance to Cypermethrin and emerging resistance to Chlorfenvinphos with SLIT. However, Chlorfenvinphos had never been used for tick control on this farm.

At Welgevind daries, the ELT method detected Cypermethrin resistant *B. decoloratus,* whilst resistance to Chlorfenvinphos was detected with the SLIT.

Resistance to Cypermethrin was detected on two of the farms using the ELT while it was detected only on one farm using the SLIT. Ektoban®, a mixture of Cymiazol and Cypermethrin, has been used for tick control for nearly ten years on these farms. According to the farmers it is no longer effective in controlling the ticks. This would indicate that the tick resistance was to one or both of the active ingredients in the mix, probably to Cypermethrin. Coetzee et al. [55] demonstrated that the development of resistance to one of the pyrethroids took 18 months after intensive application as an acaricide.

From the results of the current study, it appears that the field populations of *B. decolotatus* were more resistant to Chlorfenvinphos, moderately resistant to Cypermethrin and less resistant to Amitraz. There was also a good agreement between the high field burdens of *B. decoloratus* observed on cattle at Brycedale dairy, where the tick showed resistance with ELT and RET tests.

In summary, resistance to Amitraz was detected at Brycedale using the RET and the ELT with an indication of emerging resistance using the SLIT methods. Resistance to Chlorfenvinphos was detected at this farm using the SLIT, the RET and the ELT and at Welgevind dairy using the SLIT. Resistance to Cypermethrin was detected at Welgevind with the ELT and at Sunny Grove with the SLIT methods. While there was general agreement between the results from SLIT, RET and ELT, some refinement of the techniques and further sampling is needed for direct comparison of the three tests. In addition, the ELT and the RET need to be refined by using the Adult Immersion Test-Discriminating Dose (AIT-DD) test.

4. General Discussion

Resistance Status of Ticks Collected from the Communal and Commercial Farms

The *in vitro* larval tests indicated that the B. decoloraius population in the study areas had developed a high degree of resistance to both Cypermethrin and Chlorfenvinphos with higher levels of resistance than seen in A. hebraeum, R. appendiculatus and R. evertsi evertsi. Previous studies [11, 15, 37, 56] support this finding as the development of resistance in one-host ticks is usually faster than in two- and three-host ticks. The rate of development of tick resistance against acaricides is also linked to the degree of dominance of the resistant alleles, the frequency of acaricide application and the strength of the acaricide used [11, 53].

Our results from the commercial farms are similar to that of the National Tick Resistance Survey, in South Africa, as both surveys demonstrated high resistance in B. decoloraius to pyrethroids (Fourie, pers. comm., 2001). The pyrethroid acaricides have a long residual activity [57] and are consequently very effective in controlling ticks when compared with Amitraz or Chlorfenvinphos. The pyrethroids rapidly eliminate all susceptible members of the target tick population, resulting in a higher incidence of inter-breeding between resistant members. This soon leads to a higher proportion of genetically resistant offsprings and an overall increase in the resistance in the tick population [38].

Multiple resistances to more than one chemical group of acaricides were reported at some of the commercial farms. Multiple resistances at Brycedale to all three actives tested were extremely worrying, as dipping was no longer controlling the B. decoloratus population. At Middelfontein, B. decoloratus was resistant to both Amitraz and Cypermethrin. In the National Tick Resistance Survey, multiple resistances by B. decoloratus to Chlorfenvinphos and Cypermethrin (Fourie, pers. comm., 2001) were reported from the same farm. One must be careful not to change the dipping program from Cypermethrin to another pyrethroid since the use of alternative pyrethroids might accelerate the selection process, leading to an even more severe pyrethroid resistance [58].

In the study areas, few A. hebraeum and R. evertsi evertsi were collected, however, R. appendiculatus was more numerous in the communal farming areas, probably as a result of more scrub and bush in these areas which protects the free-living immature. More tick resistant Zebu type cattle [59] were common in the traditional grazing areas which, might lead to decreased tick burdens. In Uganda, however, Kaiser, Sutherst, & Bourne [60] found that indigenous Zebu cattle carried heavier burdens of R. appendiculatus and concluded that the resistance of indigenous Zebu to this species was not as strong as to other tick species.

Previous DDT-resistance in the field tick population might have resulted in the pyrethroid resistance in East London. This link between the use of the pyrethroid acaricides for tick control and cross-resistance to DDT [61] shows that once ticks have acquired resistance to an acaricide then the ability is retained long after the acaricide has been replaced [62].

Comparison of the Different Resistance Testing Methods Used during this Study (SLIT, RET and ELT

An increased awareness of acaricide resistance has led to the multiplication of a range of bioassay methods used at different laboratories. The *in vitro* methods include the "gauze-bag" technique [51], the "teabag" method [63] and the "pipette" bioassay method [64]. At present, however, the most widely used bioassay methods are the AIT (49], the SLIT (23] and the LPT [65]. During this study we used the SLIT and ELT and the RET methods which are both part of the AIT. All these test methods have their advantages and disadvantages.

The SLIT has the advantage that it uses unfed larvae, which are more easily standardized, and the mortality of the larvae can be recorded easily [35]. The larvae are also treated identically which leads to more statistically credible results. One disadvantage of this method, however, is that it magnifies the factors of resistance (FOR). In addition, the exposure of tick larvae for ten minutes in an emulsion of a commercial acaricide is not a satisfactory imitation of the field situation [66].

The ELT and the RET use commercial acaricides at the recommended field concentration to immerse the female ticks [49]. The big advantage of both the ELT and the RET over the SLIT is that the adult tests can be interpreted earlier, i.e. seven days with the ELT and 42 days with the RET. One disadvantage of both the ELT and RET is that sufficient numbers of folly engorged female ticks are sometimes not available to do the test. Another disadvantage of the ELT and RET is that female ticks may have already started to lay eggs before they reach the laboratory. Since the adult tests are not yet standardized so any comparison of results from different laboratories would be difficult (Kenip, pers. Comm., 2001). It is also not clear how much of the percentage resistance (% R) in this study was due to loss of control through acaricide resistance in the field. In the present study, however, if the %R was above 80% then resistance was considered to be present.

Although the three test methods could not be compared statistically, the ELT and the RET in most cases showed similar acaricide resistance results which differed from the SLIT. Stendel [67] also noted that the AIT mimicked the field conditions better than the SLIT and he also noted that there was poor correlation between the larval and adult test results. Malan [68] and Nari [69] also reported that data from engorged female ticks did not coincide with similar data from the larval tests.

The acaricide resistance tests differ from screening tests, which are undertaken to select new acahcides because with the former the acaricide is used as a standard to check any changes in the ticks, whilst in the screening tests, the ticks are used to evaluate the chemicals. Naturally there would be advantages in having the same test methods for screening acaricides and detecting acaricide resistance [70].

CONCLUSIONS

Larvae were obtained from engorged adult female A. hebraeum, B. decolaratus, R. appendiculatus and R. evertsi evertsi, from six communal and six commercial farms in five districts of the Eastern Cape and North-West Provinces, South Africa. The larvae were then tested for resistance against the three most important acaricide groups used in these areas, namely, the formamidines (Amitraz), the organophosphates (Chlorfenvinphos) and the pyrethroids (Cypermethrin).

Our results support the hypothesis that single-host ticks develop resistance much faster than multi-host ticks. This trend was recorded on all the farms where single-and multi-host ticks co-existed. It was concluded that the use of acaricides at high frequencies and high concentrations was one of the underlying causes of the development of tick resistance in the study area. The high dipping frequency and over use of the acaricides removed all the susceptible ticks from the population leaving only resistant males to mate with resistant females, leading to a population of homogenous resistant ticks.

The efficacy of the different acaricides in controlling the tick populations at the communal and commercial farms also varied considerably. On the communal farms, Amitraz was highly effective in controlling both single- and multi-host ticks, while Cypermethrin and Chlorfenvinphos still controlled most of the multi-host ticks.

The populations of B. decoloratus on the majority of the commercial farms were equally resistant to Amitraz, Cypermethrin and Chlorfenvinphos. The A. hebraeum, R. appendiculatus and R. evertsi evertsi populations on the commercial farms were, however, still susceptible to the acaricides used. Resistant strains of B. decoloratus were common on both the communal and the commercial farms. Cypermethrin and Chlorfenvinphos were, however, still effective in controlling multi-host ticks. In general, higher level of tick resistance to Amitraz was observed on commercial farms than on the communal farms.

A comparison of the three *in vitro* acaricide resistance testing methods (SLIT, RET and ELT) was done on the B, decoloratus populations from three dailies with known acaricide resistance problems. The ELT was able to give results after seven days and it was suggested that it could be used to screen acaricide resistance. The ELT detected resistance at Bryceclale dairy to Amitraz and Chlorfenvinphos. The RET, which detects resistance within 42 days, also detected severe acaricide resistance at the Brycedale dairy, but not at the other two dairies. The SLIT, which detects resistance after 60 days found OP resistance as well as emerging resistance to Amitraz at Brycedale and pyrethroid resistance on Sunny Grove. The SLIT also indicated that the tick population at Brycedale was susceptible to Cypermethrin.

In general there was a good correlation between the RET and the ELT, while in many cases there was poor correlation between the SLIT and the two AIT methods (RET and ELT). From this study it appears that the ELT was a good method to detect resistance within seven days, as opposed to the 42 days required for the RET and the 60 days for the SLIT. The ELT and the RET could possibly be used as screening methods to detect acaricide resistance on farms while the SLIT would remain the test of choice for national surveys. In addition the ELT is less costly and does not require sophisticated equipment for field testing, compared with other *in vitro* test methods. This method, however, still needs to be validated and standardized for use in South Africa and the rest of Africa where tick control is important.

Presently, in both communal and commercial farming sectors, cattle were kept free of ticks by regular application of acaricides, which is usually done irrespective of tick burdens. To delay acaricide resistance, this approach needs to be re-evaluated and a shift towards less intensive and more tactical tick control needs to be implemented. Strategic tick control with the application of acaricides only at critical times of the year and the use of locally adapted Zebu cattle should be considered to maintain endemic stability to the TBDs. The findings in this study would benefit planning strategies to develop resistance-management programs worldwide, particularly in tropical countries aspiring to start acaricide resistance testing programs.

REFERENCES

[1] Bram, R.A. (1983). Tick-borne livestock diseases and their vectors: The global problem. In: *Ticks and Tick-home Diseases, FAO Animal Production and Health.* Food and Agriculture Organization, Rome, pp. 7-11.

[2] De Castro, J.J. (1997). Sustainable tick and tickbome disease control in livestock improvement in developing countries. *Veterinary Parasitology* 71: 77-97

[3] McCosker, J. (1979). Global aspects of the management and control of ticks of veterinary importance. *Recent Advances in Acarology*, 2: 45-53.

[4] TEEL, P.O. (1985). Ticks. In: *Livestock Entomology*. Eds. Williams RJHL, Hall, R.D., Broce, A.B. & Scholl, PJ. Wiley-interscience. New York, Chichester, Brisbane, Toronto, Singapore, pp. 129-149.

[5] Jongejan, F. & Uilenberg, G. (1994). Ticks and control methods. *Revue Scientiflque et Technique de Office International des Epizooties,* 13: 1201-1226.

[6] Dipeolu, O. O. & Ndungu, J.N. (1991). Acaricidal activity of Kupetaba, a ground mixture of natural products against R. appendiculatus. *Veterinary Parasitology,* 38: 327-338.

[7] Bekker, P.M., (1960). The history of dipping. *Veld,* 20: 1-5.

[8] Du Toit, R., Graf, H. & Bekker, P.M. (1941). Resistance to arsenic as displayed by the single host blue tick B. decoloratm (Koch) in a localized area of the Union of South Africa. *Journal of the South African Medical Association,* 12:50-58.

[9] Whitnall, A. B. M. & Bradford, B. (1947). An arsenic resistant tick and its control with ganimahexane dips. *Bulletin Entomological Research*, 38: 353-372.

[10] Whitehead, G. B. (1959). Pyrethrum resistance conferred by resistance to DDT in the blue tick. *Nature,* 184: 378 -379.

[11] Whitehead, G. B. & Baker, J.A.F. (1961). Acaricide resistance in the red-legged tick Rhipicephalus evertsi (Neumann). *Bulletin of Entomological Research,* 51: 755-764.

[12] Matthewson, M. D. & Baker, J. A. F. (1975). Arsenic resistance in multi-host ticks in the South Africa and Swaziland. *Journal of the South African Veterinary Association,* 46: 341-344.

[13] Baker, J. A. F. & Shaw, R. D. (1965). Toxaphene and lindaiie resistance in R. appendiculatm, the brown ear tick of equatorial Africa and South Africa. Journal of the South African Veterinary? *Medical Association*, 36: 321-330.

[14] Baker, J. A. K, Jordaan, J. O & Robertson, W. D. (1979). Ixodicidal resistance in Boophilus microplus (Canestrini) in South Africa and the Transkei. *Journal of the South African Veterinary Association*, 50, 296-301.

[15] Nolan, J. (1990). Acaricide resistance in single and multi-host ticks and strategies for control. *Parasitologia,* 32: 145-153.

[16] Norval, R.A.I., Barrett J.C., Perry B.D. & Mukhebi A.W. (1992). Economics, epidemiology and ecology: A multidisciplinary approach to the planning appraisal of ticks and tick-borne disease control in southern Afiica. In: Fivaz B, Petney T. and Horak IG. (eds). *Tick Vector Biology, Medical and Veterinary Aspects.* Springer-Verlag, Berlin. Pp. 35-55.

[17] Baker, J.A.F. (1982). Some thoughts on resistance to ixodicides by ticks in South Africa. *Symposium on ectoparasites of cattle, 15 to 16 March 1982.* South African Bureau of Standards, Pretoria, South Africa. Pp. 53-67.

[18] Hoogstraal, H., (1956). African Ixodoidea. I. *Ticks of the Sudan (with special reference to Equatoria province and with preliminary reviews of the genera Boophilus, Margaropus and Hyalomma*) Pp.1101. Washington DC., Department of Navy, Bureau of Medicine and Surgery

[19] Walker, J.B., Keirans, I.E. & Horak, I.G. (2000). *The genus Rhipicephalus (Acari, Ixodidae): a guide to the brown ticks of the world.* Cambridge University Press, Cambridge.

[20] Walker, A.R., Bouattour, A., Camicas, J.L., Estrada-Penã, A., Horak, I.G., Latif, A.A., Pegram, R.G. and Preston, P.M., 2003. *Ticks of domestic animals in Africa: a guide to identification of species.* Pp. 1-219.

[21] Pegram, R.G., Clifford, C.M., Walker, J.B. and Keirans, J.E., (1987a). Clarification of the Rhipicephalus sanguineus group (Acari, Ixodoidea, Ixodidae). I. R. sulcatus Neumann 1980 and R. turanicus Pomarantsev, 1936. *Syst. Parasitol.* 10, 3-26.

[22] Pegram, R.G., Walker, J.B., Clifford, C.M. and Keirans, J.E., (1987b). Comparison of populations of the Rhipicephalus simus group: R. simus, R. praetextatus, and R. muhsamae (Acari: Ixodidae). J. *Med. Entomol.* 24, 666-682.

[23] Shaw, R.D. (1966). Culture of an organophosphoras resistant strain of BoopMlus microphis (Can.) and an assessment of its resistance spectrum. *Bulletin of Entomological Research*, 56: 389-405.

[24] Shaw, R.D., Cook, M. & Carson, R.E. (1968). Developments in the resistance status of the southern cattle tick to organopuosphonis and carbamate insecticides. *Journal of Economic Entomology*, 61:1590-1594.

[25] Wilson, I.T. (1980). Resistance testing techniques. In*: Ticks and tick-borne diseases. Australian Development Assistance Course*. Queensland Agricultural College, Lawes, Queensland, Australia.

[26] Abbott, W.S. (1925). A method for computing the effectiveness of insecticides. *Journal of Economic Entomology*, 18: 265-267.

[27] Coetzee, B. B., Stanford, G. D. & Davis, D. A. T. (1987b). The resistance spectrum shown by a fenvalerate-resistant strain of blue tick (B. decoloratus) to a range of ixodicides. *Onderstepoort Journal of Veterinary?* Research, 54: 79-82.

[28] Nolan, J., Roulston, WJ. & Wharton, R.H. (1977). Resistance to pyrethroids in a DDT-resistant strain ofBoophilus microphts. *Pesticide science*, 8: 484-486.

[29] Whitehead, G. B. (1956). DDT resistance in the blue tick, B. decoloratm (Koch). *Journal of the South African Veterinary Medical Association*, 27: 117-120.

[30] Jonnson, N.N. (1997). Control of cattle ticks (Boophilus microphis) on Queensland daily farms. *Australian Veterinary Journal*,, 15: 802-807.

[31] Spickett, A.M. (1998). Acaricides and resistance. Facts, figures and tables concerning tick control. Laboratory Diagnostic Series. Module: Veterinary Ectoparasitology and Protozoology. *Acaricides and Resistance*, 1-13.

[32] Baker, J. A. F, Miles, J. O., Robertson, W. D., Stanford, G. D. & Taylor, R J. (1978). The current status of resistance to organophosphorous ixodicides by the blue tick, B. decoloratm (Koch) in the South Africa and Transkei. *Journal of the South African Veterinary Association*, 49: 327-333.

[33] Nolan, J. (1981). Current developments in resistance to Amidine and pyrethroid tikicides in Australia. In: *Tick biology and control. Proceedings of an International Conference held in Grahamstown,* 27-29 January 1981. Editors, G.B. Whitehead and J.D. Gibson. Tick Research Unit, Rhodes University, Grahamstown, South Africa, pp. 109-114.

[34] Taylor, R.J. & Oberem, P. (1995). Some characteristics of an Amitraz resistant strain of B. decoloratus originating from South Africa. In: *Proceedings of the Second International Conference on Tick-borne Pathogens at the Host-Vector Interface*. Eds. L. Coons & M. Rothschild, p. 62, August, 28 to September 1, 1995, Kniger National Park, South Africa.

[35] Wharton, R.H. & Roulston, WJ. (1970). Resistance of ticks to chemicals. *Annual Review of Entomology*, 15: 381 -404.

[36] Pegram, R.G., Perry, B.D., Mussisi, F.L. & Mwanaumo, B. (1986). Ecology and phenology of ticks in Zambia: Seasonal dynamics on cattle. *Experimental and Applied Acarology*, 2: 25-45.

[37] Kunz, S.E. & Kemp, D.H. (1994). Insecticides and acaricides: resistance and environmental impact. *Revue Scientifique et Technique de Office International des Epizootics,* 13: 1249-1286.

[38] Solomon, K.R. (1983). Acaricide Resistance in Ticks. *Advances in Veterinary Science and Comparative Medicine,* 21: 273-296.

[39] Walker, J.B. (1991). A review of the Ixodid ticks (Acari:, Ixodidae) occurring in South Africa, *Onderstepoort Journal of Veterinary Research,* 58: 81 -105.

[40] Theiler, G. (1943). Notes on the ticks off domestic stock from Portuguese East Africa. *Estacao Anti-Malarica de Lourenco Marques*.55pp.

[41] Norval, R.A.I (1994). Vectors: Ticks. In: Coetzer, J.A.W. Thomson, G.R. and Tiistin, R.C. (Eds), *Infectious Diseases of Livestock with special reference to southern Africa, Volume 1*. Oxford University Press, Cape Town. 3-24.

[42] Horak, I.G., Potgieter, F.T., Walker, IB., De Vos, V. & Boomker, J. (1983). The Ixodid tick burdens of various large ruminant species in South African nature reserves. *Onderstepoort Journal Veterinary Research*, 50: 221-228.

[43] Lounsbury, C.P. (1899). The bout tick, A. hebmeum Koch, its life history and habitats. *Agricultural Journal of Cape Good Hope*, 15: 728-743.

[44] Rechev, Y. (1982). Dynamics of tick populations (Acari: Ixodidae) in the Eastern Cape Province of South Africa. *Journal of Medical Entomology*, 19: 679 — 700.

[45] Horak, 1G. (1982). Parasites of domestics and wild animals in South Africa. XV. The seasonal prevalence of ectoparasites on impala and cattle in the northern Transvaal. *Onderstepoort Journal of Veterinary Research*, 49: 85-93.

[46] Theiler, G. (1962). The Ixodidae parasites of vertebrates in Africa south of the Sahara, Project S 9958. *Report to the Director of Veterinary Services*. Onderstepoort, South Africa,

[47] Horak, I.G., Stoltsz, W.H. & Heyne, H. (2000). Short courses in the identification of southern African ticks. *Department of Veterinary Tropical Diseases, Faculty of Veterinary Science, University of Pretoria*. Pretoria, South Africa.

[48] Yeoman, G.H. & Walker, IB. (1967). The Ixodid Ticks of Tanzania. *A Study of the Zoogeography of the Ixodidae of an East African Country*. London: Commonwealth Institute of Entomology.

[49] Drummond, R.O, Ernest, S.E., Trevino, J.L., Gladney, WJ. & Graham, O.H. (1973). Boophilus anmjlatm and Boophilus microplus: laboratory tests for insecticides. *Journal of Economic Entomology*, 66: 130-133.

[50] Stone, B.F. & Webber, E.G. (1960). Cattle ticks, Boophilus microplus, resistant to DDT, BHC and dieldrin. *Australian Journal of Agricultural Research,* 11: 105-119.

[51] Graham, O.K. & Drummond, O.K. 1964. Laboratory screening of insecticides for the prevention of reproduction of Boophilus ticks. *Journal of Economic Entomology* 9 57:335-339.

[52] Stone, B.F. (1957). Resistance to DDT in the cattle tick, Boophilus microplus (Canestrini). *Australian Journal of Agricutural Research*, 8: 424-431.

[53] Stone, B.F. (1972). The genetics of resistance by ticks to acaricides. *Australian Veterinary Journal*, 48: 345-350.

[54] Shaw, R. D., Thompson, G. E. & Baker, J. A. F. (1967). Resistance to cholinesterase-inhibitors in the blue tick, B. decoloratus, in South Africa. *Veterinary Record*, 81: 548-549.

[55] Coetzee, B. B., Stanford, G. D. & Davis, D. A. T. (1987a). Resistance by the blue tick (B. decoloratus) to the pyrethroid, fenvalerate. Onderstepoort *Journal of Veterinary Research*, 54: 83-86.

[56] Purchase, H.S. (1955). Some thoughts on ticks and their practical control (Part I). *Bulletin Epizootiological Diseases in Africa*, 3: 226-230.

[57] Adams, H.R. (1995). *Veterinary Pharmacology and Therapeutics Seventh edition*, Iowa State University Press, USA.

[58] Beugnet, F. & Chardonnet, L. (1995). Tick resistance in New Caledonia. *Veterinary Parasitology*, 56: 325-338.

[59] Baker, M.K. & Ducasse, F.B.W. (1967). Tick infestation of livestock in Natal. *Journal of the South African Veterinary Medical Association*, 38:447-453.

[60] Kasier, M.N., Sutherst, R.W. & Bourne, A.S. (1982). Relationship between ticks and zebu cattle in southern Uganda. *Tropical Animal Health and Production,* 14: 63-74.

[61] Nolan, J., Wilson, J.T., Green, P.E. & Bird, P.E. (1989). Pyrethroid resistance in field samples in the cattle tick (Boophilus microplus). *Australian Veterinary Journal*, 66: 179-182.

[62] Roulston, W.J. (1980). The history of acaricide resistance in Australia. In: *Ticks and tick-borne diseases. Australian Development Assistance Course.* Queensland Agricultural College, Lawes, Queensland, Australia.

[63] Fiedler, O.G.H. (1968). A new biological method for evaluating the efficiency of acaricides against ticks. *Journal of the South African Medical Association.* 39: 84-87.

[64] Kigaye, M.K. & Matthysse, J.G. (1973). Testing acaricide susceptibility of the brown dog tick. Rhipicephalus sanguinem (Latreille, 1806), with a disposal pipette method. *Bulletin Epizootiological Disease of Africa*, 21: 429-435.

[65] Stone, B.F. & Haydock, K.P. (1962). A method for measuring the acaricide susceptibility of the cattle tick Boophilus microplus (Can.). *Bulletin of Entomological Research,* 53: 563-578.

[66] Lourens, J.H.M.& Shaw, R.D. (1975). *Research on tick-borne cattle diseases and tick control. Kenya, Acaricides and acahcide resistance*. A preliminary report.

[67] Stendel, W. (1980). The relevance of different test methods for the evaluation of tick controlling substances. *Journal of the South African Veterinary Association*, 31: 147-152.

[68] Malan, J.R. (1973). Examples of laboratory dipping of engorged females. Section Dip Research, *Veterinary Research Institute*, Onderstepoort, 19pp.

[69] Nari, A. J. (1981). *The evaluation of in vitro methods of testing for acaiicidal resistance in Boophilus microphts (Canestrini, 1887).* MSc Thesis, University of Pretoria, Pretoria, SouthAfrica.

[70] Busvine, J.R. (1977). International standardization of tests for detecting and measuring resistance of pests to pesticides. In: Crop Protection Agents-their biological evaluation. *Proceedings of the international conference on the evaluation of biological activity, held at Wageningen, The Netherlands, 16^-18* April, 1975.* Edited by N.R. McFarlane. Academic Press. London-New York-Sail Francisco. A subsidary of Harcourt Brace Jovanovich, publisher. Page 557-567.

In: Ticks: Disease, Management and Control
Editor: Moges Woldemeskel
ISBN: 978-1-62081-136-8

Chapter 8

EXPERIENCES ON THE CONTROL OF CATTLE TICK RHIPICEPHALUS (BOOPHILUS) MICROPLUS IN MEXICO

Carlos R. Bautista-Garfias[*,1] ***and Francisco Martínez-Ibañez***[2]

[1]Centro Nacional de Investigación Disciplinaria en Parasitología Veterinaria (CENID-PAVET), Instituto Nacional de Investigaciones Forestales, Agrícolas y Pecuarias (INIFAP), Jiutepec, Morelos, México

[2]Centro Nacional de Sanidad, Inocuidad y Calidad Agroalimentaria (SENASICA), Secretaria de Agricultura. Ganadería, Desarrollo Rural, Pesca y Alimentación (SAGARPA) Jiutepec, Morelos, México

ABSTRACT

Different approaches have been used in Mexico during several years for the control of the cattle tick *Rhipicephalus (Boophilus) microplus,* the vector of two very important diseases of cattle: babesiosis, caused by the haemoprotozoans *Babesia bovis* and *B. bigemina* and anaplasmosis caused by the haemotropic rickettsia *Anaplasma marginale.* The most frequently used methods in Mexico include strategic acaricidal baths, which when used improperly have generated tick populations resistant to some acaricides, use of anti-tick commercial vaccines, and the more recently proposed use of entomopathogenic fungus for biological control of ticks. These methods utilized in Mexico have shown various degrees of success. In this chapter the efficacy of the aforementioned methods of tick control and their drawbacks are discussed. Furthermore, the tendencies for the use of alternative new control procedures, for example the use of the fly *Megaselia scalaris* as a biologic control method as well as the practice of integrated control strategies is described.

[*] Correspondence: E-mail: bautista.carlos@inifap.gob.mx or carlosramon_bautista@mail.com.

INTRODUCTION

Ticks are second only to mosquitoes in the number of diseases vectored [1]. In this context, several piroplasm protozoans (Phylum Apicomplexa) are specific to ticks. Sexual reproduction takes place in the tick (definitive host). Transmission is both transstadial and transovarial [2]. The southern cattle tick *Riphicephalus (Boophilus) microplus* (Canestrini, 1888) is a recurrent vector of protozoan blood parasites of the genus *Babesia,* which causes babesiosis, a disease affecting cattle around the world, as well as the red blood cell bacterium *Anaplasma* spp., the causative agent of the bovine anaplasmosis [3,4].

Riphicephalus (Boophilus) microplus (formerly *Boophilus microplus*) [3] is considered to be the most important tick parasite of livestock in the world. *R. microplus* is a hard tick that can be found on many hosts including cattle, buffalo, horses, donkeys, goats, sheep, deer, pigs, dogs and some wild animals. Heavy tick burdens on animals can decrease production and damage hides and skins. *R. microplus* as previously indicated also transmits babesiosis caused by *Babesia bigemina* and *Babesia bovis* and anaplasmosis caused by *Anaplasma marginale* [4].

Babesiosis or "cattle fever" was eradicated from the United States between 1906 and 1943, by eliminating its vectors, *R. microplus* and *R. annulatus*. Before its eradication, babesiosis costed the U.S. an estimated $130.5 million (equivalent of $3 billion in current value) in direct and indirect annual losses. *R. microplus* and *R. annulatus* still exist in Mexico with the distribution of *R. microplus* being wider than that of *R. annulatus*. This has led to the establishment and maintenance of a permanent quarantine zone along the Mexican border to prevent their introduction into the U.S.A.

Bovine parasites, *B. bigemina* and *B. bovis* have been implicated as the most common agents of cattle babesiosis in Mexico [5] rendering 75% of cattle in Mexico (a little more than 23 million of heads) [6] at risk of acquiring babesiosis [7].

With respect to its geographical distribution, *R. microplus* can be found worldwide in subtropical and tropical regions. This tick is endemic in the Indian region, much of tropical and subtropical Asia, northeastern Australia, Madagascar, southeastern Africa, the Caribbean, and many countries in South and Central America including Mexico. Although it has been eradicated from the U.S., it can be sometimes found in Texas or California, in a buffer quarantine zone along the Mexican border [8].

R. (Boophilus) microplus is a one-host tick; all stages are spent on a single animal. The eggs hatch in the environment and the larvae crawl up grass or other plants to find a host. They may also be blown by the wind. In summer, *R. microplus* can survive for as long as 3 to 4 months without feeding. In cooler temperatures, the tick may live without food for up to six months. Ticks that do not find a host eventually die of starvation [4].

Newly attached seed ticks (larvae) are usually found on the softer skin inside the thigh, flanks, and forelegs. They may also be seen on the abdomen and brisket. After feeding, the larvae molt twice, to become nymphs and then adults. Each developmental stage (larva, nymph and adult) feeds only once, but the feeding takes place over several days. Adult male ticks become sexually mature after feeding, and mate with feeding females. An adult female tick that has fed and mated detaches from the host and deposits a single batch of many eggs in the environment. Typically, these eggs are placed in crevices or debris, or under stones. The female tick dies after oviposition. Ticks in the subgenus *Boophilus* have a life cycle that can be completed in 3 to 4 weeks; this characteristic can result in a heavy tick burden on animals [4].

At present, tick control in the world relies mainly in genetic selection of resistant animals and in the use of chemical acaricides in spite of their disadvantages and limitations [9, 10]. Other approaches to control ticks comprise vaccination with commercial anti-tick vaccines [10, 11, 12], application of potential biocontrol agents such as entomopathogenic fungi [13, 14, 15] entomopathogenic nematodes [16, 17], bacteria [18, 19], predaceous dipterans [20], and the use of crude plant extracts [21, 22]. In recent years, integrated pest management (IPM) has been carried out to control *Rhipicephalus (Boophilus) microplus* [10]. IPM involves the systematic application of two or more technologies to control pest populations which adversely affect the host species [23, 24]. Its aim is to achieve pest or parasite control in a more sustainable, environmentally compatible and cost-effective manner than is achievable with a single, stand-alone technology [10]. In this connection, it has been suggested that the use of chemical acaricides must be limited to reduce the risk of chemical resistance [25, 26].

OCCURENCE OF *RHIPICHEPHALUS (BOOPHILUS) MICROPLUS* IN MEXICO: AN OVERVIEW

Rhipicephalus (Boophilus) microplus is endemic to Mexico and causes severe economic damage to cattle industry [27]. Therefore, an official national campaign against *Rhipicephalus* (*Boophilus*) spp., (Campaña Nacional contra la Garrapata *Boophilus* spp.) established to control these arthropods [28, 29].

National Campaign against *Boophilus* spp. (NCB)

At present (April 2011) and in accordance with the Dirección General de Sanidad Animal (DGSA) from Secretaría de Agricultura, Ganadería, Desarollo Rural, Pesca y Alimentación (SAGARPA) of the Mexican government, the NCB handles three areas:

1) *Free area* (FA), it encompasses 94.4 millions of hectares (47.88% of Mexico). The FA includes part of the following states: Baja California Sur, Chihuahua, Coahuila, Nuevo Leon, Tamaulipas, Durango, Zacatecas, San Luis Potosi, Guanajuato, Hidalgo, Estado de Mexico, Puebla and the whole territories of Baja California, Sonora, Aguascalientes, Tlaxcala and Distrito Federal.
2) *Eradication area* (EA), which covers 1.1 millions of hectares (0.57% of Mexico). The EA embraces part of the states of Sinaloa and Coahuila.
3) *Control area* (CA), which covers 101.6 millions of hectares (51.5% of Mexico) (Figure 1). The CA includes zones found mainly in the states located in the Gulf of Mexico and Pacific coast and in the southern part of the country.

Figure 1. Free, control and buffer areas of the Mexican National Campaign against *Rhipicephalus (Boophilus) microplus* ticks (November 2010). The buffer area is a free area which is susceptible to tick infestation.

Control of *R. (Boophilus) microplus*: Approaches Used in Mexico

As indicated previously, there is a need of using new technologies different from the use of acaricides to reduce the growing problem of resistance developed by *R.* (*Boophilus*) *microplus* against the applied chemicals. The control methods which have been used in the last few years in México are summarized in Table 1.

Biological and Chemical Methods of Tick Control

Various methods are tested and applied to control ticks in different parts of the world including Mexico. These methods and approaches are briefly given below:

Use of Chemical Acaricides

In México, as in other countries, the main method employed for the control of *R.* (*Boophilus*) *microplus* is the use of chemical acaricides, in spite of its limitations and disadvantages. At present, the chemicals authorized in Mexico for the control of ticks include: Acaricides: organophosphates, pyrethroids, amidines, and phenylpyrazolones; Endectocides: ivermectin, doramectin, and moxidectin; growth regulators (Source: DGA, SAGARPA, Mexican government). However, the phenomenon of resistance of ticks to chemical acaricides has been growing in recent years. For example, 1717 out of 1920 tick samples (89.43%) from 22 states tested at SENASICA from 2001 to 2005 were positive for resistance to one or more acaricides. Additionally, 622 out of 655 samples (94.96%) from 16 states analyzed at SENASICA from January to November 2010 were positive to one or more acaricides (DGSA, SAGARPA, Mexican government, 2011). (Figure 2)

Table 1. Methods used in Mexico for the control of *Rhipicephalus (Boophilus) microplus* ticks

Method	Comments	Reference(s)
Chemical: Ixodicides, endectocides and growth regulators*	The most common method used; however, resistance of ticks to chemical substances is a growing problem	[29,30,31,32,33,34,35]
Anti-tick vaccines	Limited field studies	[36]
Entomopathogenic fungi: *Metahrrizium anisopliae*	Repeated treatment with Ma34 strain showed efficacy of 40-91% in naturally infested cattle	[37]
	In the wet and dry season trials, the efficacy of M. anisopliae was 67.7% and 100%, respectively.	[38]
Isaria fumosorosea	Experimental	[39]
Bacteria:		
Staphycoccocus saprophyticus	Experimental	[18]
Bacillus thuringiensis	Experimental	[19]
Arthropods:		
Megaselia scalaris	Experimental	[20]
Combined:		
Anti-tick vaccine and ixodicide	Limited studies. The use of the vaccine reduced the use of ixodicides	[40,41]

* Chemicals authorized in Mexico for the control of ticks are Acaricides (organophosphates, pyrethroids, amidines, and phenylpyrazolones), Endectocides (ivermectin, doramectin, and moxidectin), and growth regulators (Source: DGA, SAGARPA, Mexican government).

Use of Anti-Tick Vaccines

The results obtained with commercial anti-tick vaccines used in the last few years suggest that these have the potential as an improved method for tick control of cattle [11, 12, 42]; however, antigens from ticks so far described do not induce the necessary protection in cattle upon tick exposure to curb the use of acaricides [43]. In this context, limited studies on the use of anti-tick vaccines have been carried out in Mexico [36, 40, 41].

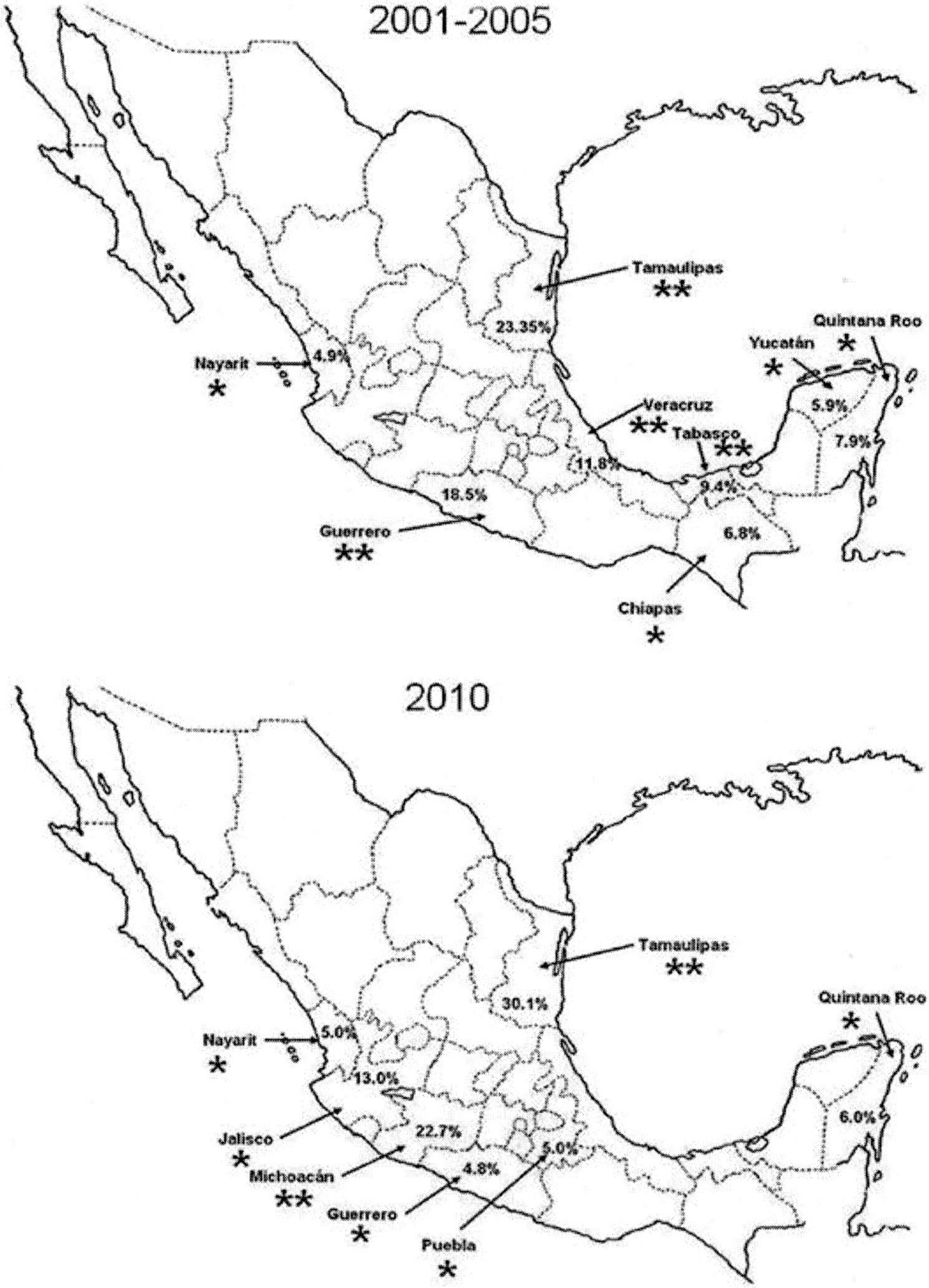

Figure 2. States which showed percentages of *Riphicephalus* (*Boophilus*) spp. resistant to one or more acaricides higher than 4.5% with respect to the total number of resistant cases examined by SENASICA during 2001-2005 (top) and from January to November 2010 (bottom). * States with multi-resistance (organophosphates/pyrethroids and amidines). ** States with more than 90 cases of multi-resistance. (Source: SENASICA, SAGARPA, Mexican government).

Use of Entomopathogenous Fungi

In recent years, some entomopathogenic fungi have been evaluated for the control of *R. (Boophilus) microplus* ticks in Mexico. The fungi species tested so far are *Metarhizium anisopliae* [37, 38, 44] and *Isaria fumosorosea* [39]. The entomopathogenic fungi caused different degrees of mortality in the referred ticks.

Use of Bacteria

The demonstration that under certain circumstances *Staphylococcus saprophyticus* is able to produce a lethal infection in fully engorged *R. (Boophilus) microplus* ticks [18] has suggested that it may be considered as a candidate for the biological control of the cattle tick. Furthermore, in other study it was also observed that four strains of *Bacillus thuringiensis* showed a harmful effect on a strain of *R. (Boophilus) microplus* resistant to acaricides [19].

Use of Predaceous Arthropods

In a recent study, it was experimentally demonstrated that the predatory dipteran *Megaselia scalaris* might be a potential agent for biological control of *R. (Boophilus) microplus* ticks [20].

Integrated Control Assays

Limited studies of this approach have been carried out in Mexico. In one of the studies, both vaccination with anti-tick vaccine and the acaricide amidine were used. It was observed that vaccination reduced the use of acaricides [40]. Recently Martínez-Ibañez, 2011 unpublished data), strategic control programs were established in several cattle ranches from three geographical areas: Santa Isabel, in Morelos State, El mirador in Nuevo León State and three ranches, Yahalum, Los Limones and San José, in Campeche State. The acaricide treatment regimen was determined in accordance with the highest peak of *Rhipicephalus* (*Boophilus*) *microplus* ticks during one year in a specific geographical area and the techniques used by SENASICA to determine susceptibility to acaricides [28, 45, 46], was employed to choose the appropriate acaricide [47]. Additionally, a commercial anti-tick vaccine was used in one of the ranches. In all ranches the ranchers were trained to count ticks and to properly apply the acaricides. In a ranch at Santa Isabel, two acaricide treatments were applied in one year. In El Mirador, the anti-tick vaccine and two acaricide treatments were used; while in the ranches at Yashalum, Los Limones, and San José, a monthly treatment during one year was applied (Figure 3).

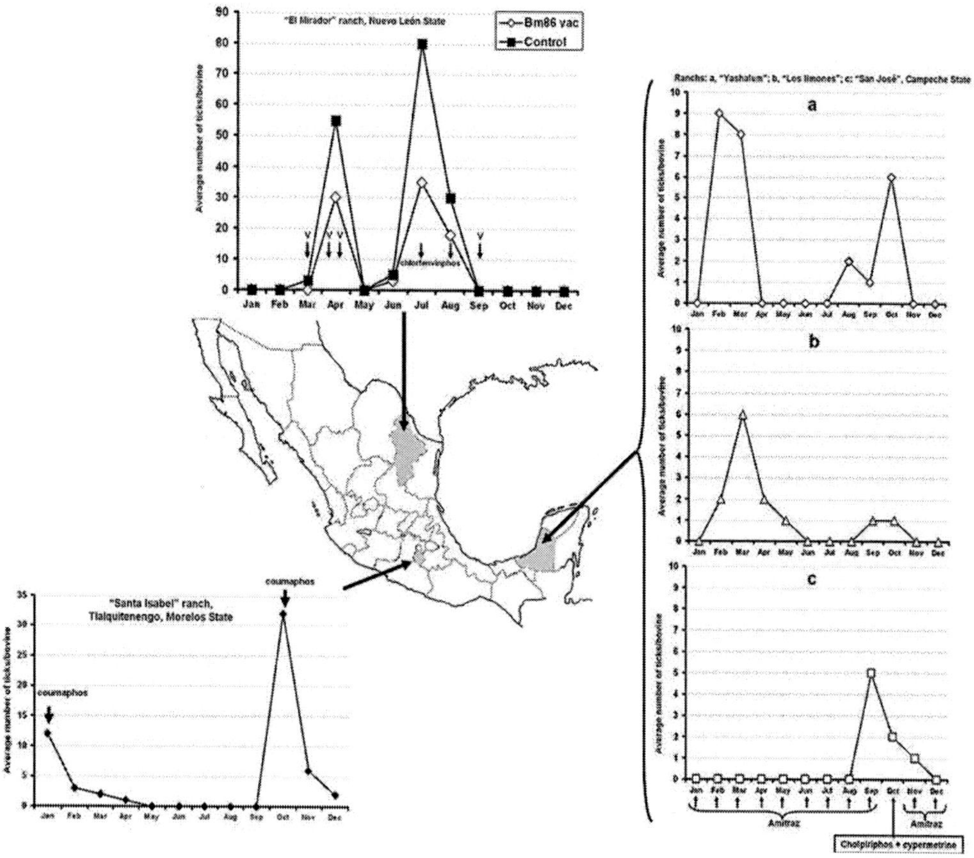

Figure 3. Location of ranches treated (arrows) against *Rhipicephalus (Boophilus) microplus* ticks in three different geographical areas in Mexico (clockwise direction, beginning bottom left). In the ranch at Santa Isabel, Morelos state, bovines were treated with coumaphos twice because ticks were susceptible to this ixodicide (as previously assessed). In the ranch at El Mirador, Nuevo Leon State, one group of animals was vaccinated with a commercial vaccine (V) containing the Bm86 antigen of *Rhipicephalus (Boophilus) microplus* at 0, 4 and 7 weeks and at 6 months after the initial vaccination. Besides, in July and August chlofenvinphos ixodicide was used. Another group served as nontreated control. In ranchs at Yashalum, Los Limones and San José, Campeche State, Amitraz was used every month, except in October, when a mixture containing chlorpyrifos and cypermethrin was used.

CONCLUSIONS

In México, *R.* (*Boophilus*) *microplus* is the most economically important tick species affecting cattle. The resistance of *R.* (*Boophilus*) *microplus* to chemical acaricides is a growing problem. In spite of the existence of a National Campaign for the control of this tick species, it has not been controlled as expected and still poses a problem in the livestock production. Our experience in Mexican indicates that comprehension of the biology of *R.* (*Boophilus*) *microplus* in a particular geographical (ecological) area, the breeds of the infested cattle and the susceptibility of the ticks to acaricides are required to improve the use of acaricides and to alleviate the development of resistance to these chemicals. The use of various new methods for the control of *R.* (*Boophilus*) *microplus* is a viable alternative; however, they have not been adequately evaluated in the field.

REFERENCES

[1] Hoogstraal, H. (1985). Argasid and nuttalliellid ticks as parasites and vectors. *Adv. Parasitol*, 24:135-238.

[2] Klompen, H. (2005). Ticks, the Ixodida. *In: Marquardt W.C. et al. (editors) Biology of disease vectors 2nd edition, Amsterdam, Elsevier Academic Press* Pp.45-55.

[3] Horak, I.G., Camicas, J.L., & Keirans, J.E. (2002). The Argasidae, Ixodidae and Nuttalliellidae (Acari: Ixodida): a world list of valid tick names. *Exp. Appl. Acarol.* 28:27-54.

[4] Bautista Garfias, C.R. (2006). *Entomología Veterinaria Esencial. México, D.F. Ed. INIFAP.*

[5] Kjemtrup, A. M., & Conrad, P.A. (2000) Human babesiosis: an emerging tick-borne disease. *Int. J. Parasitol.* 30: 1323-1337.

[6] INEGI (2009). Estados Unidos Mexicanos. Censo Agropecuario 2007. VIII Censo Agrícola, Ganadero y Forestal. *Aguascalientes, Ags*., Mexico.

[7] Bautista, C.R., Alvarez, J.A., Mosqueda, J.J., Falcon, A., Ramos, J.A., Rojas, C., Figueroa, J.V. & Ku, M. (2008). Enhancement of the Mexican bovine babesiosis vaccine efficacy by using *Lactobacillus casei.* Ann. N. Y. Acad. Sci.1149:126-30.

[8] Estrada-Peña, A., Bouattour, A., Camicas, J.L., Guglielmone, A., Horak, I., Jongejan, F., Latif, A., Pegram, R. & Walker, A.R. (2006). The known distribution and ecological preferences of the tick subgenus *Boophilus* (Acari: Ixodidae) in Africa and Latin America. *Exp. Appl. Acarol.* 38:219-235.

[9] Brossard, M. (1998). The use of vaccines and genetically resistant animals in tick control. *Rev. sci. tech. Off. Int. Epiz.* 17:188-199.

[10] Willadsen, P. (2006). Tick control: Thoughts on a research agenda. *Vet. Parasitol.*, 138:161-168.

[11] Willadsen, P. (2004). Anti-tick vaccines. *Parasitology,* 129:S367-S387.

[12] De la Fuente, J., Almazan, C., Canales, M., Perez de la Lastra, J.M., Kocan, K.M. & Willadsen, P. (2007). A ten-year review of commercial vaccine performance for control of tick infestations on cattle. *Anim. Health Res. Rev.* 8:23-28.

[13] Kaaya, G.P. & Hassan, S. (2000). Entomogenous fungi as promising biopesticides for tick control. *Exp. Appl. Acarol.* 24:913-926.

[14] Campos R.A., Boldo, J.T., Pimentel, I.C., Dalfovo, V., Araújo, W.L., Azevedo, J.L., Vainstein, M.H. & Barros N.M. (2010). Endophytic and entomopathogenic strains *of Beauveria* sp to control the bovine tick *Rhipicephalus (Boophilus) microplus. Gene. Mol. Res*., 9:1421-1430.

[15] Ángel-Sahagún, C.A., Lezama-Gutiérrez, R., Molina-Ochoa, J., Pescador-Rubio, A., Skodah, S.R., Cruz-Vázquez, C., Lorenzonie, A.G., Galindo-Velasco, E., Fragoso-Sánchez, H. & Fosteri, J.E. (2010). Virulence of Mexican isolates of entomopathogenic fungi (Hypocreales: Clavicipitaceae) upon *Rhipicephalus* (*Boophilus) microplus* (Acari:

Ixodidae) larvae and the efficacy of conidia formulations to reduce larval tick density under field conditions. *Veterinary Parasitology* 170, 278-286.

[16] De Oliveira Monteiroa, C.M., Furlong, J., De Azevedo Prata, M.C., Soares, A.E., De Paula Batista, E.S. & Dolinski, C. (2010). Evaluation of the action of *Heterorhabditis bacteriophora* (Rhabditida: Heterorhabditidae) isolate HP88 on the biology of engorged females of *Rhipicephalus (Boophilus) microplus* (Acari: Ixodidae). *Vet. Parasitol.* 170:355

[17] Barbiéri de Carvalho, L., Furlong, J., De Azevedo Prata, M.C., Dos Reis, E.S., De Paula Batista, E.S., Pasqualini Faza, A. & Cerqueira Leite, R. (2010). Evaluation *in vitro* of the infection times of engorged females of *Rhipicephalus (Boophilus) microplus* by the entomopathogenic nematode *Steinernema glaseri* CCA strain. *Cienc. Rural,* 40:939-943.

[18] Miranda-Miranda, E., Cossio-Bayugar, R., Quezada-Delgado, M.R., Sachman-Ruiz, B., & Reynaud, E. (2010). *Staphylococcus saprophyticus* is a pathogen of the cattle tick *Rhipicephalus (Boophilus) microplus*. *Biocont. Sci. Technol.*, 20:1055-1067.

[19] Fernández-Ruvalcaba, M., Peña-Chora, M., Romo-Martínez, A., Hernández-Velázquez, V., la Parra, A.B. & La Rosa, D.P. (2010). Evaluation of *Bacillus thuringiensis* pathogenicity for a strain of the tick, *Rhipicephalus microplus*, resistant to chemical pesticides. *J. Ins. Sci.,* 10:186 available online: insectscience.org/10.186.

[20] Miranda-Miranda, E., Cossio-Bayugar, R., Martínez-Ibañez, F. & Bautista-Garfias, C.R. (2010). *Megaselia scalaris* reared on *Rhipicephalus (Boophilus) microplus* laboratory cultures. *Med. Vet. Entomol.* 25:344-347.

[21] Habeeb, S.M. (2010). Ethno-veterinary and medical knowledge of crude plant extracts and its methods of application (traditional and modern) for tick control. *World Appl. Sci. J.*, 11:1047-1054.

[22] De Souza Chagas, A.C., Sanchez Georgetti, C., Olivo De Carvalho, C., De Sena Oliveira, M.C., Rodrigues, R.A., Foglio, M.A., De Magalhaes P.M. (2011). *In vitro* activity of *Artemisia annua* L (asteraceae) extracts against *Rhipicephalus (Boophilus) microplus*. *Rev. Bras. Parasitol. Vet., Jaboticabal,* 20:31-35.

[23] Bram, R.A. (1994). Integrated control of ectoparasites. *Rev. sci. tech. Off. Int. Epiz.,* 13:1357-1365.

[24] De Castro, J.J. (1997). Sustainable tick and tick borne disease control in livestock improvement in developing countries. *Vet. Parasitol.*, 71:77-97.

[25] Kunz, S.E. & Kemp, D.H. (1994). Insecticides and acaricides: resistance and environmental impact. *Rev. sci. tech. Off. int. Epiz.*, 13:1249-1286.

[26] Jonsson, N.N., Mayer, D.G & Green, P.E. (2000). Possible risk factors on Queensland dairy farms for acaricide resistance in cattle tick (*Boophilus microplus*). *Vet. Parasitol.*, 88:79-92.

[27] Li, A.Y., Davey, R.B. & George J.E. (2005). Carbaryl resistance in Mexican strains of the southern cattle tick (Acari: Ixodidae). *J. Econ. Entomol.*, 98:552-556.

[28] Norma Oficial Mexicana NOM-019-ZOO-1994 (1994). Campaña nacional contra la garrapata *Boophilus* spp. *http://cefppmich.org.mx/cgi-bin/wp-content/uploads/2009/10/nom-019-zoo-1994-garrapata-boophilus.pdf*

[29] SENASICA (2011). Campaña Nacional contra la Garrapata *Boophilus* spp. *http://www.senasica.gob.mx/?id=803.*

[30] George, J.E., Pound, J.M. & Davey, R.B. (2004). Chemical control of ticks on cattle and the resistance of these parasites to acaricides. *Parasitology*, 129:S353-S366.

[31] Wharton, R.H. & Roulston, W.J. (1970). Resistance of ticks to chemicals. *Ann. Rev. Entomol.*, 15:381-404.

[32] Aguirre, E.J. & Santamaría, V.M. (1986). Purificación y caracterización toxicologica de garrapatas *Boophilus microplus* resistentes a ixodicidas organofosforados y organoclorados. *VII Reunión Annual. Asociación Mexicana de Parasitología Veterinaria A.C.* Cd. Victoria, Tamaulipas, México.

[33] Miller, R.J., Davet, R.B. & George, J.E. (1999). Characterization of pyrethroid resistance and susceptibility to coumaphos in Mexican *Boophilus microplus* (Acari:Ixodidae). *J. Med. Entomol.*, 36:533-538.

[34] Li, A.Y., Davey, R.B., Miller, R.J. & George, J.E. (2004). Detection and characterization of amitraz resistance in the southern cattle tick, *Boophilus microplus* (Acari: Ixodidae). *J. Med. Entomol.*, 41:193-200.

[35] Pérez-Cogollo, L.C., Rodriguez-Vivas, R.I., Ramírez-Cruz, G.T. & Miller, R.J. (2010). First report of the cattle tick *Rhipicephalus microplus* resistant to ivermectin in Mexico. *Vet. Parasitol.* 168:165-169.

[36] Fragoso, H., Rad, P.H., Ortiz, M., Rodrigues, M., Redondo M., Herrera, L. & De la Fuente, J. (1998). Protection against *Boophilus annulatus* infestations in cattle vaccinated with the *B. microplus* Bm86-cantaining vaccine Gavac. *Vaccine.* 16:1990-1992.

[37] Alonso-Díaz, M.A., García, L., Galindo-Velasco, E., Lezama-Gutiérrez, R., Angel-Sahagún, C.A., Rodríguez-Vivas, R.I., Fragoso-Sánchez, H. (2007). Evaluation of *Metarhizium anisopliae* (Hyphomycetes) for the control of *Boophilus microplus* (Acari: Ixodidae) on naturally infested cattle in the Mexican tropics. *Vet. Parasitol.* 147:336-340.

[38] Ojeda-Chi, M.M., Rodriguez-Vivas, R.I., Galindo-Velasco, E. & Lezama-Gutierrez, R. (2010). Laboratory and field evaluation of *Metarhizium anisopliae* (Deuteromycotina: Hyphomycetes) for the control of *Rhipicephalus microplus* (Acari: Ixodidae) in the Mexican tropics. *Vet. Parasitol.* 170:348-354.

[39] Ángel-Sahagún, C.A., Lezama-Gutiérrez, R., Molina-Ochoa, J., Pescador-Rubio, A., Skodah, S.R., Cruz-Vázquez, C., Lorenzonie, A.G., Galindo-Velasco, E., Fragoso-Sánchez, H. & Fosteri, J.E. (2010). Virulence of Mexican isolates of entomopathogenic fungi (Hypocreales: Clavicipitaceae) upon *Rhipicephalus (Boophilus) microplus* (Acari: Ixodidae) larvae and the efficacy of conidia formulations to reduce larval tick density under field conditions. *Vet. Parasitol.* 170:278-286.

[40] Redondo, M., Fragoso, H., Ortiz, M., Montero, C., Lona, J., Medellin, J.A., Fria, R., Hernandez, V., Franco, R., Machado, H., Rodriguez, M. & De la Fuente, J. (1999). Integrated control of acaricide-resistant *Boophilus microplus* populations on grazing cattle in Mexico using vaccination with Gavac and Amidine treatments. *Exp. Appl. Acarol.* 23:841-849.

[41] Martínez, I.F. (2011). La dinámica poblacional una herramienta en el control de la garrapata *Boophilus microplus* en ganado bovino. *XXXV Congreso Nacional de Buiatría. León, Guanajuato, México.* Pp. 150-156.

[42] Vargas, M., Montero, C., Sanchez, D., Perez, D., Valdes, M., Alfonso, A., Joglar, M., Machado, H., Rodriguez, E., Mendez, L., Lleorant, R., Suarez, M., Fernandez, E.,

Estrada, M.P., Rodriguez-Mallon, A. & Farnos, O. (2010). Two initial vaccinations with the Bm86-based Gavac^{plus} vaccine against *Rhipicephalus (Boophilus) microplus* induce similar reproductive suppression to three initial vaccinations under production conditions. *BMC Vet. Res.* 6:43.

[43] Parizi, L.F., Poh, P.C., Masuda, A. & Da Silva Vaz Junior, I. (2009). New approaches toward anti-*Rhipicephalus* (*Boophilus*) *microplus* tick vaccine. *Rev. Bras. Parasitol. Vet.* 18:1-7.

[44] Ojeda-Chi, M.M., Rodríguez-Vivas, R.I., Galindo-Velasco, E., Lezama-Gutiérrez, C. & Cruz-Vázquez, C. (2011). Control of *Rhipicephalus microplus* (Acari: Ixodidae) using the entomopathogenic fungi *Metarhizium anisopliae* (Hypocreales: Clavicipitaceae). *Rev. Mex. Cienc. Pecu.* 2:177-192.

[45] Stone, B.F. & Haydock, K.P. (1962). A method for measuring the acaricide-susceptibility of the cattle tick *Boophilus microplus* (Canestrini). *Bull. Entomol. Res.* 53:563-578.

[46] Drummond, R.O., Ernst, S.E., Trevino, J.L., Gladney, W.J. & Graham, O.H. (1973). *Boophilus annulatus* and *B. microplus*: laboratory tests of insecticides. *J. Econ. Entomol.* 66:130-133.

[47] Norma Oficial Mexicana NOM-006-ZOO-1993 (1993). Requisitos de efectividad biológica para los ixodicidas de uso en bovinos y métodos de prueba. *http://148.206.53.231/bdcdrom/GAM06/GAMV15/root/docs/NOM-343.PDF*

In: Ticks: Disease, Management and Control
Editor: Moges Woldemeskel
ISBN: 978-1-62081-136-8

Chapter 9

BIOCONTROL OF THE CATTLE TICK *RHIPICEPHALUS (BOOPHILUS) MICROPLUS* BY THE ACARICIDAL FUNGUS *METARHIZIUM ANISOPLIAE*

Walter Orlando Beys-da-Silva*[*], *Lucelia Santi*, *Marilene Henning Vainstein and Augusto Schrank
Centro de Biotecnologia, Universidade Federal do Rio Grande do Sul, P.O. Box 15005, CEP 91501-970, Porto Alegre, RS, Brazil

ABSTRACT

The tick *Rhipicephalus (Boophilus) microplus* is a bovine ectoparasite that causes severe economic losses in herds in tropical and subtropical areas due to parasitism and disease transmission. Tick control is currently based mainly on the use of chemicals. However, alternative strategies are required due to the ability of ticks to develop resistance, the demands of consumers for chemical-free foods, and the environmental impact of chemical acaricides. In accordance with this scenario, the use of fungi to biocontrol ticks is an efficient and important tool. The entomopathogenic and acaricidal fungus *Metarhizium anisopliae* is the most studied fungus for the biocontrol of *R. microplus*. Various studies on different aspects of the use of *M. anisopliae* to control ticks and other pests have been published during the decade. Such knowledge is crucial in the search for more efficient strains and for optimizing new formulations for large-scale use of this efficient, economic and environmentally safer form of tick control. Here we focus on the biocontrol of the tick *R. microplus* based on the use of *M. anisopliae*, including the different fungal isolates, their virulence, the biochemical and molecular aspects of the pathogenesis and the control strategies.

[*] E-mail address: walterbeys@cbiot.ufrgs.br / walterbeys@yahoo.com.br.

INTRODUCTION

The tick *Rhipicephalus (Boophilus) microplus* is an obligatory haematophagous ectoparasite that mainly infests cattle. This arthropod is considered one of the most harmful ectoparasites of domestic and wild animals [1]. The parasite occurs in tropical and subtropical areas and probably originated from Asia [2]. *R. microplus* causes considerable economic losses, and has therefore become one of the main targets of control and eradication programs in commercial herds worldwide. During the parasitic phase, the tick is able to engorge large amounts of blood from its host, significantly impairing milk, meat and leather production. Additionally, during the feeding phase, ticks may cause trauma by damaging tissues, leading to toxicity triggered by the inoculation of substances present in their saliva [3]. They may also act as a vector for diseases such as cattle sadness [3, 4, 5] and cause systemic haemostatic changes during the infection [6]. The losses associated with this parasite as well as efforts to control it have a significant economic impact [7]. Similar to other pest arthropods, this organism is controlled by using chemicals, in strategies that ultimately promote resistance, especially in regions with a high incidence of ticks, such as Brazil, a country with the largest commercial herd in the world. Alternatives to chemical control like the use of the pathogenic fungus *Metarhizium anisopliae*, which has proven efficient in tick control both in laboratory experiments and in the field, could be an interesting commercial alternative in the light of all the intrinsic advantages it affords [1, 8, 9, 10]. For several decades, this fungus has had commercial applications in the control of several pests with medical relevance and agricultural importance [11 - 14], underlining its potential use to control one of the most economically important pest affecting herds worldwide. The potential application of *M. anisopliae* in the control strategies against the bovine tick *R. microplus* in terms of the advantages afforded, methodologies and molecular characteristics of the biocontrol process are addressed below.

The Tick *Rhipicephalus microplus*

Ticks were among the first arachnids to thrive in the world [15]. These arachnids belong to the subclass of mites, super order *Parasitiformes*, order *Ixodida*, and are found in all continents (America, Asia, Europe, Oceania and Africa). Around 870 tick species have been described to date, and these are classified in two main families, *Ixodidae* and *Argasidae*. *R. microplus*, commonly referred to as the cattle tick, belongs to the family *Ixodidae*, which includes the hard ticks. This term is used in allusion to the scutum, a sclerotic hard dorsal shield that covers the idiosome (the tick body) completely in males, and partially in females. The cuticle is dark grey, and the scutum is not fully developed [16, 17].

The bovine tick is distributed between latitudes 32° N and 32° S, that is, in tropical and subtropical regions, which are important areas of cattle production (Latin America and Oceania) [18]. It has been suggested that *R. microplus* is native to Asia and was introduced to other regions via the traffic of parasitized Asian cattle, mainly the zebu (*Bos indicus*). This supports the observation that European cattle breeds (*Bos taurus*) are more susceptible to infestation by *R. microplus*, and explains why Asian breeds are more resistant, in the light of a more consistent balance in the population dynamics of the parasite due to coevolution [18].

The bovine tick is a monoxous parasite, which uses a single host throughout its parasitic life cycle. The only natural host of *R. microplus* is cattle, although it may occasionally be found on other mammals (especially other ruminants) living in the same infested area, and only when there is a high parasitic burden in the environment. Considerable rates of rejection are observed in hosts other than bovines, especially during the attachment of infesting larvae [16, 18].

The life cycle of *R. microplus* is characterized by a parasitic phase and a free-living phase. The parasitic phase lasts 3 weeks on average, starting when infesting larvae, which position themselves on the upper structures of grasses, detect the host. Larvae may survive for over 8 months without feeding, in scenarios in which hosts are not available [16, 18, 19, 20]. Once on the host, larvae migrate to the attachment site, preferably where the temperature of the skin is higher, its thickness is lower, and where it is difficult for the infested cattle to mechanically remove them [16, 18, 20]. Females increase their blood meal intake considerably, and their weight may increase to 1,400% of the weight at initial attachment to the host. Despite ingesting much smaller blood meals compared to females at all life stages, males are also haematophagous, and blood meal intake plays an important role in spermatogenesis. As a rule, on the 22nd day of the attachment, females detach from the host and drop to the ground, starting the free-living cycle [18, 19, 21]. As soon as they are on the ground, the fully engorged female searches for a humid site protected from sunlight for oviposition. Each female is able to lay around 3,000 eggs. It is important to stress that about 95% of ticks in a given population are found in the environment, either as larvae or adult females [16, 18, 19, 20]. Climatic conditions influence free-living stages of *R. microplus*. At low temperatures the time elapsing between detachment and the emergence of infesting larvae may exceed 6 months [18, 21].

The tick *R. microplus* is the major ectoparasite affecting livestock in America, Asia, Africa and Oceania [22], and is considered the most important parasite affecting domestic animals in Latin America. Almost all cattle herds in Brazil are at risk of infestation with *R. microplus* [21]. Infestation by ticks seems to be more important in regions in which European breeds and their crosses that are highly susceptible to tick infestation and its deleterious effects predominate. They are also important in milk herds in which Holstein and Jersey are the main breeds throughout Brazil and other countries [21].

R. microplus has various deleterious effects on the parasitized animals, mainly due to the sucking of large volumes of blood, which leads to anaemia, anorexia, weight loss, apathy, decreased milk and meat production and death in extreme cases [23]. One tick alone may suck on average 2 to 3 mL of blood per day from its host, causing a reduction in gain of live weight for each tick of around 0.9 g [23]. Moreover, during the feeding process the tick may cause trauma and injury due to laceration of skin and tissues, causes discomfort and pruritus, promotes inflammation, secondary infection, and myiasis and induces toxic effects by substances present in the saliva [3, 16, 18, 21]. Tick saliva modulates - a series of reactions in the host aimed at guaranteeing haematophagy [24]. One such example is the occurrence of important systemic changes in haemostasis during tick infestation in cattle [6]. Moreover, this tick may act as a vector of diseases such as cattle sadness caused by protozoa of the genus *Babesia* and rickettsia of the genus *Anaplasma* [3, 4, 5, 25]. The parasite may change the host's blood biochemical parameters as well as renal and digestive functions [6, 26, 27]. Grisi et al. [7] reported that economic losses in Brazil caused by *R. microplus* might reach US$ 2 billion per year, manifested as weight loss, costs of ectoparasiticides, decreased milk and meat productions, lowered leather quality as well as treatment costs of the secondary lesions such as myiasis at the sites of attachment.

Since *R. microplus* causes both direct and indirect damages to its host, poses serious economic problems and warrants effective control measures. From the beginning of the 20th century onwards, measures based on the use of chemicals have comprised the main strategies in the control of *R. microplus*. The most important acaricides used nowadays are organophosphorous compounds, pyrethroids, formamidines (amitraz), macrocyclic lactones (avermectins), fluazuron (growth/chitin incorporation inhibitors) and fipronil [21, 28]. However, the emergence of resistance to active principles present in acaricides is relatively common [1, 29, 30, 31]. Despite being the most efficient, chemical control methods are the object of frequent criticism due to cost-related issues and possible chemical residues in milk and meat used for human consumption. Alternative control measures are thus receiving increasing attention [32, 33].

An alternative approach to controlling *R. microplus* lies in the development of vaccines. Two anti-tick vaccines have been produced and commercialized. Both are antigen-based vaccines that use Bm86, a protein present in the tick gut [33]. Under the commercial names of TickGard® and GAVAC®, these two vaccines have been relatively successful in the countries they were originally developed, Australia and Cuba, respectively [33]. Nevertheless, the results of the inoculation of cattle with these vaccines in other regions have not been promising. Compared to the traditional chemical control methods, these vaccines give limited protection, and have significantly higher associated costs. Such drawbacks have raised questions regarding the viability and use of vaccines as a parasite control strategy. Further studies are being conducted to develop vaccines against *R. microplus*, though the results obtained so far have shown that this strategy still suffers from low efficiency and high cost issues [22].

Another interesting, viable and promising choice is the biological control, based on the use of natural enemies of a given target pest. In this context, arthropod-pathogenic microorganisms like the filamentous fungus *M. anisopliae* deserve special attention as one of the most studied biocontrol agents used worldwide in the control strategies against numerous pests [11-14, 32]. Furthermore, in the light of the substantial advancements in discovering more about the molecular characteristics of *M. anisopliae* infection of the cattle tick [34 - 37], along with current data addressing the use of this strategy in laboratory and field experiments alike, this fungus may be considered as having great promise in biological control strategies [9, 32, 38].

BIOCONTROL OF TICKS

The use of microorganisms for the biological control of arthropod pests and disease vectors was first proposed in the middle of the 19th century; nevertheless, the full potential and the many advantages of this practice have only recently been applied at a commercial scale [37]. Since the beginning of the 20th century, investigators have documented numerous candidates for the biocontrol of ticks, including pathogens, parasitoids and predators of ticks [39]. Nevertheless, strategies to control ticks are still based on the use of chemicals. Factors hampering the wider use of biocontrol strategies are the laborious work of isolating and identifying potential biocontrol agents and the development of suitable formulations to maximize the efficiency of the bioproducts [37]. With the advance in studies and consequently in the knowledge about the biological control of arthropod pests, mainly microorganisms, this alternative is becoming more attractive commercially. A recent study reported that the biopesticide market is growing and is expected to reach over one billion US dollars over the next 5 years [40]. The success of the commercial application of microorganisms, especially filamentous fungi, for protection against agricultural arthropod pests around the beginning of the last century reinforces the potential use of this strategy to control ticks. As indicated by Samish et al. [39], biological control is an increasingly attractive approach for tick management because of: (1) increasing concerns about environmental safety and human health (e.g. the gradual increase in the use of chemical insecticides is stimulating the growing market for 'organic' food); (2) the increasing cost of chemical control; and (3) the increasing resistance of ticks to pesticides. During the last 10 years much progress has been made in research and field trials to control ticks, especially the cattle tick, mainly using *R. microplus* and the filamentous fungus *M. anisopliae* as a model.

The Biocontrol Agent *Metarhizium anisopliae*

M. anisopliae (Metschnikoff) was first isolated from a cereal pest and used to control the beetle *Anisoplia austriaca* by the Russian microbiologist Metchnikoff in 1879 [40], and was described and classified by Sorokin in 1883 as *M. anisopliae*. This Ascomycete is one of the best-characterized arthropod-pathogenic fungi and also one of the most widely applied in biological control programs against several pests worldwide. In Brazil it has been extensively used to control insect pests such as spittlebugs (*Mahanarva posticata*) in sugarcane plantations since the 1970s. The area treated with *M. anisopliae* in 2008 is estimated to be approximately one million hectares [41], and several commercial products are available [14, 42]. *M. anisopliae* is widespread in nature, being found in the soil, in the rhizosphere and on plant roots [43], and in arthropod cadavers as saprophytes or parasitizing a broad range of insects and ticks. This fungus grows vegetatively as hyphae, producing mycelia and conidia that are the infective propagules on arthropod hosts and in culture media. Although the overall host range of *M. anisopliae* is broad, individual strains may be more specific. As recently reviewed [37], strains such as E6, CARO14, CG47 and CG97 caused 90–100% mortality within 4 days in bioassays using the tick *R. microplus*, as did other strains listed in Table 1.

Commercial aspects of Biocontrol: Conidia Production, Formulation and Methods of Application

Commercially, asexual spores (conidia) are used as mycopesticides in different formulations and methods of application, and in combination with other biocontrol agents, such as *Beauveria bassiana* and chemical acaricides [41, 44-48]. In nature, conidia are also the infective form and the infection cycle begins with the contact and attachment of conidia to the surface of a susceptible host, triggering mechanical and enzymatic processes. Therefore, the efficient production of conidia is a crucial step for the successful large-scale use of *M. anisopliae* in the biocontrol of pests in the field.

Table 1. Acaricidal fungus for tick control

Ticks	Country	Biocontrol agent	Strain	Experiment	Developmental cycle	Reference
R. appendiculatus *A. variegatum*	Kenya	*M. anisopliae* *B. bassiana*		field/laboratory	larvae, nymph, adult	[71]
I. scapularis	USA	*M. anisopliae*		laboratory	larvae, engorged female	[151]
R. (B.) microplus	Brazil	*M. anisopliae*	E9	field	engorged female	[26]
R. (B.) microplus	Brazil	*M. anisopliae*	E6S1	laboratory	engorged female	[8]
A. variegatum *R. apendiculatus* *B. decoloratus*	Namibia	*M. anisopliae* *B. bassiana*		field	larvae, adult	[47]
R. appendiculatus *A. variegatum*	Namibia	*M. anisopliae* *B. bassiana*		field	larvae, nymph, adult	[58]
R. (B.) microplus	Brazil	*M. anisopliae* *M. flavoviride*	CG291	laboratory	engorged female	[152]
R. sanguineus	Israel	*M. anisopliae* *M. flavovirirde* *B. bassiana* *P. fumosoroseus*	108	laboratory	larvae, nymph, unfed and engorged female	[153]
R. (B.) microplus	Brazil	*M. anisopliae* *B. bassiana*		laboratory	engorged female	[154]
R. annulatus *R. sanguineus* *H. excavatum*	Israel	*M. anisopliae* *M. flavoviride* *B. bassiana* *P. fumosoroseus* *V. lecanii*	7, 43, 108	laboratory	egg, larvae, nymph, adult	[155]
I. scapularis	USA	*M. anisopliae*		laboratory	unfed adult	[156]
R. (B.) microplus	Brazil	*M. anisopliae*	Ma01, Ma02, Ma04	laboratory	egg, larvae, engorged female	[157]
I. scapularis	USA	*M. anisopliae*		field/laboratory	nymph, larvae, adult	[158]
D. variabilis *R. sanguineus* *I. scapularis*	Usa	*M. anisopliae* *B. bassiana*		laboratory	nymph, adult	[159]

Ticks	Country	Biocontrol agent	Strain	Experiment	Developmental cycle	Reference
A. variegatum	Kenya	*M. anisopliae* *B. bassiana*	ICIPE 7	field/laboratory	engorged female	[73]
R. sanguineus	Brazil	*M. anisopliae* *B. bassiana*	ESALQ 959	laboratory	nymph, unfed adult	[66]
R. (B.) microplus	USA	*M. anisopliae*	ARSEF3297	laboratory	engorged female	[63]
R. (B.) microplus *R. sanguineus*	USA	*M. anisopliae* *S. lamellicola* *P. farinosus*	ARSEF3297	laboratory	engorged female, larvae, egg	[59]
R. (B.) microplus	Trinidad and Tobago	*M. anisopliae*	ARSEF3297, IMI386697	field/laboratory	engorged female	[160]
I. scapularis	USA	*M. anisopliae*	ESC1	field/laboratory	Nymph	[161]
I. scapularis	USA	*M. anisopliae*		field/laboratory	Larvae	[162]
R. (B.) microplus	Brazil	*M. anisopliae*	ESALQ 959	laboratory	Larvae	[30]
R. (B.) microplus	Brazil	*M. anisopliae*	E9, 319	laboratory	egg, larvae, engorged female	[163]
R. (B.) microplus	Brazil	*M. anisopliae* *B. bassiana* *Pe. corylophilum* *F. oxysporum*	ESALQ 959	laboratory	hemocytes (engorged female)	[164]
R. (B.) microplus	Brazil	*M. anisopliae* *Penicillium sp.*	ESALQ 959	laboratory	hemocytes (engorged female)	[165]
A. variegatum	Kenya	*M. anisopliae* *B. bassiana*	ICIPE 07	field	unfed adult	[166]
R. (B.) microplus	Brazil	*M. anisopliae*	ESALQ 959	field	engorged female	[167]
A. cajennense	Brazil	*M. anisopliae* *B. bassiana*	ESALQ-1037	laboratory	Nymph	[168]
R. (B.) microplus	Mexico	*M. anisopliae*	Ma34	field	engorged female	[72]
R. (B.) annulatus	Iran	*M. anisopliae* *B. bassiana* *L. psalliotae*	IRAN 437C	laboratory	engorged female, larvae, egg	[169]

Table 1. (Continued)

Ticks	Country	Biocontrol agent	Strain	Experiment	Developmental cycle	Reference
I. scapularis *A. americanum*	USA	*M. anisopliae*	2575	laboratory	cell lines	[170]
R. (B.) microplus	Australia	*M. anisopliae*		laboratory	engorged female	[49]
R. sanguineus	Brazil	*M. anisopliae* *B. bassiana*	Ma959 Bb986		engorged female	[45]
R. (B.) microplus	Australia	*M. anisopliae*	ARIM16, ARIM10	field	unfed and engorged female	[9]
R. (B.) microplus	Brazil	*M. anisopliae*	ESALQ 959	field	engorged female	[48]
R. evertsi evertsi	Namibia	*M. anisopliae*	RS2	laboratory	larvae, unfed adult	[68]
H. excavatum *R. annulatus* *R. sanguineus*	Israel	*M. anisopliae* *M. acridum*		laboratory	Egg	[171]
A. variegatum	Kenya	*M. anisopliae*	ICIPE 07	field	engorged female	[46]
R. (B.) microplus	Mexico	*M. anisopliae*	Ma34, Ma14	field/laboratory	engorged female, larvae	[10]
R. (B.) microplus	Brazil	*M. anisopliae*	E6,CARO14, CG47, CG97	laboratory	engorged female	[37]
Argas persicus	Iran	*M. anisopliae*	V245, 685, 715C	laboratory	egg, larvae, unfed and engorged female	[172]
A. variegatum	Kenya	*M. anisopliae*	ICIPE 07	field	unfed adult	[50]
R. pulchellus	Kenya	*M. anisopliae*	ICIPE 60	laboratory	unfed adult and nymph	[173]
R. annulatus *R. sanguineus* *H. excavatum*	Israel	*M. anisopliae*	7	laboratory	Egg	[99]
I. scapularis	USA	*M. anisopliae* *B. bassiana*	F52	field	Nymph	[174]
I. scapularis	USA	*M. anisopliae*	F52	field	Nymph	[175]
R. (B.) microplus	Mexico	*M. anisopliae* *P. fumosoroseus*	Ma 14	field/laboratory	Larvae	[44]

Ticks	Country	Biocontrol agent	Strain	Experiment	Developmental cycle	Reference
H. anatolicum anatolicum	China	*M. anisopliae* *B. bassiana* *P. lilacinus*	M.aAT26	laboratory	engorged female	[176]
R. (B.) microplus	Brazil	*M. anisopliae*	E9	field/laboratory	engorged female, larvae	[31]
A. variegatum *R. appendiculatus* *R. evertsi evertsi*	Namibia	*M. anisopliae*	RS2	laboratory	egg, larvae, nymph, adult	[60]
R. (B.) decoloratus *R. evertsi evertsi*	Namibia	*M. anisopliae*	NA1	field	engorged female	[61]

Abbreviations: Tick genera: *R.* = *Rhipicephalus*; *A.*= *Amblyomma*; *I.*= *Ixodes*; *B.*= *Boophilus*; *H.*= *Hyalomma*; *D.*= *Dermacentor*; Fungus genera: *M.*= *Metarhizium*; *B.*= *Beauveria*; *P.*= *Paecilomyces*; *L.*= *Lecanicillium*; *F.*= *Fusarium*; *V.*= *Verticillium*; *Pe.*= *Penicillium*; *S.*= *Simplicillium*.

In Brazil, several industries use rice as a substrate for *M. anisopliae* conidial production [42], this being both efficient [14, 49, 50] and providing the best results in UV stress studies [51] with important implications for application in the field. However, other substrates such as potatoes, wheat flour and corn have been used [42, 52]. Selection of the optimum substrate for commercial conidial production needs to consider the local availability and cost of the substrate. Opportunities to reuse the substrate for conidial production may help to reduce costs. Recent results showed that rice can be reused as a substrate in two consecutive production cycles [14].

Several factors may limit the utilization of microbial biocontrol agents, including the time required to cause target mortality, which is longer than that required for conventional pesticides [14, 37]. Additionally, adverse environmental conditions, such as high and low temperatures, low humidity and ultraviolet radiation, that may affect the microbial agent should be considered [53, 54]. The development of efficient formulations containing appropriate carriers for host infection is fundamental for the successful utilization of arthropod-pathogenic fungus on a large scale [41, 55]. For example, vegetable oils may be used as a component in formulations to protect spores from ultraviolet radiation, probably by absorbing the UV [38, 56, 57]. Oil was shown to increase the effectiveness and virulence of fungal pathogens against arthropod pests when compared to aqueous formulations [14, 38, 44, 46, 58-61], especially in low humidity climates [62]. This may be due to the greater adhesion of the conidia to the surface of the arthropod [45, 62]. Oil reacts well with hydrophobic *M. anisopliae* conidia [63], has low volatility, does not affect germination [64], and enhances the effectiveness of the pathogen in natural conditions [65]. Studies have also demonstrated satisfactory results in formulations containing a polymerized cellulose gel [45, 66]. With the aim of protecting the conidia in the environment, some formulations containing sunscreens have been investigated under laboratory conditions [67, 68] and in the field [69]. Hedimbi et al. [68] observed that inclusion of a sunscreen in an *M. anisopliae* oil formulation improved survival, and had no effect on viability and virulence. Several *M. anisopliae*-based formulations have been developed to improve the biocontrol of the cattle tick *R. microplus* and other ticks (Table 2).

Different methods of application of *M. anisopliae* as a mycoacaricide can be used for tick control. Treatment of the environment is a good alternative for reducing levels of infestation. While *in vitro* tests are promising, trials that mimic field conditions tend to be less successful. Therefore, studies that evaluate the effect of mycoacaricides in the field are important. Pastures are the main source of bovine tick infestations in the field and thus should be a major target for control [70]. Although in some cases field or pen trials showed little effect on adult ticks, other parameters such as egg laying and hatching were reduced [10, 26, 71, 72, 73]. Garcia et al. [31] evaluated the effect of *M. anisopliae* on off-host *R. microplus* from tick-infested pasture on which cattle were grazing. Kaaya et al. [61] evaluated the effect of oil formulation of *M. anisopliae* for the control of two tick species on Afrikana bulls grazing in paddocks for a period of one year. The fungus reduced the on-host tick populations by 83% within three months of the experiment. *M. anisopliae* may be applied to livestock grazing large fields in the same way as in sugarcane plantations, which have been treated by aircraft spraying with great success for many decades in Brazil and Australia [74, 75].

Table 2. Formulations using *M. anisopliae* for tick control

Tick	Country	Formulations tested *	Reference
A. variegatum *R. apendiculatus* *B. decoloratus*	Namibia	15% peanut oil + 1% Tween-80	[47]
R. appendiculatus *A. variegatum*	Namibia	15% peanut oil + Tween 80 (1%)	[58]
R. (B.) microplus	USA	10% coconut oil AEO + 0.2% Tween 80 10% liquid paraffin AEO + 0.2% Tween 80	[63]
R. (B.) microplus	Trinidad and Tobago	2% emulsifiable adjuvant oil Newman's Cropspray 11-E	[160]
R. sanguineus	Brazil	polymerized gel (cellulose stearate) emulsion (calcium phenylsulfonate and nonyl phenol ethoxylated) polymerized gel + emulsion	[45, 66]
A. variegatum	Kenya	15% peanut oil + 1% emulsogen-1900T	[73, 166]
R. (B.) microplus	Australia	10% vegetable oil	[49]
R. (B.) microplus	Australia	10% cocacide	[9]
R. evertsi evertsi	Namibia	3% sunscreens + 20% olive oil + 0.05% Triton X-100	[68]
A. variegatum	Kenya	49.5% corn oil + 1% Tween 80	[46]
R. pulchellus	Kenya	49.5% corn oil + 1% Tween 80	[50, 173]
R. (B.) microplus	Mexico	0.1% Tween 80 + 1% agricultural surfactant Wheat bran (50 kg/ha)**	[44]
A. variegatum *R. appendiculatus* *R. evertsi evertsi*	Namibia	20% olive oil + 0.05% Triton X-100	[60]
R. (B.) decoloratus *R. evertsi evertsi*	Namibia	20% sunflower oil + 0.05% Triton X-100	[61]

Abreviations: Tick genera: R. = *Rhipicephalus*; A.= *Amblyomma*; B.= *Boophilus*.
* All formulation containing spore suspension in water.
** Solid formulation Infection by *M. anisopliae*.

While tick biocontrol methods that are completely free of chemical acaricides are most desirable, they may occasionally be combined with low quantities of chemical acaricides. This would help to reduce environmental pollution and the rate of development of resistance [71]. Therefore, compatible chemical products can be used with pathogenic fungi to increase control efficiency [76]. As reported previously [57], the compatibility of chemical and biological products results in higher efficacy in the treatment of some chemical-resistant or less susceptible arthropods, as observed in the use of *M. anisopliae* and deltamethrin against *R. microplus* [30, 48]. Acaricides may act in combination with arthropod-pathogenic agents, with sub-lethal doses facilitating the infectious process by debilitating or stressing the arthropod rendering them more susceptible to the action of pathogens [38, 77]. The combination of *M. anisopliae* with other arthropod-pathogenic microorganisms has also been tested, providing a clear indication that under certain conditions, a cocktail of pathogenic fungi may induce high mortality in ticks (Table 2). In this context, higher mortality induced by a combination of *B. bassiana* and *M. anisopliae* might be attributed to the arrays of mycotoxins produced by each fungal species, which become more lethal when combined [73].

Host infection by *M. anisopliae* starts with contact between the conidia and the host surface followed by adhesion and, under specific conditions of humidity and temperature, spore germination. Specific events define the entire infection cycle, as follows: conidia adhesion and germination over the host surface (cuticle); development of specific structures (germ tube and appressorium); penetration through the host cuticle; death of the host and intense fungal growth; and fungal emergence through the host cuticle followed by production of conidia that could initiate another infection cycle in a new host target (Figure 1). The system is highly conserved in nature and the invasion of arthropod tissues resembles that occurring in fungal plant pathogens, including the formation of germ tubes, appressoria and penetration pegs [37, 78].

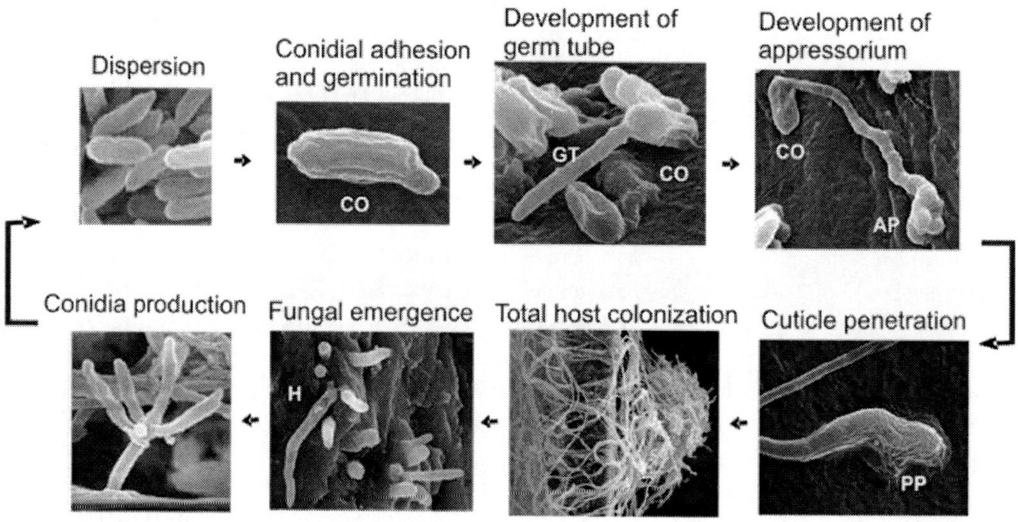

Figure 1. Scanning electron microscopy showing the infection cycle of *M. anisopliae* in the host *R. microplus*. The infection process starts with conidial adhesion and germination, and ends with conidial production followed by dispersal into the environment. CO – conidia; GT – germ tube; AP – appressorium; PP – penetration peg; H – hyphae.

M. anisopliae infects susceptible hosts via direct penetration through the cuticle. When conidia find the cuticle of a susceptible host, the infection process starts and both the pathogen and host trigger the expression of genes related to infection structures and hydrolytic enzymes in the fungus and defense mechanisms in the host [37]. Initial infection involves a combination of two processes: the mechanical pressure exerted by appressoria, the modified extremities of the hyphae, and degradation by hydrolytic enzymes like lipases, proteases and chitinases (Figure 2) [35, 36, 57, 79-88].

When favorable conditions are encountered, fungal spores interact hydrophobically with the host's cuticle [89]. This stage is considered a limiting step in terms of infection success, and the role played by hydrophobins [90, 91], adhesins and other constituents like GAPDH (glyceraldehyde 3-phosphate dehydrogenase) are extremely relevant [92, 93]. Spore adhesion is mediated by the hydrophobic interaction between polar proteins (hydrophobins) and the host's lipid layer [90, 91]. An adhesin (MAD1) is involved in spore adhesion, and the germination and virulence of null mutants for the gene codifying this protein have been observed to diminish [92]. Other components that play a role in the adhesion process include the enzyme GAPDH, which was shown to be involved in the binding of conidia to the host surface, indicating new roles for this well-known glycolytic enzyme both physiologically and during the host interaction [93]. The surface structure, topology and chemical composition of arthropod cuticle may also affect spore adhesion [94, 95]. *M. anisopliae* is able to recognize specific signals from the host, probably epicuticle lipids, and specificity is a result of these signals [95]. After contact with and adsorption to the host's tegument, the spore germinates and the fungus develops, marking the onset of the infection process.

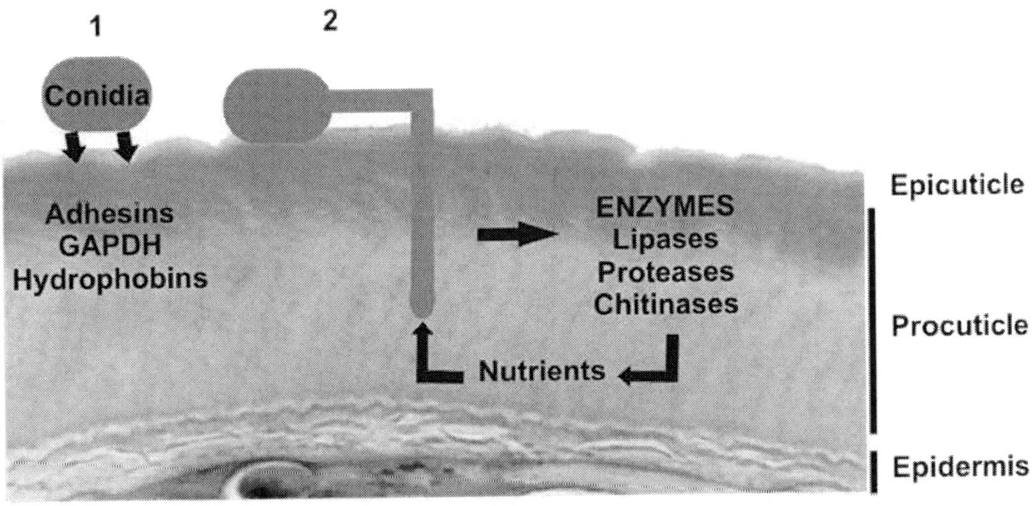

Figure 2. Adhesion to and penetration of *R. microplus* cuticle by *M. anisopliae*. 1- Adhesion of conidia to host cuticle mediated by different components. 2- During the penetration of the host's cuticle the fungus secretes several hydrolytic enzymes whose roles include nutrient acquisition and degradation of host tegument components. A microscopic section of *R. microplus* cuticle is shown.

The Host Target *R. microplus*

The host cuticle is the main barrier against infection. In arthropods it is a complex structure formed by two layers: the epicuticle and procuticle (Figure 3). The procuticle comprises the majority of the cuticle, and is composed essentially of proteins and chitin [96]. Therefore, it is the procuticle that is the substrate for specific proteases and chitinases. However, for the fungus to reach this tegument layer it must first cross the epicuticle, the outer layer formed mainly by esterified lipids, which are specific substrates for lipases.

The epicuticle is thin (0.1–3 μm) and is formed by at least two distinct sublayers: cuticulin, the innermost layer composed of lipoproteins, polymerized lipids and polyphenols, and an outer waxy layer produced by secretions from the dermal glands and is composed of an oriented layer of esterified lipids and fatty acids (mostly long-chain ones) as well as alcohols [95, 97, 98, 99]. The outer epicuticular waxy layer of ticks has several functions, such as the prevention of water loss, protection from physicochemical hazards, and the production of sex pheromone [99-102]. The epicuticular lipid composition of ticks shows much variation, even among ticks from the same family [103] and among various ages of the same tick species [99, 100, 102, 104].

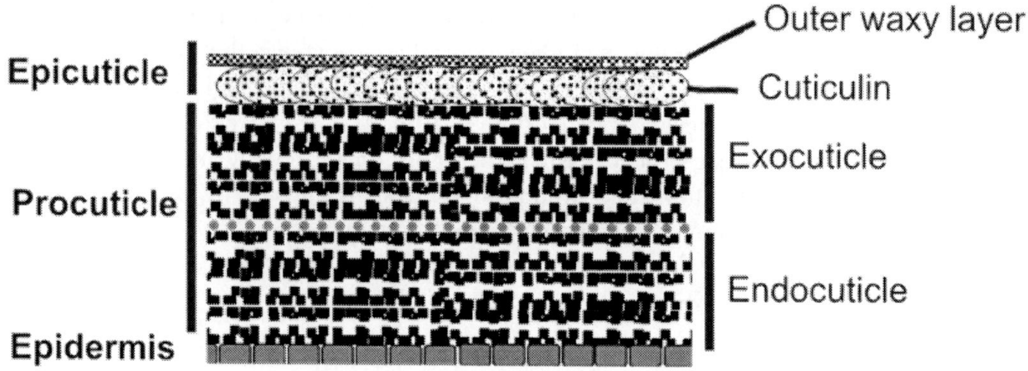

Figure 3. Basic constitution of arthropod tegument.

However, the deposition of additional surface wax over the course of tick development is a general trend. For example, the amount of surface wax may increase three times during the engorgement of adult *R. microplus* [99, 100].

The Host Infection Process: Adhesion, Penetration, Colonization and Host Death

Enzymes like lipases, proteases, esterases and N-acetylglycosidases have been detected on the surface of ungerminated spores [34, 86]. These enzymes help in adhesion and the initial acquisition of nutrients, leading to surface changes on the host cuticle [105]. These spores germinate on the host, producing a germ tube that, at one of its ends, is modified so as to produce the appressorium. *M. anisopliae* may develop the appressorium *in vitro* as well as *in vivo*, in a process influenced by topography and the presence of hard hydrophobic surfaces [80, 106]. The appressorium stores sugars, and is responsible for the generation of the osmotic pressure required for the successful penetration of the host [107]. At least four different signals affect the development of appressorium: (1) the polar and non-polar moieties of the host's cuticle, (2) the nutrient levels, (3) the hydrophobic surface, and (4) the signals generated by the host's scutum, which remain unknown [37, 99, 108]. Ment et al. [99] reported the effects of lipids from tick epicuticle on conidial germination and the development of appressoria in two subspecies of *M. anisopliae* that have different levels of virulence toward ticks. The germination of conidia of acaricidal fungi is selectively affected by the lipids present in tick epicuticles, which may explain the variability in the control of different hosts by these fungi. In this sense, the cuticular lipid profile of susceptible and resistant hosts was found to differ in some cases, either stimulating or inhibiting conidial germination and appressorium formation [99, 109]. Ment et al. [99] also showed that lipids obtained from the ear secretions, skin and hair of a calf can stimulate the germination of *M. anisopliae* conidia. This is an important observation, since lipid substances in a host are inseparable components of the environment of feeding ticks, and may be a highly relevant factor in terms of the anti-tick efficacy of the fungus against the host [99]. The primary role of the appressoria seems to be the establishment of a nutritional relationship between the germinated conidia and the host [110]. Once the appressorium is formed, the penetration pegs form through a change in the cell wall of the appressorium section that is in contact with the host, being thinner and more prominent [80, 96]. After crossing the cuticle, the hyphae thicken and the blastospores, responsible for fungal nutrition by degradation of carbon sources from the haemolymph, are formed. With the development of the fungus, toxins and secondary metabolites, such as citric and oxalic acids, are secreted [89].

Fungal penetration of the host takes place as a combination of mechanical pressure and enzyme degradation [79, 85, 111]. At this stage of the pathogenic process, the enzymes that are essential for cuticle degradation including lipases, esterases, chitinases and proteases are produced [34, 36, 85, 87, 88, 112]. It has been postulated that lipases are formed before proteases since the epicuticle, the outer lipid layer that comprises arthropod cuticle, has to be crossed first [79]. During infection of the cattle tick *R. microplus*, the secretion rates of these enzymes present on the spore surface increase within the first few hours of infection. Moreover, lipolytic activity plays a fundamental role in *R. microplus* infection [87]. Lipolytic activity has also been proposed as a 'virulence index', enabling the selection of more efficient strains from those available naturally, as recently reported for *M. anisopliae* [113]. Chitinolytic activity begins after proteolytic activity due to the physical structure of the arthropod cuticle, since chitin fibers act as anchor sites for some proteins, forming a protein matrix [89].

The colonization process starts after penetration. The fungus colonizes the inner cavity of the host and produces secondary metabolites such as organic acids and toxins known as destruxins (DTX), which intensify the pathogenic process. Several types of DTX have been described: A, B, C, D and E, and another 27 variants [114], which cause muscle paralysis and inhibit haematocyte and Malpighian tubule function, in addition to suppressing the host immune system [89, 115, 116, 117]. The destruxins detected in *M. anisopliae* may vary as a function of the substrate used by the fungus. Colonization times also differ, depending on the host and environmental conditions [114, 118].

The symptoms of fungal pathogenicity on the host include lowered sensitivity, uncoordinated movements and paralysis, ultimately leading to death [57]. After the host's death, hyphae invade the inner organs and, once nutrients have been exhausted, grow out of the tegument. Under appropriate environmental conditions, olive green conidia are produced. These spores may be spread by the wind, enabling the infection of other susceptible hosts [57].

M. anisopliae Enzymes and Infection

The wide genetic variability of this fungal species results in the emergence of several strains with different degrees of virulence, specificity, conidia production and resistance to ultraviolet light [51, 57, 111]. *Metarhizium* strains pathogenic to a particular host species may not present the same characteristics in terms of pathogenicity as those infecting other arthropods [119]. An accepted definition of virulence is the "degree of pathogenicity between a group or different species" [120], and the variation in fungal strain virulence against different arthropods may be related to the production of enzymes, among other factors [121].

For the hyphae to cross the cuticles of arthropods, host recognition and response seem to be critical for the sequence of events that culminates in efficient infection. Several pathogenicity genes have been implicated in this process [122, 123]. These genes may be divided into groups: (1) those that codify receptors that detect the presence of a host; (2) those that codify enzymes that facilitate penetration; (3) those whose products inactivate the host defense system; and (4) those that codify toxins that are able to cause disease [122, 124].

Several factors have been suggested as possible determinants of *M. anisopliae* pathogenicity, including toxin production and hydrolytic enzyme secretion [96]. Some authors have also suggested that the differential expression of superoxide dismutase enzymes (SODs), which act as a defense system against reactive oxygen species [125, 126], and the presence of mycovirus with a dsRNA genome [127, 128] may reduce the virulence of *M. anisopliae*.

The possibility of a correlation between the virulence shown by arthropod-pathogenic fungi and the production of enzymes that degrade arthropod cuticle has been widely investigated. Lipases and esterases are involved in the infection process, since lipids are constituents of the host's cuticle [36, 79, 85, 87, 88, 96, 129]. *M. anisopliae* is able to produce lipases induced by different lipids that are used as a carbon source [85, 88].

Lipases

Fully engorged females of *R. microplus* exhibit an increase of 215% in the total lipid content at this stage, from 20 to 63 μg per tick [97]. Lipids are probably the first point of contact between arthropod and pathogenic microorganisms, reinforcing the importance of lipases in initial recognition stages. Of the lipids in the cuticle, the major components are acids and alcohols that are generally combined as esters [97], thus forming the substrate for the action of lipases. Cholesteryl stearate, a lipid characteristic of animal cells and present in tick cuticle, induced higher levels of lipase activity in comparison to olive oil, mainly when combined with chitin. This substrate combination also led to a better growth of *M. anisopliae*, suggesting that these specific constituents, which are present in the host tegument, are more easily recognized and used by the fungus, as reflected by the higher mycelial mass produced [88].

A *M. anisopliae* spore surface lipase (MASSL) that is strongly bound to the fungal spore surface with a molecular mass of approximately 66 kDa was purified and characterized [36]. MASSL is able to degrade a variety of substrates, which could be an important adaptive mechanism by which new host environments with different compositions could be conquered by the pathogen [36]. This enzyme was the first lipolytic enzyme to be characterized in *M. anisopliae*, and its presence on the conidial surface reinforces the possibility of its participation in the initial stages of the infection process. During the course of tick exposure to spores (6 to 120 h) lipase activity increased around 10 times, and bands of lipase activity were detected in zymograms of ticks treated with *M. anisopliae* spores. The treatment of spores with lipase activity inhibitor, ebelactone B, completely inhibited lipolytic activity, and prevented the infection of *R. microplus*. Spores treated with the inhibitor did not germinate on the tick, although this effect was not observed in the culture medium. Therefore, lipolytic activity has a pivotal role in infection of the tick *R. microplus*, and is one of the key factors in tick infection [87].

Proteases

M. anisopliae produces a diverse array of proteases, especially subtilisin-like serine proteases Pr1 (Pr1A-K), which form an important family of proteases involved in fungal infection [108, 122, 130, 131]. The different isoforms of subtilisins are the result of gene duplication events. It is likely that they are associated with pathogenicity, increasing adaptability to different environments and improving host specificity. Gene duplication may also have occurred with genes encoding for other hydrolases involved in this process [130, 132]. Pr1A has been proven to play a role in the entomopathogenicity of *M. anisopliae* [133, 134]. A second class of proteases that show high *in vitro* output when *M. anisopliae* is cultured on a host's cuticle comprises the serine proteases associated with trypsins (Pr2) [135, 136]. Two of these Pr2 isoforms seem to be linked to the appressorium, suggesting that they are available during the first stages of cuticle colonization [136]. Although the gene codifying for one of these serine proteases has been cloned and characterized, the role played by Pr2 in pathogenicity has not been fully clarified. Nevertheless, it is possible that Pr2 is relevant in the induction or activation of Pr1A [137]. Other proteases are likewise secreted by *M. anisopliae*, including those of the type cysteine-protease (Pr4, which may be an isoform of Pr2) [138], carboxypeptidases [139] and metalloproteases [140]. No other fungal species has been reported to have such a diverse array of proteases as *M. anisopliae* [141], despite the fact that they are fundamental to transposing the arthropod cuticle, colonizing and digesting tissues, and surviving in diverse niches in the environment during saprophytic existence, and may also influence virulence and/or host specificity [141].

The identification of protease inhibitors produced by arthropods shows the importance of these enzymes for *M. anisopliae* infection in different hosts. The tick *R. microplus* produces two different protease inhibitors that are able to inhibit Pr1 activity in *M. anisopliae*: BmSI-6 and BmSI-7, which could be involved in the defense mechanisms of *R. microplus* against fungal infection [142]. Different proteases were detected in proteomic studies (carboxypeptidase, subtilisins Pr1A and Pr1J, and elastase-like serine protease) under conditions of infection, and are probably secreted during the infection of *R. microplus* by *M. anisopliae* [34, 112]. The presence of different proteases, especially Pr1, in different infection conditions suggests their involvement in the recognition of different hosts [112].

Chitinases

Chitinases comprise another class of hydrolytic enzymes that, apart from being involved in the morphogenesis of this fungus, may also play a role in its pathogenicity, as virulence factors. It is believed that chitinases may be associated with virulence in the same way as in phytopathogenic, mycopathogenic and entomopathogenic fungi, [35, 83, 84, 143-147]. It is likely that these enzymes develop a synergistic relationship with proteolytic enzymes, thereby solubilizing the host's cuticle during penetration [79, 146, 148, 149].

In studies on the chitinolytic complex secreted by *M. anisopliae* var. *anisopliae* under induction-promoting conditions, at least three chitinolytic enzymes [82] and at least six chitinolytic enzymes [35] were present. One of these, a 30-kDa chitinase (CHIT30), exhibits endo- and exo-chitinase activity. This ability may allow this chitinase to degrade chitin more efficiently and probably plays an important role in the penetration of the tick cuticle. Using an immunolocalization protocol, da Silva et al. [35] demonstrated the presence of CHIT30 in *M. anisopliae* infection during penetration of the cattle tick *R. microplus*, suggesting the importance of this enzyme for pathogenicity. Different chitinases were also detected in proteomic studies and are probably involved in the infection process of *R. microplus* and other arthropods [34, 112, 150]. These chitinases were detected with other enzymes *in vitro* under conditions of infection, suggesting that a synergistic effect between pathogenicity factors and hydrolase diversity may be important in this relationship [34]. This leads to the conclusion that chitinases have an important role in pathogenicity.

CONCLUSION

Advances in biocontrol research, mainly in the fungus *M. anisopliae*, and the need for alternative, efficient and environmentally safe strategies of production make the biocontrol of pests even more commercially attractive. In this context, several studies discussed in this chapter have demonstrated the efficiency of *M. anisopliae* in control of the tick *R. microplus*, the major ectoparasite affecting livestock worldwide. *M. anisopliae* has been applied commercially and successfully for the control of different pests for many decades. Today, it could also be used for the control of the cattle tick *R. microplus,* either totally or partially replacing chemical acaricides, and thus reducing the problems associated with their use.

REFERENCES

[1] Samish, M., & Rehacek, J. (1999). Pathogens and predators of ticks and their potential in biological control *Annu. Rev. Entomol., 44,* 159-182.

[2] Hoogstral, H. (1985). Argasid and nuttallied ticks as parasites and vectors. *Adv. Parasitol., 1*, 135-238.

[3] Leal, A. T., Freitas, D. R. J., & Vaz Jr, I. S. (2003). Perspectivas para o controle do carrapato bovino. *Acta Sci. Vet.*, *31,* 1-11.

[4] Angus, B. M. (1996). The history of the cattle tick *Boophilus microplus* in Australia and achievements in its control. *Int. J. Parasitol., 26*, 1341-1355.

[5] Estrada-Peña, A., García, Z., & Sánchez, H.F. (2006). The distribution and ecological preferences of *Boophilus microplus* (Acari: Ixodidae) in Mexico. *Exp. Appl. Acarol.*, *38*, 307-316.

[6] Reck Jr, J., Berger, M., Terra, R. M. S., Marks, F. S., Da Silva Vaz Jr, I., Guimarães, J. A., & Termignoni, C. (2009). Systemic alterations of bovine hemostasis due to *Rhipicephalus (Boophilus) microplus* infestation. *Res. Vet. Sci., 86,* 56-62.

[7] Grisi, L., Massard, C., Moya Borja, G. E., & Pereira, J. B. (2002). Impacto econômico das principais ectoparasitoses em bovinos no Brasil. *A Hora Vet.*, *21*, 8-10.

[8] Frazzon, A. P. G., Vaz Jr, I. V. S., Masuda, A., Schrank A., & Vainstein, M. H. (2000). *In vitro* assessment of *Metarhizium anisopliae* isolates to control the cattle tick *Boophilus microplus*. *Vet. Parasitol., 93*, 117-125.

[9] Leemon, D. M., Turner, L. B., & Jonsson, N. N. (2008). Pen studies on the control of cattle tick *Rhipicephalus* (*Boophilus*) *microplus* with *Metarhizium anisopliae* (Sorokin). *Vet. Parasitol., 156,* 248-260.

[10] Ojeda-Chi, M. M., Rodriguez-Vivas, R. I., Galindo-Velasco, E., & Lezama-Gutiérrrez, R. (2010). Laboratory and field evaluation of *Metarhizium anisopliae* (Deuteromycotina: Hyphomycetes) for the control of *Rhipicephalus microplus* (Acari: Ixodidae) in the Mexican tropics. *Vet. Parasitol.,* 170, 348-354.

[11] Quesada-Moraga, E., Santos-Quirós, R., Valverde-García, P., & Santiago-Álvarez, C. (2004). Virulence, horizontal transmission, and sub lethal reproductive effects of *Metarhizium anisopliae* (Anamorphic fungi) on the German cockroach (Blattodea: Blattellidae). *J. Invertebr. Pathol., 87,* 51-58.

[12] Lazzarini, G. M. J., Rocha, L. F. N., & Luz, C. (2006). Impact of moisture on *in vitro* germination of *Metarhizium anisopliae* and *Beauveria bassiana* and their activity on *Triatoma infestans. Mycol. Res., 110,* 485-492.

[13] Scholte, E. J., Takken, W., & Knols, B. G. J. (2007). Infection of adult *Aedes aegypti* and *Ae. albopictus* mosquitoes with the entomopathogenic fungus *Metarhizium anisopliae. Acta Trop., 102,* 151-158.

[14] Santi, L., Silva, L. A. D., Silva, W. O. B., Correa, A. P. F., Rangel, D. E. N., Schrank, A., & Vainstein, M. H. (2011). Virulence of the entomopathogenic fungus *Metarhizium anisopliae* using soybean oil formulation for control of the cotton stainer bug, *Dysdercus peruvianus. World J. Microbiol. Biotechnol., in press.*

[15] Mans, B. J., & Neitz, A. W. (2004). Adaptation of ticks to a blood-feeding environment: evolution from a functional perspective. *Insect Biochem. Mol. Biol., 34,* 1-17.

[16] Guimarães, J. H., Tucci, E. C., & Barros-Battesti, D. M. (2001). *Ectoparasitos de importância veterinária. São Paulo, Editora Plêiade.* Pp. 218.

[17] Pereira, M. C. (2009). Introdução ao *R.* (*B.*) *microplus. In Pereira, M. C., Labruna, M. B., Szabó, M. J. P., Klafke, G. M. (editors) Rhipicephalus (Boophilus) microplus – Biologia, controle e resistência. São Paulo, MedVet Livros.* Pp. 1-6.

[18] Pereira, M. C., & Labruna, M. B. (2009). *Rhipicephalus* (*Boophilus*) *microplus. In Pereira, M. C., Labruna, M. B., Szabó, M. J. P., Klafke, G. M. (editors) Rhipicephalus (Boophilus) microplus – Biologia, controle e resistência. São Paulo, MedVet Livros.* Pp. 15-56.

[19] Balashov, Y. S. (1972). *Blood sucking ticks (Ixodoidea) - vectors of diseases of man and animals, vol. 8. Entomological Society of America, Miscell. Publ.* Pp160-376.

[20] Gonzales, J. C., Silva, N. R., & Wagner, E. M. (1974). O ciclo parasitário de *Boophilus microplus* (Can. 1887) em bovinos estabulados. *Arq. Fac. Vet. UFRGS, 2,* 25-34.

[21] Martins, J. R. (2004). Carrapato bovino: epidemiologia e controle. *In Pereira Neto, O. A. (editor) Cadernos de Pecuária - Boas Práticas de Produção. Porto Alegre, Novartis Saúde Animal.* Pp. 125-136.

[22] Parizi, L. F., Pohl, P. C., Masuda, A., & Vaz da Silva Jr., I. (2009). New approaches toward anti-*Rhipicephalus (Boophilus) microplus* tick vaccine. *Rev. Bras. Parasitol. Vet., 18,* 1-7.

[23] Jonsson, N. N. (2006). The productivity effects of cattle tick (*Boophilus microplus*) infestation on cattle, with particular reference to *Bos indicus* cattle and their crosses. *Vet. Parasitol., 137,* 1-10.

[24] Wikel, S. K. (1996). Immunology of the tick-host interface. *In Wikel, S. K. (editor) The immunology of host-ectoparasitic arthropod relationships. Wallingford, CAB International.* Pp. 204-231.

[25] Jonsson, N. N., Bock, R. E., & Jorgensen, W. K. (2008). Productivity and health effects of anaplasmosis and babesiosis on *Bos indicus* cattle and their crosses, and the effects of differing intensity of tick control in Australia. *Vet. Parasitol.* 155:1-9.

[26] Correia, A. C. B., Fiorin, A. C., Monteiro, A. C., & Veríssimo, C. J. (1998). Effects of *Metarhizium anisopliae* on the tick *Boophilus microplus* (Acari: Ixodidae) in stabled cattle. *J. Invertebr. Pathol., 71*, 189-191.

[27] Gonzalez-Acuña, D., & Guglielmone, A. A. (2005). Ticks (Acari: Ixodoidea: Argasidae: Ixodidae) of Chile. *Exp. Appl. Acarol., 35*, 147-163.

[28] Labruna, M. B. (2009). Combate contra o *R.* (*B.*) *microplus*. *In Pereira, M. C., Labruna, M. B., Szabó, M. J. P., & Klafke, G. M. (editors) Rhipicephalus (Boophilus) microplus – Biologia, controle e resistência. São Paulo, MedVet Livros.* Pp. 65-80.

[29] Soares, V. E., Silveira, D. M., Nunes, T. L. S., Oliveira, G. P., Barbosa, O. F., & da Costa, A.J. (2001). *In vitro* analysis of the action of acaricides on *Boophilus microplus* strain take from dairy cattle from the north-east area of Sao Paulo State. *Semina-Ciências Agrárias, 22,* 71-75.

[30] Bahiense, T. C., Fernandes, E. K. K, & Bittencourt, V. R. E. P. (2006). Compatibility of the fungus *Metarhizium anisopliae* and deltamethrin to control a resistant strain of *Boophilus microplus* tick. *Vet. Parasitol., 141,* 319-324.

[31] Garcia, M. V., Monteiro, A. C., Szabó, M. P. J., Mochi, D. A., Simi, L. D., Carvalho, W. M., Tsuruta, S. A., & Barbosa, J. C. (2011). Effect of *Metarhizium anisopliae* fungus on off-host *Rhipicephalus (Boophilus) microplus* from tick-infested pasture under cattle grazing in Brazil. *Vet. Parasitol., in press.*

[32] Samish, M., Ginsberg, H., & Glazer, I. (2008). Anti-tick biological control agents: assessment and future perspectives. *In: Bowman, A. S. & Nutall, P. A. (editors) Ticks – Biology, Disease and control. New York, Cambridge University Press.* Pp. 447-469.

[33] Willadsen, P. (2008). Anti-tick vaccines. *In: Bowman, A. S. & Nutall, P. A. (editors) Ticks – Biology, Disease and control. New York, Cambridge University Press.* Pp. 424-446.

[34] Santi, L., Silva, W. O. B., Pinto, A. F. M., Schrank, A., & Vainstein, M. H. (2009). Differential immunoproteomics allows the identification of *Metarhizium anisopliae* proteins related with *Rhipicephalus microplus* infection. *Res. Microbiol., 160,* 824-828.

[35] Da Silva, M. V., Santi, L., Staats, C. C., Costa, A. M., Colodel, E. M., Driemeier, D., Vainstein, M. H., & Schrank, A. (2005). Cuticle-induced endo/exoacting chitinase CHIT30 from *Metarhizium anisopliae* is encoded by an ortholog of the chi3 gene. *Res. Microbiol., 156,* 382-392.

[36] Silva, W. O. B., Santi, L., Berger, M., Pinto, A. F. M., Guimaraes, J. A., Schrank, A., & Vainstein, M. H. (2009). Characterization of a spore surface lipase from the biocontrol agent *Metarhizium anisopliae*. *Proc. Biochem., 44,* 829-834.

[37] Schrank, A., & Vainstein, M. H. (2011) *Metarhizium anisopliae* enzymes and toxins. *Toxicon, 56*, 1267-1274.

[38] Fernandes, E. K. K., & Bittencourt, V. R. E. P. (2008). Entomopathogenic fungi against South American tick species. *Exp. Appl. Acarol. 46,* 71-93.

[39] Samish, M., Ginsberg, H., & Glazer, I. (2004). Biological control of ticks. *Parasitol., 129,* S389-S403.

[40] Thakore, Y. (2006). The biopesticide market for global agricultural use. *Ind. Biotechnol., 2,* 194-209.

[41] Faria, M. R, & Wraight, S. P. (2007). Mycoinsecticides and mycoacaricides: a comprehensive list with worldwide coverage and international classification of formulation types. *Biol. Control, 43,* 237-256.

[42] Li, Z., Alves, S. B., Roberts, D. W., Fan, M., Delalibera, I. Jr, Tang, J., Lopes, R. B., Faria, M., & Rangel, D. E. N. (2010) Biological control of insects in Brazil and China: history, current programs and reasons for their successes using entomopathogenic fungi. *Biocontrol Sci. Technol., 20,* 117-136.

[43] St Leger, R.J. (2008). Studies on adaptation of *Metarhizium anisopliae* to life in the soil. *J. Invertebr. Pathol., 98,* 271-276.

[44] Ángel-Sahagúna, C. A., Lezama-Gutiérrezb, R., Molina-Ochoab, J., Pescador-Rubioc, A., Skodah, S. R., Cruz-Vázquezd, C., Lorenzonie, A.G., Galindo-Velascof, E., Fragoso-Sánchezg, H., & Fosteri, J. E. (2010). Virulence of Mexican isolates of entomopathogenic fungi (Hypocreales:Clavicipitaceae) *upon Rhipicephalus (Boophilus) microplus* (Acari: Ixodidae) larvae and the efficacy of conidia formulations to reduce larval tick density under field conditions. *Vet. Parasitol., 170,* 278-286.

[45] Reis, R. C. S., Fernandes, E. K. K., & Bittencourt, V. R. E. P. (2008). Fungal formulations to control *Rhipicephalus sanguineus* engorged females. *Ann. N.Y. Acad. Sci., 1149,* 239-241.

[46] Nchu, F., Maniania, N. K., Toure, A., Hassanali, A., & Eloff, J. N. (2009). The use of a semiochemical bait to enhance exposure of *Amblyomma variegatum* (Acari: Ixodidae) to *Metarhizium anisopliae* (Ascomycota: Hypocreales). *Vet. Parasitol., 160,* 279-284.

[47] Kaaya, G. P. (2000). Laboratory and field evaluation of entomogenous fungi for tick control. *Ann. N. Y. Acad. Sci., 916,* 559-564.

[48] Bahiense, T. C., Fernandes, E. K., Angelo Ida, C., Perinotto, W. M., & Bittencourt, V. R. (2008). Performance of *Metarhizium anisopliae* and its combination with deltamethrin against a pyrethroid-resistant strain of *Boophilus microplus* in a stall test. *Ann. N. Y. Acad. Sci., 1149,* 242-245.

[49] Leemon, D. M., & Jonsson, N. N. (2008). Laboratory studies on Australian isolates of *Metarhizium anisopliae* as a biopesticide for the cattle tick *Boophilus microplus. J. Invertebr. Pathol., 97,* 40-49.

[50] Nchu, F., Maniania, N. K., Hassanali, A., & Eloff, K. N. (2010a). Performance of a *Metarhizium anisopliae*-treated semiochemical-baited trap in reducing *Amblyomma variegatum* populations in the field. *Vet. Parasitol., 169,* 367-372.

[51] Rangel, D. E. N., Braga, G. U., Flint, S. D., Anderson, A. J., & Roberts, D. W. (2004). Variations in UV-B tolerance and germination speed of *Metarhizium anisopliae* conidia produced on insects and artificial substrates. *J.Invertebr. Pathol., 87,* 77-83.

[52] Jenkins, N. E. (1998). Development of mass production technology for aerial conidia for use as mycospesticides. *Biocontrol News Inform., 19,* 21-31.

[53] Rangel, D. E., Alston, D. G., & Roberts, D. W. (2008a). Effects of physical and nutritional stress conditions during mycelial growth on conidial germination speed, adhesion to host cuticle, and virulence of *Metarhizium anisopliae*, an entomopathogenic fungus. *Mycol. Res., 112,* 1355-1361.

[54] Rangel, D. E., Anderson, A. J., & Roberts, D. W. (2008b). Evaluating physical and nutritional stress during mycelial growth as inducers of tolerance to heat and UV-B radiation in *Metarhizium anisopliae* conidia. *Mycol Res*. 2008b Nov;112(Pt 11):1362-72.

[55] Goettel, M. S., & Roberts, D. W. (1992). Mass production, formulation and field application of entomopathogenic fungi. *In Lomer, C. J., & Prior, C. (editors) Biological control of locusts and grasshoppers. USA, CAB International.* Pp 230-238.

[56] Moore, D. P., Bridge, D., Higgins, P. M., Bateman, R., & Proir, C. (1992). Ultraviolet radiation damage to *Metarhizium flavoviride* conidia and the protection given by vegetable and mineral oils and chemical sunscreens. *Ann. Appl. Biol., 122,* 605-616.

[57] Alves, S. B. (1998). *Controle microbiano de insetos, 2nd edition. Piracicaba, FEALQ.* Pp 1163.

[58] Kaaya, G. P., & Hassan, S. (2000). Entomogenous fungi as promising biopesticides for tick control. *Exp. Appl. Acarol., 24*, 913-926.

[59] Polar, P., Kairo, M. T., Peterkin, D., Moore, D., Pegram, R., & John, S. A. (2005b). Assessment of fungal isolates for development of a myco-acaricide for cattle tick control. *Vector Borne Zoonotic Dis., 5*, 276-284.

[60] Hedimbi, M., Kaaya, G. P., & Chinsembu, K. C. (2011). Mortalities induced by entomopathogenic fungus *Metarhizium anisopliae* to different ticks of economic importance using two formulations. *Int. Res. J. Microbiol., 2,* 141-145.

[61] Kaaya, G. P., Samish, M., Hedimbi, M., Gindin, G., & Glazer, I. (2011). Control of tick populations by spraying *Metarhizium anisopliae* conidia on cattle under field conditions. *Exp. Appl. Acarol.,* in press.

[62] Bateman, R. P., Carey, M., Moore, D., & Prior, C. (1993). The enhanced infectivity of *Metarhizium flavoviride* in oil formulations to desert locust at low humidities. *Ann. Appl. Biol., 122,* 145-152.

[63] Polar, P., Kairo, M. T., Moore, D., Pegram, R., & John, A. S. (2005a). Comparison of water, oils and emulsifiable adjuvant oils as formulating agents for *Metarhizium anisopliae* for use in control of *Boophilus microplus*. *Mycopathol., 160,* 151-157.

[64] Batta, Y.A. (2003). Production and testing of novel formulations of the entomopathogenic fungus *Metarhizium anisopliae* (Metschinkoff) Sorokin (Deuteromycotina:Hyphomycetes). *Crop Prot., 22,* 415-422.

[65] Albernaz, D. A. S., Tai, M. H. H., & Luz, C. (2009). Enhanced ovicidal activity of an oil formulation of the fungus *Metarhizium anisopliae* on the mosquito *Aedes aegypti*. *Med. Vet. Entomol., 23,* 141-147.

[66] Reis, R. C. S., Melo, D. R., Perinotto, W. M., & Bittencourt, V. R. E. P. (2005). *In vitro* pathogenicity of fungic formulation on nymphs and adults of *Rhipicephalus sanguineus* (Latreile, 1806) (Acari:Ixodidae). *Rev. Bras. Parasitol. Vet., 14,* 101-105.

[67] Hunt, T., Moore, D., Higgins, P. M., & Prior, C. (1994). Effect of sunscreen irradiance and resting periods on the germination of *Metarhizium flavoviride* conidia. *Entomophaga, 39*, 313-322.

[68] Hedimbi, M., Kaaya, G. P., Singh, S., Chimwamurombe, P. M., Gindin, G., Glazer, I., & Samish, M. (2008). Protection of *Metarhizium anisopliae* conidia from ultra-violet radiation and their pathogenicity to *Rhipicephalus evertsi evertsi* ticks. *Exp. Appl. Acarol., 46,* 149-156.

[69] Shah, P. A., Douro-Kpindou, O. –K., Sidibe, A., Daffe, C. O., & van der Pauuw, H., et al. (1998). Effects of the sunscreen oxybenzone on field efficacy and persistence of *Metarhizium flavoviride* conidia against *Kraussella amabile* (Orthoptera: Acrididae) in Mali, West Africa. *Biocontrol Sci.Technol., 8*, 357-364.

[70] Cordovés, C.O. (1997). *Carrapato: Controle ou Erradicação. 2th edition. Porto Alegre, Agropecuária.* Pp 176.

[71] Kaaya, G. P, Mwangi, E. N., & Ouna, E. A. (1996). Prospects for biological control of liverstock tick *Rhipecephalus appendiculatus* and *Amblyomma variegatum*, using the entomogenous for *Beauveria bassiana* and *Metarhizium anisopliae. J. Invertebr. Pathol., 67*, 15-20.

[72] Alonso-Díaz, M. A., García, L., Galindo-Velasco, E., Lezama-Gutierrez, R., Angel-Sahagún, C. A., Rodríguez-Vivas, R. I., & Fragoso-Sánchez, H. (2007). Evaluation of *Metarhizium anisopliae* (Hyphomycetes) for the control of *Boophilus microplus* (Acari: Ixodidae) on naturally infested cattle in the Mexican tropics. *Vet. Parasitol., 147*, 336-40.

[73] Maranga, R. O., Kaaya, G. P., Mueke, J. M., & Hassanali, A. (2005). Effects of combining the fungi *Beauveria bassiana* and *Metarhizium anisopliae* on the mortality of the tick *Amblyomma variegatum* (Ixodidae) in relation to seasonal changes. *Mycopathol., 159,* 527-532.

[74] Batista Filho, A., Almeida, J. E. M., Santos, A. S., Machado, L. A., & Alves, S. B. (2003). Eficiência de isolados de *Metarhizium anisopliae* no controle de cigarrinha-da-raiz da cana-de-açúcar *Mahanarva fimbriolata* (Hom.: Cercopidae). *Arq. Inst. Biol., 70*, 309-314.

[75] Milner, R. J., Samson, P., & Morton, R. (2003). Persistence of conidia of *Metarhizium anisopliae* in Sugarcane Fields: Effect of isolate and formulation on persistence over 3.5 years. *Biocontrol Sci. Technol., 13*, 507-516.

[76] Moino Jr., A. M., & Alves, S. B. (1998). Efeito de imidacloprid e fipronil sobre *Beauveria bassiana* e *Metarhizium anisopliae* no comportamento de limpeza de *Heterotermes tenuis*. *Ann. Soc. Entomol. Bras., 27*, 611-619.

[77] Batista Filho, A., Almeida, J. E. M., & Lamas, C. (2001). Effect of thiamethoxam on entomopathogenic microorganisms. *Neotrop. Entomol., 30*, 437-447.

[78] Shah, P. A., & Pell, J. K. (2003). Entomopathogenic fungi as biological control agents. *Appl. Microbiol. Biotechnol., 61,* 413-423.

[79] St Leger, R. J., Cooper, R. M., & Charnley, A. K. (1986). Cuticle-degrading enzymes of entomopathogenic fungi: Synthesis in culture on cuticle. *J. Invertebr. Pathol., 48,* 85-95.

[80] St Leger, R. J., Goettel, M. S., Roberts, D. W., & Staples, R. C. (1991b). Prepenetration events during infection of host cuticle by *Metarhizium anisopliae. J. Invertebr. Pathol., 58,* 168-179.

[81] Goettel, M. S., St Leger, R. J., Rizzo, N. W., Staples, R. C., & Roberts, D. W. (1989). Ultrastructural localization of a cuticle-degrading protease produced by the entomopathogenic fungus *Metarhizium anisopliae* during penetration of host (*Manduca sexta*) cuticle. *J. Gen. Microbiol., 135,* 2233-2239.

[82] Pinto, A. S., Barreto, C. C., Schrank, A., Ulhoa, C. J., & Vainstein, M. H. (1997). Purification and characterization of an extracellular chitinase from the entomopathogen *Metarhizium anisopliae. Can. J. Microbiol., 43,* 322-327.

[83] Krieger de Moraes, C., Schrank, A., & Vainstein, M. H. (2003). Regulation of extracellular chitinases and proteases in the entomopathogen and acaricide *Metarhizium anisopliae*. *Curr. Microbiol., 46,* 205-210.

[84] Barreto, C. C., Staats, C. C., Schrank, A., & Vainstein, M. H. (2004). Distribution of chitinases in the entomopathogen *Metarhizium anisopliae* and effect of N-acetylglucosamine in protein secretion. *Curr. Microbiol., 48*, 102-107.

[85] Silva, W. O. B., Mitidieri, S., Schrank, A., & Vainstein, M. H. (2005). Production and extraction of an extracellular lipase from the entomopathogenic fungus *Metarhizium anisopliae*. *Proc. Biochem., 40,* 321-326.

[86] Santi, L., Silva, W. O. B., Berger, M., Guimaraes, J. A., Schrank, A., & Vainstein, M. H. (2010a). Conidial surface proteins of *Metarhizium anisopliae*: source of activities related with toxic effects, host penetration and pathogenesis. *Toxicon, 55,* 874-880.

[87] Beys da Silva, W. O., Santi, L., Schrank, A., & Vainstein, M. H. (2010a). *Metarhizium anisopliae* lipolytic activity plays a pivotal role in *Rhipicephalus (Boophilus) microplus* infection. *Fungal Biol., 114,* 10-15.

[88] Beys da Silva, W. O., Santi, L., Corrêa, A. P., Silva, L. A., Bresciani, F. R., Schrank, A., & Vainstein, M. H. (2010b). The entomopathogen *Metarhizium anisopliae* can modulate the secretion of lipolytic enzymes in response to different substrates including components of arthropod cuticle. *Fungal Biol., 114*, 911-916.

[89] Bidochka, M. J., St Leger, R. J., & Roberts, W. D. (1997). Mechanisms of Deuteromicete fungal infections in grasshoppers and locusts: an overview. *Mem. Entomol. Soc. Canada.,* 171: 213-224.

[90] St Leger, R. J., Staples, R. C., & Roberts, D. W. (1992). Cloning and regulatory analysis of starvation-stress gene, ssgA, encoding a hydrophobin-like protein from the entomopathogenic fungus, *Metarhizium anisopliae*. *Gene, 120,* 119-124.

[91] Fang, W., Pei, Y., & Bidochka, M. J. (2007). A regulator of a G protein signaling (RGS) gene, cag8, from the insect-pathogenic fungus *Metarhizium anisopliae* is involved in conidiation, virulence and hydrophobin synthesis. *Microbiol., 153*, 1017-1025.

[92] Wang, C., & St Leger, R. J. (2007a). The MAD1 adhesin of *Metarhizium anisopliae* links adhesion with blastospore production and virulence to insects, and the MAD2 adhesin enables attachment to plants. *Eukaryot. Cell, 6,* 808-816.

[93] Broetto, L., Da Silva, W. O., Bailão, A. M., De Almeida Soares, C., Vainstein, M. H., & Schrank, A. (2010). Glyceraldehyde-3-phosphate dehydrogenase of the entomopathogenic fungus *Metarhizium anisopliae*: cell-surface localization and role in host adhesion. *FEMS Microbiol. Lett., 312,* 101-109.

[94] Lord, J. C., & Howard, R. W. (2004). A proposed role for the cuticular fatty amides of *Liposcelis bostrychophila* (Psocoptera: Liposcelidae) in preventing adhesion of entomopathogenic fungi with dry-conidia. *Mycopathol., 158,* 211-217.

[95] Pedrini, N., Crespo, R., & Juarez, M. P. (2007). Biochemistry of insect epicuticle degradation by enomopathogenic fungi. *Comp. Biochem. Physiol. part C, 146,* 124-137.

[96] Clarkson, J. M., & Charnley, A. K. (1996). New insights into the mechanisms of fungal pathogenesis in insects. *Trends Microbiol., 4*, 197-204.

[97] Hackman, R. H., & Filshie, B. K. (1982). The tick cuticle. *In Obenchain, F.D., & Galun, R. (editors) Physiology of ticks – Current themes in tropical science, vol. Ist edition. New York, Pergamon Press*. Pp 450.

[98] Sonenshine, D. E. (1991). *Biology of ticks, vol. 1, NewYork, Oxford University Press.* Pp. 472.

[99] Ment, D., Gindin, G., Soroker, V., Glazer, I., Rot, A., & Samish, M. (2010). *Metarhizium anisopliae* conidial responses to lipids from tick cuticle and tick mammalian host surface. *J. Invertebr. Pathol.* 103: 132-139.

[100] Cherry, L. M. (1969). The production of cuticle wax by engorged females of the cattle tick, *Boophilus Microplus* (Canestrini). *J. Exp. Biol.* 50: 705-709.

[101] Hamilton, J. G. C., Sonenshine, D. E., & Lusby, W. R. (1989). Cholesteryl oleate: mounting sex pheromones of the hard tick *Dermacentor variabilis* (Say) (Acari: Ixodidae). *J. Insect Physio.* 35: 873-879.

[102] Yoder, J. A., Theriot, G. C., & Rivers, D. B. (1996). Venom from *Nasonia vitripennis* alters water loss from the flesh fly, *Sarcophaga bullata. Entomol. Exp. Appl. 81,* 235-238.

[103] Hunt, L.M. (1986). Differentiation between three species of *Amblyomma* ticks (Acari:Ixodidae) by analysis of cuticular hydrocarbons. *Ann. Trop. Med. Parasitol., 80,* 245-249.

[104] Davis, M. T. B. (1974). Critical temperature and changes in cuticular lipids in the rabbit tick, *Haemaphysalis leporispalustris. J. Insect Physiol., 20,* 1087-1100.

[105] St Leger, R. J., Butt, T. M., Staples, R. C., & Roberts, D. W. (1990). Second messenger involvement in differentiation of the entomopathogenic fungus *Metarhizium anisopliae. J. Gen. Microbiol., 136,* 1779-1789.

[106] Milner, R. J. (2000). Current status of *Metarhizium* as a mycoinsecticide in Australia. *Biocontrol:* 21: 47-50.

[107] Wang, C., & St Leger, R. J. (2007b). The *Metarhizium anisopliae* perilipin homolog MPL1 regulates lipid metabolism, appressorial turgor pressure, and virulence. *J. Biol. Chem.* 28: 21110-21115.

[108] Wang, C., & St Leger, R. J. (2005). Developmental and transcriptional responses to host and nonhost cuticles by the specific locust pathogen *Metarhizium anisopliae* var. *acridum. Eukaryot. Cell.* 4: 937-947.

[109] Golebiowski, M., Malinski, E., Bogus, M.I., Kumirska, J., & Stepnowski, P. (2008). The cuticular fatty acids of *Calliphora vicina, Dendrolimus pini* and *Galleria mellonella* larvae and their role in resistance to fungal infection. *Insect Biochem. Mol. Biol.* 38: 619-627.

[110] St Leger, R. J., Roberts, D. W., & Staples, R. C. (1989). Novel GTP-binding proteins in plasma membranes of the fungus *Metarhizium anisopliae. Biochem. Biophys. Res. Commun.* 164: 562-566.

[111] Arruda, W., Lubeck, I., Schrank, A., & Vainstein, M. H. (2005). Morphological alterations of *Metarhizium anisopliae* structures during the penetration events leading to the infection of the cattle tick *Boophilus microplus. Exp. Appl. Acarol., 37,* 231-244.

[112] Santi, L., Silva, W. O. B., Pinto, A. F. M., Schrank, A., & Vainstein, M. H. (2010b). *Metarhizium anisopliae* host-pathogen interaction: differential immunoproteomics reveals proteins involved in the infection. *Fungal Biol., 114,* 312-319.

[113] Mustafa, U., & Kaur, G. (2009). Extracellular enzyme production in *Metarhizium anisopliae* isolates. *Folia Microbiol., 54,* 499-504.

[114] Wang, C., Skrobek, A., & Butt, T. M. (2003). Investigations on the destruxin production of the entomopathogenic fungus *Metarhizium anisopliae*. *J. Invertebr. Pathol., 85,* 168-174.

[115] Melo, I. S., & Azevedo, J. L. (1998). *Controle biológico, volume I. Brasília, EMBRAPA.* Pp 262.

[116] Kershaw, M. J., Moorhouse, E. R., Bateman, R., Reynolds, S. E., & Charnley, A. K. (1999). The role of destruxins in the pathogenicity of *Metarhizium anisopliae* for three species of insect. *J. Invertebr. Pathol., 74*, 213-223.

[117] Pal, S., St Leger, R. J., & Wu, L. P. (2007). Fungal peptide destruxin A plays a specific role in suppressing the innate immune response in *Drosophila melanogaster*. *J. Biol. Chem., 282,* 8969-8977.

[118] Hsiao, Y. M., & Ko, J. L. (2001). Determination of destruxins, cyclic peptide toxins, produced by different strains of *Metarhizium anisopliae* and their mutants induced by ethyl methane sulfonate and ultraviolet using HPLC method. *Toxicon, 39*, 837-841.

[119] Huxham, I. M., Samuels, K. D. L., Heale, J. B., & McCorkindale, N. J. (1989). *In vivo* and *in vitro* assays for pathogenicity of wild-type and mutant strains of *Metarhizium anisopliae* for three insect species. *J. Invertebr. Pathol., 53*, 143-151.

[120] Shapiro-Ilan, D. I., Fuxa, J. R., Lacey, L. A., Onstad, D. W., & Kaya, H. K. (2005). Definitions of pathogenicity and virulence in invertebrate pathology. *J. Invertebr. Pathol., 88,* 1-7.

[121] Bidochka, M. J., & Khachatourians, G. G. (1990). Identification of *Beauveria bassiana* extracellular protease as a virulence factor in pathogenicity toward the migratory grasshopper, *Melanoplus sanguimipes*. *J. Invertebr. Pathol., 56*, 362-370.

[122] Freimoser, F. M., Screen, S., Bagga, S., Hu, G., & St Leger, R. J. (2003). Expressed sequence tag (EST) analysis of two subspecies of *Metarhizium anisopliae* reveals a plethora of secreted proteins with potential activity in insect hosts. *Microbiol., 149,* 239-247.

[123] Dutra, V., Nakazato, L., Broetto, L., Schrank, I. S., Vainstein, M. H., & Schrank, A. (2004). Application of representational difference analysis to identify sequence tags expressed by *Metarhizium anisopliae* during the infection of the tick *Boophilus microplus* cuticle. *Res. Microbiol., 155*, 245-251.

[124] Screen, S. E., Hu, G., & St Leger, R. J. (2001). Transformants of *Metarhizium anisopliae* sf. *anisopliae* overexpressing chitinase from *Metarhizium anisopliae* sf. *acridum* show early induction of native chitinase but are not altered in pathogenicity to *Manduca sexta*. *J. Invertebr. Pathol., 78,* 260-266.

[125] Schrank, A., Bassanesi, M. C., Pinto Jr, H., Costa, S. V., Bogo, M., & Silva, M. S. N. (1993). Superoxide dismutases in the entomopathogenic fungus *Metarhizium anisopliae*. *Ciência e Cultura, 45,* 200-205.

[126] Bittencourt, S. E. T., Castro, L. A., Farias, S. E., Bao, S. N., Schrank, A., & Vainstein, M. H. (2004). Purification and ultrastructural localization of a copper–zinc superoxide dismutase (CuZnSOD) from the entomopathogenic and acaricide fungus *Metarhizium anisopliae*. *Res. Microbiol., 155*, 681-687.

[127] Giménez-Pecci, M. P., Bogo, M. R., Santi, L., Moraes, C. K., Corrêa, C. T., Vainstein, M. H., & Schrank, A. (2002). Characterization of mycoviruses and analyses of chitinases secretion in the biocontrol fungus *Metarhizium anisopliae*. *Curr. Microbiol., 45*, 334-339.

[128] Tiago, P. V., Fungaro, M. H., Faria, M. R., & Furlaneto, M. C. (2004). Effects of double-stranded RNA in *Metarhizium anisopliae* var. *acridum* and *Paecilomyces fumosoroseus* on protease activities, conidia production, and virulence. *Can. J. Microbiol., 50,* 335-339.

[129] St Leger, R. J., Cooper, R. M., & Charnley, A. K. (1991a). Characterization of chitinase and chitobiase produced by the entomopathogenic fungus *Metarhizium anisopliae. J. Invertebr. Pathol., 58,* 415-426.

[130] Bagga, S., Hu, G., Screen, S. E., & St Leger, R. J. (2004). Reconstructing the diversification of subtilisins in the pathogenic fungus *Metarhizium anisopliae. Gene, 324*, 159-169.

[131] Gao, Q., Jin, K., Ying, S. H., Zhang, Y., Xiao, G., Shang, Y., & Duan, Z., et al. (2011). Genome sequencing and comparative transcriptomics of the model entomopathogenic fungi *Metarhizium anisopliae* and *M. acridum. PLoS Genet., 7*, e1001264.

[132] Hu, G., & St Leger, R. J. (2004). A phylogenomic approach to reconstructing the diversification of serine proteases in fungi. *J. Evol. Biol., 17*, 1204-1214.

[133] St Leger, R. L., Joshi, L., Bidochka, M. J., & Roberts, D. W. (1996). Construction of an improved mycoinseticide overexpressing a toxic protease. *PNAS, 93,* 6349-6354.

[134] Joshi, L., St Leger, R. J., & Roberts, D. W. (1997). Isolation of a cDNA encoding a novel subtilisin-like protease (Pr1B) from the entomopathogenic fungus *Metarhizium anisopliae* using differential display-RT-PCR. *Gene, 197*, 1-8.

[135] Gillespie, J. P., Baterman, R., & Chanrley, K. (1998). Role of cuticle-degrading porteases in the virulence of *Metarhizium anisopliae* spp. For the desert locust, *Schistocerca gregaria. J. Invertebr. Pathol., 71*, 128-137.

[136] St Leger, R. J., Bidochka, M. J., & Roberts, D. W. (1994b). Characterization of a novel carboxypeptidase produced by the entomopathogenic fungus *Metarhizium anisopliae. Arch. Biochem. Biophys., 314,* 392-398.

[137] Smithson, S. L., Paterson, I. C., Bailey, A. M., Screen, S. E., Hunt, B. A., Cobb, B. D., Cooper, R. M., Charnley, A. K., & Clarkson, J. M. (1995). Cloning and characterization of a gene encoding a cuticle-degrading protease from the insect pathogenic fungus *Metarhizium anisopliae. Gene, 166:* 161-165.

[138] Cole, S. C. J., Charnley, A. K., & Cooper, R. M. (1993). Purification and partial characterisation of a novel trypsin-like cysteine proteinase from *Metarhizium anisopliae. FEMS Microbiol. Lett., 113*, 189-196.

[139] Joshi, L., & St Leger, R. J. (1999). Cloning, expression, and substrate specificity of MeCPA, a zinc carboxypeptidase that is secreted into infected tissues by the fungal entomopathogen *Metarhizium anisopliae. J. Biol. Chem., 274*, 9803-9811.

[140] St Leger, R. J., Bidochka, M. J., & Roberts, D.W. (1994a). Isoforms of the cuticle-degrading Pr1 proteinase and production of a metalloproteinase by *Metarhizium anisopliae. Arch. Biochem. Biophys., 313,* 1-7.

[141] Freimoser, F. M., Hu, G., & St Leger R. J. (2005). Variation in gene expression patterns as the insect pathogen *Metarhizium anisopliae* adapts to different host cuticles or nutrient deprivation *in vitro. Microbiol., 151*, 361-371.

[142] Sasaki, S. D., Lima, C. A., Lovato, D. V., Juliano, M. A., Torquato, R. J. S., & Tanaka, A. S. (2008). BmSI-7, a novel subtilisin inhibitor from *Boophilus microplus*, with activity toward Pr1 proteases from the fungus *Metarhizium anisopliae. Exp. Parasitol., 118,* 214-220.

[143] Hassan, A., & Charnley, A. K. (1989). Ultrastructural study of the penetration by *Metarhizium anisopliae* throug dimilin-affected cuticle of *Manduca sexta*. *J. Invertebr. Pathol., 54*, 117-124.

[144] Ulhoa, C. J., & Peberdy, J. F. (1991). Regulation of chitinase synthesis in *Trichoderma harzianum*. *J. Gen. Microbiol., 137,* 2163-2169.

[145] Ulhoa, C. J., & Peberdy, J. F. (1993). Effect of carbon source on chitobiase production by *Trichoderma harzianum*. *Mycol. Res., 97,* 45-48.

[146] St Leger, R. J., Joshi, L., & Roberts, D. W. (1998). Ambient pH is a major determinant in the expression of cuticle-degrading enzymes and hydrophobin by *Metarhizium anisopliae*. *Appl. Environ. Microbiol., 64,* 709-713.

[147] Kishmoto, K., Nishizawa, Y., Tabei, Y., Hibi, T., Nakajima, M., & Akutsu, K. (2002). Detailed analysis of rice chitinase gene expression in transgenic cucumber plants showing different levels of disease resistance to gray mold (*Botrytis cinerea*). *Plant Sci., 162*, 655-662.

[148] St Leger, R. J., Charnley, A. K., & Cooper, R. M. (1987). Characterization of cuticle-degrading proteases produced by the entomopathogen *Metarhizium anisopliae*. *Arch. Biochem. Biophys., 253,* 221-232.

[149] St Leger, R. J., Cooper, R. M., & Charnley, A. K. (1993). Analysis of aminopeptidase and dipeptidylpeptidase IV from the entomopathogenic fungus *Metarhizium anisopliae*. *J. Gen. Microbiol., 139,* 237-243.

[150] Murad, A. M., Noronha, E. F., Miller, R. N. G., Costa, F. T., Pereira, C. D., Mehta, A., Caldas, R. A., & Franco, O. L. (2008). Proteomic analysis of *Metarhizium anisopliae* secretion in the presence of the insect pest *Callosobruchus maculatus*. *Microbiol., 154*, 3766-3774.

[151] Zhioua, E., Bowing, M., Jonhson, P. W., Ginsesberg, H. S., & Lebrum, R. A. (1997). Pathogenicity of the entomopathogenic fungus *Metarhizium anisopliae* (Deuteromicetos) to *Ixodes scapularum* (Acari: Ixodidae).*J. Parasitol., 83,* 815-818.

[152] Onofre, S. B., Miniuk, C. M., Barros, N. M., & Azevedo, J. L. (2001). Pathogenicity of four strains of entomopathogenic fungi against the bovine tick *Boophilus microplus*. *Am. J. Vet. Res., 62,* 1478-1480.

[153] Samish, M., Gindin, G., Alekseev, E., & Glazer, I. (2001). Pathogenicity of entomopathogenic fungi to different developmental stages of *Rhipicephalus sanguineus* (Acari: Ixodidae). *J. Parasitol., 87,* 1355-1359.

[154] Costa, G. L., Sarquis, M. I. M., Moraes A. M. L., & Bittencourt, V. R. E. P. (2001). Isolation of *Beauveria bassiana* and *Metarhizium anisopliae* var. *anisopliae* from *Boophilus microplus* tick (Canestrini, 1887), in Rio de Janeiro State, Brazil. *Mycopathol., 154,* 207-209.

[155] Gindin, G., Samish, M., Zangi, G., Mishoutchenko, A., & Glazer, I. (2002). The susceptibility of different species and stages of ticks to entomopathogenic fungi. *Exp. Appl. Acarol. 28,* 283-288.

[156] Benjamin, M. A., Zhioua, E., & Ostfeld, R. S. (2002). Laboratory and field evaluation of the entomopathogenic fungus *Metarhizium anisopliae* (Deuteromycetes) for controlling questing adult *Ixodes scapularis* (Acari: Ixodidae). *J. Med. Entomol., 39,* 723-728.

[157] Fernandes, E. K., Costa, G. L., Moraes, A. M., & Bittencourt, V. R. (2004). Entomopathogenic potential of *Metarhizium anisopliae* isolated from engorged females

and tested in eggs and larvae of *Boophilus microplus* (Acari: Ixodidae). *J. Basic Microbiol., 44*, 270-274.

[158] Hornbostel, V. L., Ostfeld, R. S., Zhioua, E., & Benjamin, M. A. J. (2004). Sublethal effects of *Metarhizium anisopliae* (Deuteromycetes) on engorged larval, nymphal, and adult *Ixodes scapularis* (Acari: Ixodidae). *Med. Entomol., 41*, 922-929.

[159] Kirkland, B. H., Westwood, G. S., & Keyhani, N.O. (2004). Pathogenicity of entomopathogenic fungi *Beauveria bassiana* and Metarhizium anisopliae to Ixodidae tick species *Dermacentor variabilis, Rhipicephalus sanguineus,* and *Ixodes scapularis. J. Med. Entomol., 41*, 705-711.

[160] Polar, P., Muro, M. A., Kairo, M. T. K., Moore, D., Pegram, R., John, S. A., & Roach-Benn, C. (2005c). Thermal characteristics of *Metarhizium anisopliae* isolates important for the development of biological pesticides for the control of cattle ticks. *Vet. Parasitol., 134,* 159-167.

[161] Hornbostel, V. L., Zhioua, E., Benjamin, M. A., Ginsber, H. S., & Ostfeld, R. S. (2005a). Pathogenicity of *Metarhizium anisopliae* (Deuteromycetes) and permethrin to *Ixodes scapularis* (Acari: Ixodidae) nymphs. *Exp. Appl. Acarol., 35*, 301-316.

[162] Hornbostel, V. L., Ostfeld, R. S., & Benjamin, M. A. (2005b). Effectiveness of *Metarhizium anisopliae* (Deuteromycetes) against *Ixodes scapularis* (Acari: Ixodidae) engorging on *Peromnyscus leucopus. J. Vector Ecol., 30*, 91-101.

[163] De Melo, D. R., Reis, R. C., & Bittencourt, V. R. (2006). *In vitro* patogenicity of the fungi *Metarhizium anisopliae* (Metschnikoff, 1879) Sorokin, 1883, on the tick *Boophilus microplus* (Canestrini,1887).*Rev. Bras. Parasitol. Vet., 15,* 157-162.

[164] Da Silva, S. B., & Bittencourt, V. R. (2006). Evaluation of cellular response in engorged females of *Boophilus microplus* (Canestrini, 1887) inoculated with *Metarhizium anisopliae, Beauveria bassiana, Penicillium corylophilum* or *Fusarium oxysporum. Rev. Bras. Parasitol. Vet., 15,* 151-156.

[165] Da Silva, S. B., Savastano, G., & Bittencourt, V. R. (2006). Cellular types involved in the immune response of females engorged of *Boophilus microplus* inoculated with *Metarhizium anisopliae* and *Penicillium* sp. *Rev. Bras. Parasitol. Vet.15,* 128-131.

[166] Maranga, R. O., Hassanali, V., Kaaya, G. P., & Mueke, J. M. (2006). Performance of a prototype baited-trap in attracting and infecting the tick *Amblyomma variegatum* (Acari: Ixodidae) in field experiments. *Exp. Appl. Acarol., 38,* 211–218.

[167] Bahiense, T. C., Fernandes, E. K., Angelo Ida, C., Perinotto, W. M., & Bittencourt, V. R. (2007). Evaluation of the biological control potential of *Metarhizium anisopliae* toward *Boophilus microplus* in pen trials. *Rev. Bras. Parasitol. Vet., 16,* 243-245.

[168] Lopes, R. B., Alves, S. B., Padulla, L. F., & Pérez, C. A. (2007). Efficiency of *Beauveria bassiana* and *Metarhizium anisopliae* formulations on *Amblyomma cajennense* (Fabricius, 1787) nymphae.*Rev. Bras. Parasitol. Vet., 16,* 27-31.

[169] Pirali-Kheirabadi, K., Haddadzadeh, H., Razzaghi-Abyaneh, M., Bokaie, S., Zare, R., Ghazavi, M., & Shams-Ghahfarokhi, M. (2007). Biological control of Rhipicephalus *(Boophilus) annulatus* by different strains of *Metarhizium anisopliae, Beauveria bassiana* and *Lecanicillium psalliotae* fungi. *Parasitol. Res., 100*, 1297–1302.

[170] Kurtti, T. J., & Keyhani, N. O. (2008). Intracellular infection of tick cell lines by the entomopathogenic fungus *Metarhizium anisopliae. Microbiol., 154*, 1700-1709.

[171] Gindin, G., Ment, D., Rot, A., Glazer, I., & Samish, M. (2009). Pathogenicity of *Metarhizium anisopliae* (Hypocreales: Clavicipitaceae) to tick eggs and the effect of egg cuticular lipids on conidia development. *J. Med. Entomol., 46,* 531-538.

[172] Pourseyeda, S. H., Tavassolia, M., Bernousib, I., & Mardanic, K. (2010). *Metarhizium anisopliae* (Ascomycota: Hypocreales): An effective alternative to chemical acaricides against different developmental stages of fowl tick *Argas persicus* (Acari: Argasidae). *Vet. Parasitol., 172,* 305-310.

[173] Nchu, F., Maniania, N. K., Hassanali, A., & Eloff, K. N. (2010b). Optimizing modes of inoculation of *Rhipicephalus ticks* (Acari: Ixodidae) with a mitosporic entomopathogenic fungus in the laboratory. *Exp. Appl. Acarol. 51,* 373-382.

[174] Stafford, K. C., & Allan, S. A. (2010). Field applications of entomopathogenic fungi *Beauveria bassiana* and *Metarhizium anisopliae* F52 (Hypocreales: Clavicipitaceae) for the control of *Ixodes scapularis* (Acari: Ixodidae). *J. Med. Entomol., 47,* 1107-1115.

[175] Bharadwaj, A, & Stafford, K. C. (2010). Evaluation of *Metarhizium anisopliae* strain F52 (Hypocreales: Clavicipitaceae) for control of *Ixodes scapularis* (Acari: Ixodidae). *J.Med. Entomol., 47,* 862-867.

[176] Sun, M., Ren, Q., Guan, G., Liu, Z., Ma, M., Gou, H., Chen, Z., Li, Y., Liu, A., Niu, Q., Yang, J., Yin, H., & Luo, J. (2011). Virulence of *Beauveria bassiana, Metarhizium anisopliae and Paecilomyces lilacinus* to the engorged female *Hyalomma anatolicum anatolicum* tick (Acari:Ixodidae). *Vet. Parasitol., in press.*

In: Ticks: Disease, Management and Control
Editor: Moges Woldemeskel
ISBN: 978-1-62081-136-8

Chapter 10

DERMOSCOPY IN THE DIAGNOSIS OF TICK-BITE LESIONS IN HUMANS AND IDENTIFICATION OF THE INVOLVED TICKS

Naoki Oiso* and Akira Kawada
Department of Dermatology, Kinki University Faculty of Medicine, Osaka-Sayama, Osaka, Japan

ABSTRACT

Ticks are vectors of various protozoan and bacterial diseases. Ticks transmit infectious diseases such as Lyme disease, Japanese spotted fever, Rocky Mountain spotted fever, Colorado tick fever, and others. It is significant to check for the presence of a tick bite, because immediate and precise treatment may reduce the morbidity and the mortality due to tick-borne diseases. The management of tick-bite should not be limited to the treatment of only the tick-bite lesions. Physicians need to examine and consider the possibility of consequent hazardous tick-transmitted infectious diseases. As the distribution of ticks is dependent on the environment, it is recommended for physicians to know the types of ticks prevalent in their surroundings, the time ticks bite their victims, and the types of tick-borne disease transmitted by a given tick species prevalent in the surrounding. Recently, dermoscopy is used for detecting and examining the tick-bite lesions and the involved biting tick. It also helps to determine whether or not the tick has been completely removed from the lesion. Dermoscopy is useful for easy initial classification of the tick species, which will suggest the strategy for treatment. Now, the term entodermoscopy is proposed for the new field connecting entomology, infectiology and dermatology.

* Address correspondence to: Naoki Oiso, Department of Dermatology, Kinki University Faculty of Medicine, 377-2 Ohno-Higashi, Osaka-Sayama, Osaka 589-8511, Japan; Tel: +81 72 366 0221; Fax: +81 72 368 2120; E-mail: naoiso@med.kindai.ac.jp.

INTRODUCTION

Ticks are vectors of the causative agents of various diseases such as Lyme disease, Rocky Mountain spotted fever, and Japanese spotted fever. Dermoscopy is a useful tool in evaluating the magnified morphological structures and in taking pictures to keep the record for diagnosis [1]. Dermoscopy is applied for examining the tick-bite lesions and the captured tick. It is also useful for identification of the lesion and easy initial classification of the involved tick species [1-3]. This chapter describes the preliminary findings of case reports of tick-bite lesions in humans and the use of dermoscopy in the diagnosis of the tick-bite lesions and the biting tick species.

LIFE CYCLE OF TICKS

The life cycle of ticks consists of the egg, the six-legged larva, the eight-legged immature nymph, and the mature adult [4]. Ticks feed once during each three stage of larva, nymph and adult [5]. *Ixodes ticks, the arthropod vectors for* Lyme disease, have two-year life cycle [5, 6]. Throughout the summer season, the larvae emerge from eggs deposited on the ground, and then crawl up low vegetation and grass stems, and attach to the skin of animals or humans in contact [4, 5]. Cases of numerous tick bites are commonly caused by the larval ticks [1, 6-9]. Following the fall and winter season, larvae develop into nymphs. The nymph feeds once during late spring and summer [5]. Usually, cases of tick-bite lesions are caused by these immature nymphal ticks [6]. Nymphal ticks then develop to adult tick, which feed on their hosts once again during summer and autumn [5]. Humans are at greater risk of Lyme disease from bites of the nymphal forms of the *Ixodes* tick in the spring and summer [6], because the smaller size of the nymphal ticks allows them to stay unnoticed for a longer time than the adult ticks [5].

CLINICAL FEATURE OF LESIONS ASSOCIATED WITH TICK-BITE AND TICK REMOVAL

The prevalence of tick-bite depends on the presence of ticks in a given area and human factors. Tick-bite is related to the distribution of ticks, the seasonal tick availability, and the survival of ticks [5]. There is an increase in cases of tick-bites and Japanese spotted fever in Japan. The current increase in tick-bite and Japanese spotted fever in Japan may be associated with the recent increase in the number of deer population [10]. Furthermore, the exposure to ticks carrying the causative microorganism is associated with human behavior and outdoor activities. It comprises the time spent outdoors playing during leisure or working in endemic areas without protective clothes [5]. Two distinct clinical types are observed and illustrated; numerous tick-bites by the larval ticks and a common tick-bite caused by a nymphal or adult tick.

Numerous Tick-Bites by the Larval Ticks

Case 1 refers to an 87-year-old woman with numerous tick-bite lesions. The patient was referred to the clinic for examination in June, 2010. The patient lived in a rural area of Kawachi-Nagano, Osaka, Japan, and habitually goes into fields to gather edible wild plants. Physical examination revealed infestation with many larval ticks, which caused pruritic erythema on the trunk and extremities, especially on the abdomen, in the genital area, bilaterally on the thighs and the popliteal fossa (Figures 1a, 1b) [1].The pruritic erythema most likely is elicited by allergic reaction, which might have been sensitized by previous tick-bites.

Tick-Bite Caused by the Nymphal Tick

In case 2 a 42-year-old man with a trace of tick-bite skin lesion was examined [2].The patient was admitted to the outpatient clinic of the Department of Dermatology, Kinki University, Faculty of Medicine, Osaka-Sayama, Japan, two days after noticing a tick-bite lesion on the left thigh while working in an area with tall grasses in Gobo, Wakayama, Japan, in April 2009. Physical examination revealed a slightly pruritic erythema with purpura on the patient's thigh [2].

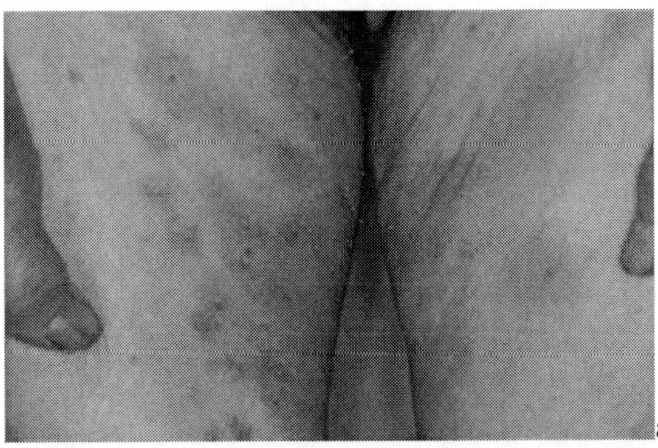
a.

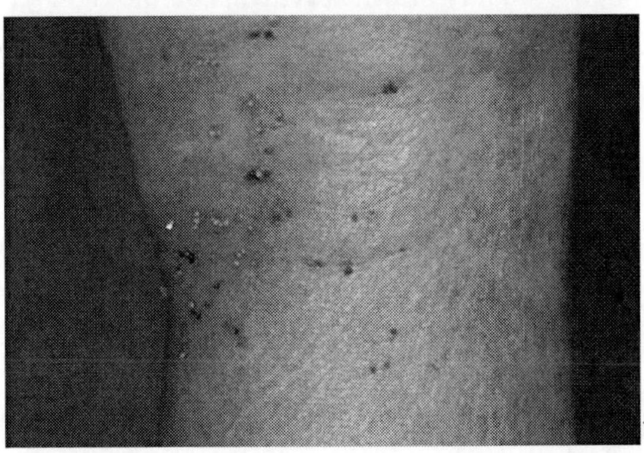
b.

Figures 1a and 1b. Case 1. Clinical appearance of numerous tick-bite lesions by the larval ticks.

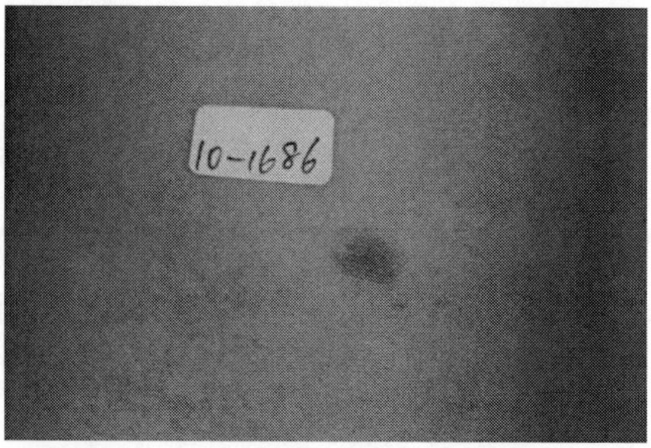

Figure 2a. Case 3. Clinical appearance of the trace of tick-bite-lesion.

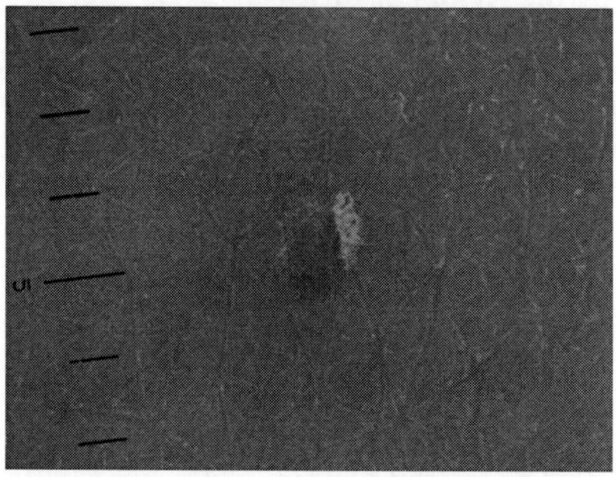

Figure 2b. Case 3. Dermoscopy without immersion gel for the trace of tick-bite lesion.

In case 3 a 61-year-old woman with a trace of tick-bite skin lesions was examined. In May 2010, the patient visited the clinic two days after removing a tick from the bite site (Figure 2a).

The patient lived in a rural area of Tondabayashi, Osaka, Japan. Physical examination revealed a 10 mm in diameter, slightly pruritic erythema on the right lateral abdomen [3].

Tick Removal

The tick removal technique should prevent the push which would induce the tick to salivate and regurgitate the infective body fluids into the skin. There is very limited experimental evidence on the validity of tick removal techniques, even though many methods of tick removal have been described [11]. Would Health Organization recommends the method using fine-tipped forceps, in which the tick is grasped closest to the point of entry into the skin to completely remove the hypostome [12, 13]. After the removal of the tick, the patient should be monitored up to 30 days for signs and symptoms of tick-borne diseases [11].

ENTODERMOSCOPY

The concept of entodermoscopy is proposed by Zalaudek *et al.* in 2008 [14]. It is a new and interesting field of science connecting entomology, infectiology and dermatology. Entodermoscopy is a fast, easy and non-invasive useful method for the clinical diagnosis of many skin infections and infestations [14].

The application of dermoscopy to the affected skin is useful to detect the trace of the lesion. Dermoscopic findings in case 2 and 3 showed a round skin lesion, 0.1 mm in diameter [2, 3]. In case 3, dermoscopy without immersion gel detected the skin hole at the bite site (Figures 2b) [3].

The use of dermoscopy in the identification of the biting tick can detect the number of legs of the biting tick, the condition of tick's hypostome, and other features of each tick species.

In case 2, dermoscopy showed an eight-legged nymphal tick that lost the hypostome. The tick had a pair of eyes at both sides of the scutum, typical feature of nymphal stage of *Amblyomma testudinarium* (Figure 3a).

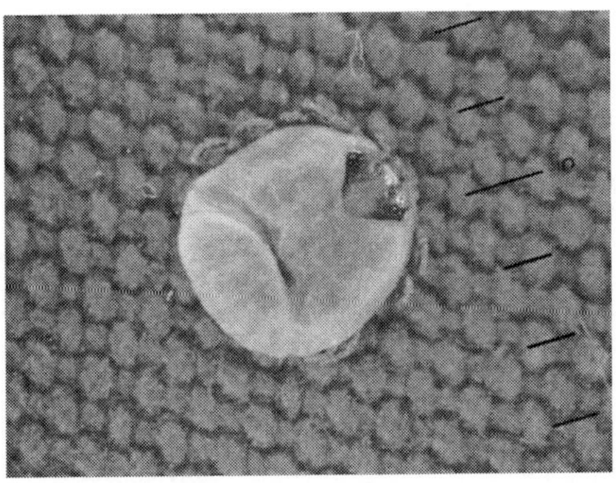

Figure 3a. Case 2. Dermoscopic appearance of the location of eyes of eight-legged nymph of *Amblyomma testudinarium*.

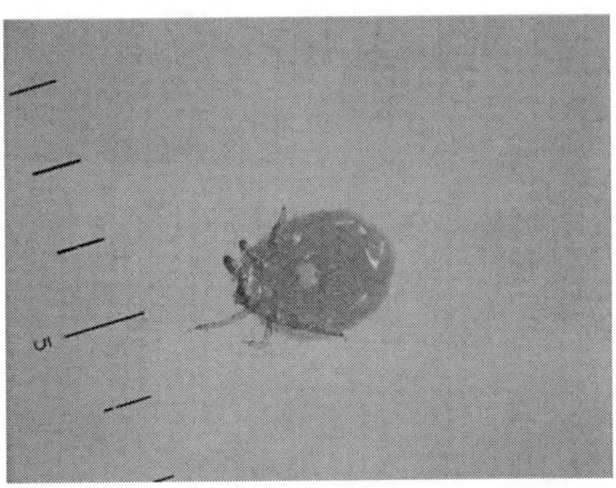

Figure 3b. Case 1. Dermoscopic appearance of six-legged larval of *Amblyomma testudinarium*.

In case 1, dermoscopy showed a six-legged larval tick consistent with the larva of *A. testudinarium* (Figure 3b).

Tick-Borne Diseases

Various tick-borne diseases which infect human beings are known. The common tick-borne diseases recorded includes Lyme disease, Rocky Mountain spotted fever, Japanese spotted fever, human monocyte ehrlichiosis, human granulocyte ehrlichiosis, tularemia, babesiosis, Colorado tick fever and relapsing fever. Each disease has its specific causative agents, tick vectors, and geographic regional distribution [15]. The three illustrated cases of tick-bite lesions in this chapter are caused by *A. testudinarium*, which sometimes carries *Rickettsia tamurae*, a rare pathogenic organism [16, 17].

Conclusion

In this chapter, the clinical feature of tick-bite lesions, which is the initial step in the diagnosis of tick-borne diseases, is illustrated. Entodermoscopy is a powerful tool to detect and examine the lesions induced by tick-bites and to identify the involved human-biting tick species. It enables to determine whether or not the tick has been removed completely, especially for tiny six-legged larval ticks. Moreover, it allows for easy initial classification of the species of the tick, which would help to determine the strategy for medical treatment. Physicians are required not only to treat the tick-bite lesions but also to prevent and treat the likely tick-borne diseases which could be transmitted during the tick-bite. Hereupon, it is worth noting that entodermoscopy opens a new opportunity and wide scope for entomologists, infectiologists and dermatologists in identification of skin lesions and associated causative agents.

References

[1] Oiso, N., Nakano, A., Yano, Y., and Kawada, A. (2010). The diagnostic usefulness of dermoscopy for identifying six-legged larval ticks. *Ticks Tick Borne Dis.*, 1, 197-198.

[2] Oiso, N., Kawara, S., Yano, Y., and Kawada, A. (2010). Diagnostic effectiveness of dermoscopy for tick bite. *J. Eur. Acad. Dermatol. Venereol.*, 24, 231-232.

[3] Matsuda, M., Oiso, N., Yano, Y., and Kawada, A. (2011). Dermoscopy for tick bite: reconfirmation of the usefulness for the initial diagnosis. *Case Rep. Dermatol.*, 3, 94-97.

[4] Fisher, E. J., Mo, J., and Lucky, A. W. (2006). Multiple pruritic papules from lone star tick larvae bites. *Arch. Dermatol.*, 142, 491-494.

[5] Bhate, C., and Schwartz, R. A. (2011). Lyme disease: Part I. Advances and perspectives. *J. Am. Acad. Dermatol.*, 64, 619-636.

[6] Bratton, R. L., Whiteside, J. W., Hovan, M. J., Engle, R. L., and Edwards, F. D. (2008). Diagnosis and treatment of Lyme disease. *Mayo Clin. Proc.*, 83, 566-571.

[7] Duckworth, P.F., Hayden, G., and Reed, C. (1985). Human infestation by *Amblyomma americanum* larvae ("seed ticks"). *South Med. J.*, 78, 751-753.

[8] Halkier-Sorensen, L., and Kragballe, K. (1989). Multiple bites from the larvae of the tick *Ixodes ricinus*. A case report. *Acta Derm. Venereol.*, 69, 268-269.

[9] Nakamura-Uchiyama, F., Komuro, Y., Yoshii, A., and Nawa, Y. (2000). *Amblyomma testudinarium* tick bite: one case of engorged adult and a case of extraordinary number of larval tick infestation. *J. Dermatol.*, 20, 774-777.

[10] Kondo, M., Nishii, M., Gabazza, E. C., Kurokawa, I., and Akachi, S. (2010). Nine cases of Japan spotted fever diagnosed at our hospital in 2008. *Int. J. Dermatol.*, 49, 430-434.

[11] Roupakias, S., Mitsakou, P., and Nimer. A. A. (2011). Tick removal. *J. Prev. Med. Hyg.*, 52, 40-44.

[12] Bhate, C., and Schwartz, R. A. (2011). Lyme disease: Part II. Management and prevention. *J. Am. Acad. Dermatol.*, 64, 639-653.

[13] Rozendaal, J. A. (1997). Vector control: methods for use by individuals and communities. Gneneva: Would Health Organization.

[14] Zalaudek, I., Giacomel, J., Cabo, H., Di Stefani, A., Ferrara, G., Hofmann-Wellenhof, R., Malvehy, J., Puig, S., Stolz, W., and Argenziano, G. (2008). Entodermoscopy: a new tool for diagnosing skin infections and infestations. *Dermatology*, 216, 14-23.

[15] Bratton, R. L., and Corey, R. (2005) Tick-borne disease. *Am. Fam. Physician*, 71, 2323-2330.

[16] Fournier, P. E., Takada, N., Fujita, H., and Raoult, D., (2006). *Rickettsia tamurae* sp. nov., isolated from *Amblyomma testudinarium* ticks. *Int. J. Syst. Evol. Microbiol.*, 56, 1673-1675.

[17] Imaoka, K., Kaneko, S., Tabara, K., Kusatake, K., and Morita, E. (2011). The first human case of *Rickettsia tamurae* infection in Japan. *Case Rep. Dermatol.*, 3, 68-73.

INDEX

B

C

D

E

I

J

K

L

M

N

O

P

T

U

V

W

Y

Z